ALGEBRAIC TOPOLOGY

A First Course

MATHEMATICS LECTURE NOTE SERIES

Volumes of the Series published from 1961 to 1973 are not officially numbered. The parenthetical numbers shown are designed to aid librarians and bibliographers to check the completeness of their holdings.

ISBN			
0-8053-5801-3	(1)	S. Lang	Algebraic Functions, 1965
0-8053-8703-X	(2)	J. Serre	Lie Algebras and Lie Groups, 1965 (4th printing, 1981)
0-8053-2327-9	(3)	P. J. Cohen	Set Theory and the Continuum Hypothesis, 1966 (5th printing, 1980)
0-8053-5808-0 0-8053-5809-9	(4)	S. Lang	Rapport sur la cohomologie des groupes, 1966
0-8053-8750-1 0-8053-8751-X	(5)	J. Serre	Algèbres de Lie semi-simples complexes, 1966
0-8053-0290-5 0-8053-0291-3	(6)	E. Artin and J. Tate	Class Field Theory, 1967 (2nd printing, 1974)
0-8053-0300-6 0-8053-0301-4	(7)	M. F. Atiyah	K-Theory, 1967
0-8053-2434-8 0-8053-2435-6	(8)	W. Feit	Characters of Finite Groups, 1967
0-8053-3555-4	(9)	Marvin J. Greenberg	Lectures on Algebraic Topology, 1966 (6th printing, with corrections, 1979)
0-8053-3757-1	(10)	Robin Hartshorne	Foundations of Projective Geometry, 1967 (4th printing, 1980)
0-0853-0660-9	(11)	H. Bass	Algebraic K-Theory, 1968
0-8053-0668-4	(12)	M. Berger and M. Berger	Perspectives in Nonlinearity: An Introduction to Nonlinear Analysis, 1968
0-8053-5208-2 0-8053-5209-0	(13)	I. Kaplansky	Rings of Operators, 1968
0-8053-6690-3 0-8053-6691-1	(14)	I. G. MacDonald	Algebraic Geometry: Introduction to Schemes, 1968
0-8053-6698-9 0-8053-6699-7	(15)	G. W. Mackey	Induced Representation of Groups and Quantum Mechanics, 1968
0-8053-7710-7 0-8053-7711-5	(16)	R. S. Palais	Foundations of Global Nonlinear Analysis, 1968
0-8053-7818-9 0-8053-7819-7	(17)	D. Passman	Permutation Groups, 1968
0-8053-8725-0	(18)	J. Serre	Abelian *l*-Adic Representations and Elliptic Curves, 1968
0-8053-0116-X	(19)	J. F. Adams	Lectures on Lie Groups, 1969
0-8053-0550-5	(20)	J. Barshay	Topics in Ring Theory, 1969
0-8053-1021-5	(21)	A. Borel	Linear Algebraic Groups, 1969
0-8053-1050-9	(22)	R. Bott	Lectures on K(X), 1969
0-8053-1430-X 0-8053-1431-8	(23)	A. Browder	Introduction to Function Algebras, 1969
		G. Choquet	Lectures on Analysis
0-8053-6955-4	(24)		Volume I. Integration and Topological Vector Spaces, 1969 (4th printing, 1980)
0-8053-6957-0	(25)		Volume II. Representation Theory, 1969 (4th printing, 1981)
0-8053-6959-7	(26)		Volume III. Infinite Dimensional Measures and Problem Solutions, 1969 (4th printing, 1981)
0-8053-2366-X 0-8053-2367-8	(27)	E. Dyer	Cohomology Theories, 1969

ALGEBRAIC TOPOLOGY

A First Course

Marvin J. Greenberg
University of California
Santa Cruz, California

John R. Harper
University of Rochester
Rochester, New York

The Advanced Book Program

CRC Press
Taylor & Francis Group
Boca Raton London New York

CRC Press is an imprint of the
Taylor & Francis Group, an **informa** business
A CHAPMAN & HALL BOOK

First published 1981 by Westview Press

Published 2018 by CRC Press
Taylor & Francis Group
6000 Broken Sound Parkway NW, Suite 300
Boca Raton, FL 33487-2742

CRC Press is an imprint of the Taylor & Francis Group, an informa business

Visit the Taylor & Francis Web site at
http://www.taylorandfrancis.com

and the CRC Press Web site at
http://www.crcpress.com

ISBN 13: 978-0-8053-3557-6 (pbk)

Library of Congress Cataloging in Publication Data

Greenberg, Marvin J.
Algebraic topology.

(Mathematics lecture note series ; 58)
"A revision of the first author's Lectures on algebraic topology"—P.
Bibliography: p.
Includes index.
1. Algebraic topology. I. Harper, John R., 1941- . II. Title. III. Series.
QA612.G7 514'.2 81-17108
ISBN 0-8053-3558-7 AACR2
ISBN 0-8053-3557-9 (pbk.)

MATHEMATICS LECTURE NOTE SERIES *(continued)*

ISBN			
0-8053-2420-8 0-8053-2421-6	(28)	R. Ellis	Lectures on Topological Dynamics, 1969
0-8053-2570-0 0-8053-2571-9	(29)	J. Fogarty	Invariant Theory, 1969
0-8053-3080-1 0-8053-3081-X	(30)	William Fulton	Algebraic Curves: An Introduction to Algebraic Geometry, 1969 (5th printing, with corrections, 1978)
0-8053-3552-8 0-8053-3553-6	(31)	M. J. Greenberg	Lectures on Forms in Many Variables. 1969
0-8053-3940-X 0-8053-3941-8	(32)	R. Hermann	Fourier Analysis on Groups and Partial Wave Analysis, 1969
0-8053-4551-5	(33)	J. F. P. Hudson	Piecewise Linear Topology, 1969
0-8053-5212-0 0-8053-5213-9	(34)	K. M. Kapp and H. Schneider	Completely O-Simple Semi-groups: An Abstract Treatment of the Lattice of Congruences, 1969
0-8053-5240-6 0-8053-5241-4	(35)	J. B. Keller and S. Antman (eds.)	Bifurcation Theory and Nonlinear Eigenvalue Problems, 1969
		O. Loos	Symmetric Spaces
0-8053-6620-2 0-8053-6621-0	(36)		Volume I. General Theory, 1969
0-8053-6622-9 0-8053-6623-7	(37)		Volume II. Compact Spaces and Classification, 1969
0-8053-7024-2 0-8053-7025-0	(38)	H. Matsumura	Commutative Algebra, 1970 (2nd Edition—cf. Vol. 56)
0-8053-7574-0 0-8053-7575-9	(39)	A. Ogg	Modular Forms and Dirichlet Series, 1969
0-8053-7812-X 0-8053-7813-8	(40)	W. Parry	Entropy and Generators in Ergodic Theory, 1969
0-8053-8350-6 0-8053-8351-4	(41)	W. Rudin	Function Theory in Polydiscs, 1969
0-8053-9100-2 0-8053-9101-0	(42)	S. Sternberg	Celestial Mechanics, Part I, 1969
0-8053-9102-9	(43)	S. Sternberg	Celestial Mechanics, Part II, 1969
0-8053-9254-8 0-8053-9255-6	(44)	M. E. Sweedler	Hopf Algebras, 1969
0-8053-3946-9 0-8053-3947-7	(45)	R. Hermann	Lectures in Mathematical Physics Volume I, 1970
0-8053-3942-6	(46)	R. Hermann	Lie Algebras and Quantum Mechanics, 1970
0-8053-8364-6 0-8053-8365-4	(47)	D. L. Russell	Optimization Theory, 1970
0-8053-7080-3 0-8053-7081-1	(48)	R. K. Miller	Nonlinear Volterra Integral Equations, 1971
0-8053-1875-5 0-8053-1876-3	(49)	J. L. Challifour	Generalized Functions and Fourier Analysis, 1972
0-8053-3952-3	(50)	R. Hermann	Lectures in Mathematical Physics Volume II, 1972
0-8053-2342-2 0-8053-2343-0	(51)	I. Kra	Automorphic Forms and Kleinian Groups, 1972
0-8053-8380-8 0-8053-8381-6	(52)	G. E. Sacks	Saturated Model Theory, 1972
0-8053-3103-4	(53)	A. M. Garsia	Martingale Inequalities: Seminar Notes on Recent Progress, 1973
0-8053-5664-3 0-8053-5666-5	(54)	T. Y. Lam	The Algebraic Theory of Quadratic Forms, 1973 (2nd printing, with revisions, 1980)
0-8053-6702-0 0-8053-6703-9	55	George W. Mackey	Unitary Group Representations in Physics, Probability, and Number Theory, 1978

MATHEMATICS LECTURE NOTE SERIES *(continued)*

ISBN			
0-8053-7026-9	56	Hideyuki Matsumura	Commutative Algebra, Second Edition, 1980 (2nd printing, 1981)
0-8053-0360-X	57	Richard Bellman	Analytic Number Theory: An Introduction, 1980
0-8053-3558-7 0-8053-3557-9	58	M. J. Greenberg and J. Harper	Algebraic Topology: A First Course, 1981

Other volumes in preparation

CONTENTS

PREFACE

Algebraic Topology is one of the major creations of twentieth-century mathematics. Its influence on other parts of mathematics, such as algebra [38], number theory [4, 49], algebraic geometry [27, 31, 50], differential geometry [26], and analysis [12, 1963–64] has been enormous. In its own right, it is a major tool for the investigation of topological spaces, especially manifolds. Its key idea is to attach algebraic structures to topological spaces and their maps in such a way that the algebra is both invariant under a variety of deformations of spaces and maps, and computable.

This book is intended as a first course, sufficiently comprehensive to enable the student either to use the subject in other fields of endeavor and/or to pursue its development and applications in more advanced texts and the literature.

Our presentation is a revision of the first author's *Lectures on Algebraic Topology*. The intent in revising was to make those additions of theory, examples, and exercises which updated, enhanced, and simplified the original exposition. The point of view and organizational principles of the earlier book have been maintained. Virtually all of the original book has been reproduced.

In the additional material, special attention has been given to calculations, with more geometry to balance all the algebra.

There are essentially four parts to this work: Sections 1–7 form Part I, elementary homotopy theory. Homotopy of paths and maps is defined, and the fundamental group is constructed. The classification of covering spaces by means of subgroups of the fundamental group is given, and, finally, the higher homotopy groups are defined inductively using loop spaces, following Hurewicz.

Sections 8–21, Part II, treat singular homology theory. This Part has been influenced by the lucid notes of E. Artin [3] and the work of Eilenberg-Steenrod [23]. The advantages of singular over simplicial homology theory are that, first, it applies to arbitrary topological spaces; second, it is obviously topologically invariant; third, once the excision theorem is proved, there is never again any need to subdivide, and, finally, it is easier to calculate once the basic formulas (19.16–19.18) have been proved. Combinatorial techniques are still very important in algebraic topology [36, 62, 70]. However, it is now recognized that algebraic topology encompasses at least three different categories—topological, differential, and piecewise linear. In this book we treat primarily the first (references for the second are [15, 17, 41–44, 51, 55, 68, 71]). The classical applications of homology theory to spheres are given in Sections 15, 16, and 18.

Sections 22–28 form Part III, the orientability and duality properties of manifolds. This part has been greatly influenced by notes of Dold, Puppe, and Milnor. No assumption of triangulability is needed in this treatment. The correct cohomology theory for the duality is that of Alexander-Cech; however, for brevity's sake, we only describe the Alexander-Cech cohomology module of a subspace A as the inductive limit over the neighborhoods of A of the singular cohomology modules. We show that this coincides with the singular cohomology module when A is a compact ANR.

Finally in Part IV we develop the basic features of the theory of products in cohomology. The applications include the Lefschetz fixed point theorem for compact oriented manifolds and an introduction to intersection theory in closed manifolds.

Each part is divided into several sections. These are the organizational units of the text. There is considerable flexibility (especially in the latter parts) in the order in which they may be studied. In Part II, many sections conclude with material which may be skimmed or skipped at first reading.

Most sections end with sets of exercises. No theoretical development depends on an exercise nor is further theoretical material given as exercises. Most exercises concern calculation and, as the subject develops, geometric applications are made. There are many cross-references among exercises. Refinements of calculations available with developments of the theory are offered. Similarly, improvements in geometric results are made in several sections. This process imitates the way the subject actually developed, and may help motivate the successive layers of abstraction through which the subject passes. Some exercises are accompanied with suggestions for their solution. These suggestions should not be taken too seriously. Most problems can be solved in different ways, and one's favorite solution may not receive widespread approval. But it is discouraging to be totally "stuck" so suggestions are offered to alleviate that condition.

Prerequisites for this book, besides the usual "mathematical maturity," are very few. In algebra, familiarity with groups, rings, modules, and their homomorphisms is required. From Section 20 on, some basic results for modules over principal ideal domains will be used. Only in Sections 29 and 30 is knowledge of the basic properties of the tensor product of two modules needed. The language of categories and functors is used throughout the book, although no theorems about categories are required. For all of this material, see Lang [35].

In point-set topology, the reader is presumed to be familiar with the basic facts about continuity, compactness, connectedness and pathwise-connectedness, product spaces, and quotient spaces. Only in the appendix to Section 26* do we require a nontrivial result, Tietze's extension theorem. Section 7 uses some elementary results about the compact-open topology on function spaces. For this material, see Dugundji [20] or Kelley [34].

I recommend the survey articles [44a, 62, and 75, pp. 227–31 and its bibliography] to the reader seeking further information on the extraordinary achievements in algebraic topology in recent years.

I thank M. Artin, H. Edwards, S. Lang, B. Mazur, V. Poenaru, H. Rosenberg, E. Spanier, and A. Vasquez; also my students Berkovits, Perry, and Webber, for helpful comments.

We are grateful to a number of people for helpful remarks concerning the revision. The comments of D. Anderson, E. Bishop, G. Carlsson, M. Freedman, T. Frankel, J. Lin, and K. Millett were helpful in deciding what to include and what to leave out. As the work developed, valuable remarks were made by M. Cohen, A. Liulevicius, R. Livesay, S. Lubkin, H. Miller, R. Mandelbaum, N. Stein, and A. Zabrodsky.

The typing of the manuscript was expertly done by S. Agostinelli, R. Colon, and M. Lind. Additional figures were drawn by D. McCumber.

Special thanks are extended to Doris, Jennifer, and Allison for not overreacting to neglect endured during preparation and assembly of this material.

Lastly, we thank Errett Bishop for suggesting that we collaborate on this book.

MARVIN J. GREENBERG
JOHN R. HARPER

*(26.17)

ALGEBRAIC TOPOLOGY

A First Course

Part 1
ELEMENTARY HOMOTOPY THEORY

Introduction to Part I

The wellspring of ideas leading to algebraic topology was the perception, developed largely in the latter half of the nineteenth century, that many properties of functions were invariant under "deformations." For example, Cauchy's theorem and the calculus of residues in complex analysis assert invariance of complex integrals with respect to continuous deformations of curves. Perhaps the true starting point was Riemann's theory of abelian integrals. It was here that the significance of the connectivity of surfaces was recognized. The interested reader is strongly encouraged to examine Felix Klein's exposition of Riemann's theory [80], during the study of algebraic topology.

It was Poincaré who first systematically attacked the problem of attaching numerical topological invariants to spaces. In his investigations, he perceived the difference between curves *deformable* to one another and curves *bounding* a larger space. The former idea led to the introduction of homotopy and the fundamental group, while the latter led to homology.

The development of these ideas into a mathematical theory is elaborate. However, the idea guiding the development is easily described. Certain functors are constructed. Thus to each topological space X is assigned a group $F(X)$, and to each map $f: X \to Y$ (a "map" of topological spaces will always mean a "continuous map" unless otherwise stated) is assigned a homomorphism $F(f): F(X) \to F(Y)$ such that

(1) If $Y = X$ and $f =$ identity, then $F(f) =$ identity,

(2) If $g: Y \to Z$, then $F(gf) = F(g)F(f)$.

Illustration: Suppose we have a diagram of topological spaces and maps

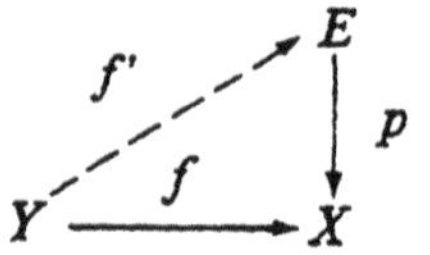

and the problem is to find f' such that $pf' = f$. Applying the functor F we see that a necessary condition for a solution to exist is that $F(f)$ send $F(Y)$ into the subgroup $F(p)(F(E))$ of $F(X)$. In certain cases later we will see this is also sufficient (6.1).

Illustration: Suppose $f : X \to Y$ is a homeomorphism. Then by functoriality $F(f^{-1})$ is inverse to $F(f)$, so that $F(f)$ is an isomorphism. Thus a *necessary* condition (but usually not sufficient) that X and Y be homeomorphic is that $F(X)$ and $F(Y)$ be isomorphic groups. This is usually the easiest way to prove that two given spaces with similar topological properties are not homeomorphic.

Illustration: Suppose $i : A \to X$ is the inclusion map of a subspace A into X and our problem is to find a map $r : X \to A$ such that ri is the identity map of A (such a map r is called a *retraction* of X onto A). By functoriality, $F(r)F(i)$ equals the identity transformation of $F(A)$, so that $F(i)$ sends $F(A)$ isomorphically onto a subgroup of $F(X)$. If we happen to know, e.g., that $F(X)$ is trivial while $F(A)$ is not, it then follows that no retraction can exist. This is the way the Brouwer Fixed Point Theorem is proved (4.11 and 15.7).

The reader may construct some more illustrations to convince himself of the fruitfulness of this point of view.

1. Arrangement of Part I

In Part I, we treat the fundamental group and the closely related notion of covering space. The geometric idea for the construction of the fundamental group functor is homotopy of paths. Roughly speaking, a homotopy of a path is a deformation leaving the end points fixed. A composition of paths may be defined when the end point of one agrees with the initial point of the other. Familiar algebraic properties, like associativity, do not hold, but do hold up to homotopy. The result is a group structure on equivalences classes, called the fundamental group. This group is not just a topological invariant, but invariant under a larger class of maps, called *homotopy equivalences*. These topics are treated in Sections 2 and 3.

In order to exploit the fundamental group, we must be able to calculate it. There are two principal routes to calculation: the Seifert-Van Kampen theorem and the use of covering spaces. The versions of the former used in this text are stated in (4.12). There are several excellent accounts available in other texts, so we do not reproduce the details. Our treatment of the fundamental group of the circle is the prototype for the theory of covering spaces. The lifting theorem for covering spaces (6.1), besides being useful, is an outstanding example of the blend of algebra and geometry that gives this subject its special flavor. Part I concludes with a brief discussion of higher homotopy groups, introduced by means of loop spaces.

2. Homotopy of Paths

Consider, in the plane, the problem of integrating a function f of a complex variable around a closed curve C, e.g., the unit circle. We have, for example,

$$\int_C z\,dz = 0$$

$$\int_C \frac{dz}{z} \neq 0$$

What is the difference? We take the point of view that C can be "shrunk to a point" within the domain of analyticity of z (i.e., the whole plane), hence integrating around C is equivalent to integrating at a point, which gives 0. On the contrary C cannot be "shrunk to a point" within the domain of $1/z$.

More precisely, let σ, τ be *paths* in a space X (i.e., maps of the unit interval I into X) with the same end points (i.e., $\sigma(0) = \tau(0) = x_0$, $\sigma(1) = \tau(1) = x_1$). We say *$\sigma$ and τ are homotopic with end points held fixed* written

$$\sigma \simeq \tau \text{ rel } (0, 1)$$

if there is a map $F : I \times I \to X$ such that

(1) $F(s, 0) = \sigma(s)$ all s

(2) $F(s, 1) = \tau(s)$ all s

(3) $F(0, t) = x_0$ all t

(4) $F(1, t) = x_1$ all t

F is called a *homotopy* from σ to τ. For each t, $s \to F(s,t)$ is a path F_t from x_0 to x_1, and $F_0 = \sigma$, $F_1 = \tau$. We write

$$F_t : \sigma \simeq \tau \qquad \text{rel } (0, 1)$$

Pictorially:

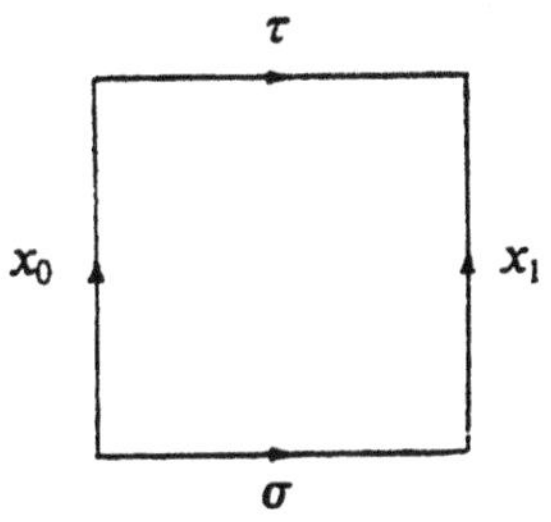

In particular if σ is a *loop* at x_0 (i.e., $x_1 = x_0$) and τ is the constant loop $\tau(s) = x_0$ for all s, and if $\sigma \simeq \tau$ rel (0, 1), we say that "σ can be shrunk to a point," or is ***homotopically trivial.***

Then the correct statement of Cauchy's Theorem is that $\int_C f(z)dz = 0$ for all loops C in the domain X of analyticity of f which are homotopically trivial (more generally, homologically trivial).

The following properties of relation $\simeq$ are easily proved:

(1) $\sigma \simeq \sigma$ rel (0, 1)

(2) $\sigma \simeq \tau$ rel (0, 1) $\Rightarrow \tau \simeq \sigma$ rel (0, 1)

(3) $\sigma \simeq \tau$ rel (0, 1) and $\tau \simeq \rho$ rel (0, 1) $\Rightarrow \sigma \simeq \rho$ rel (0, 1)

Thus we can consider the homotopy classes $[\sigma]$ of paths σ from x_0 to x_1 under the equivalence relation $\simeq$.

If σ is a path from x_0 to x_1 and τ is now taken to be a path from x_1 to x_2, we define a path $\sigma\tau$ from x_0 to x_2 by first travelling along σ, then along τ; more precisely we set

$$\sigma\tau(t) = \begin{cases} \sigma(2t) & 0 \le t \le 1/2 \\ \tau(2t-1) & 1/2 \le t \le 1 \end{cases}$$

(4) $\sigma \simeq \sigma'$ rel (0, 1) and $\tau \simeq \tau'$ rel (0, 1) $\Rightarrow \sigma\tau \simeq \sigma'\tau'$ rel (0, 1).

Proof: If $F_t : \sigma \simeq \sigma'$ rel (0, 1), $G_t : \tau \simeq \tau'$ rel (0, 1), then

$$F_t G_t : \sigma\tau \simeq \sigma'\tau' \text{ rel } (0, 1). \qquad \blacksquare$$

Thus we can multiply the *class* of σ on the right by the *class* of τ without ambiguity, always supposing the end point of σ equals the initial point of τ.

(2.1) *Theorem. Let $\pi_1(X, x_0)$ be the set of homotopy classes of loops in X at x_0. If multiplication in $\pi_1(X, x_0)$ is defined as above, $\pi_1(X, x_0)$ becomes a group, in which the neutral element is the class of the constant loop at x_0 and the inverse of a class $[\sigma]$ is the class of the loop σ^{-1} defined by*

$$\sigma^{-1}(t) = \sigma(1 - t) \qquad 0 \le t \le 1$$

(i.e., travel backwards along σ).

Proof: We will prove that $\sigma\sigma^{-1} \simeq x_0$, where now x_0 denotes also the constant loop at the point x_0. The homotopy is given by the following diagram:

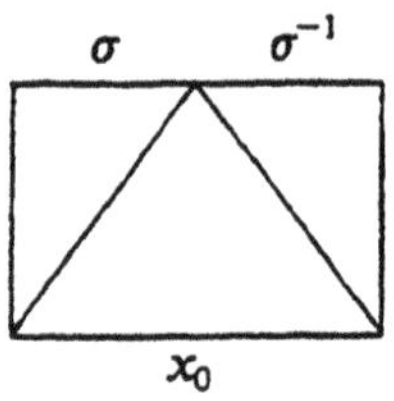

Thus, we define $F(s, t)$ by

$$F(s, t) = \begin{cases} \sigma(2s) & 0 \le 2s \le t \\ \sigma(t) & t \le 2s \le 2 - t \\ \sigma^{-1}(2s - 1) & 2 - t \le 2s \le 2 \end{cases}$$

Clearly these functions are continuous on each triangle and they agree on the intersections, hence by an elementary argument F is continuous on the whole square.

The proof that multiplication is associative (up to homotopy) can be done similarly, as can the proof that the class of x_0 is the neutral element.

$$\text{Define } F(s,t) = \begin{cases} \sigma\left(\dfrac{4s}{t+1}\right) & 0 \le s \le \frac{1}{4}(t+1) \\ \tau(4s-t-1) & \frac{1}{4}(t+1) \le s \le \frac{1}{4}(t+2) \\ \omega\left(\dfrac{4s-t-2}{2-t}\right) & \frac{1}{4}(t+2) \le s \le 1 \end{cases}$$

to establish $(\sigma\tau)\omega \simeq \sigma(\tau\omega)$ rel $(0, 1)$.

$$\text{Define } F(s,t) = \begin{cases} \sigma\left(\dfrac{2s}{t+1}\right) & 0 \le s \le \dfrac{t+1}{2} \\ x_0 & \dfrac{t+1}{2} \le s \le 1 \end{cases}$$

to establish that the constant path at x_0 is the neutral element of the fundamental group. ∎

Is there a relation between $\pi_1(X, x_0)$ and $\pi_1(X, x_1)$? There certainly is not if x_0 and x_1 lie in different path-connected components of X. However, we have the following result.

(2.2) *Proposition.* Let α *be a path from* x_0 *to* x_1. *The mapping* $[\sigma] \to [\alpha^{-1}\sigma\alpha]$ *is an isomorphism* α_* *of the group* $\pi_1(X, x_0)$ *onto* $\pi_1(X, x_1)$.

Proof. It is clearly a homomorphism, and $(\alpha^{-1})_*$ is its inverse (where α^{-1} is the path defined as in 2.1). ∎

(2.3) *Corollary. If* X *is pathwise connected, the group* $\pi_1(X, x_0)$ *is independent of the point* x_0, *up to isomorphism.*

In that case we often write simply $\pi_1(X)$ for $\pi_1(X, x_0)$ and call it *the fundamental group* of X.

We would like π_1 to be a functor from spaces to groups, but since $\pi_1(X, x_0)$ does depend on the base point x_0 in the general case, we must put the base points into our category if we are to obtain a functor. So define the category of *pointed topological spaces* to have as objects pairs (X, x_0), and as morphisms the maps $f: X \to Y$ such that $f(x_0) = y_0$. For any such f we obtain an induced homomorphism

$$f_* : \pi_1(X, x_0) \to \pi_1(Y, y_0)$$

defined by $f_*[s] = [f \circ s]$. One verifies easily that this is well-defined and a homomorphism. Moreover,

(1) $Y = X$ and $f =$ identify $\Rightarrow f_* =$ identity;

(2) Given $g : (Y, y_0) \to (Z, z_0)$, $(g f)_* = g_* f_*$.

Thus we can speak of *the fundamental group functor* from the category of pointed topological spaces to the category of groups.

3. Homotopy of Maps

Since paths are maps of I into X, we can try to replace I by any space Y and define homotopy. Thus we no longer have end points but we can substitute a subspace $A \subset Y$ for the set $\{0, 1\}$.

Given maps $f, g : Y \to X$ such that $f \mid A = g \mid A$, we say

$$f \simeq g \quad \text{rel } A$$

if there is a map $F : Y \times I \to X$ satisfying

(1) $F(y, 0) = f(y)$ all $y \in Y$

(2) $F(y, 1) = g(y)$ all $y \in Y$

(3) $F(y, t) = f(y) = g(y)$ all $y \in A, t \in I$

In case A is empty, we write simply

$$f \simeq g$$

Once again we obtain an equivalence relation.

Example 1: Let $X = Y = \mathbf{R}^n$, let f be the identity, g the constant map 0. Then

$$F(x, t) = t\,x$$

defines a homotopy from g to f.

If X is a space such that the identity map on X is homotopic to a constant map on some point in X, we say X is *contractible*.

(3.1) *Exercise.* X is contractible if and only if for any space Y any two maps of Y into X are homotopic. A contractible space is pathwise connected.

Example 2: Every convex subset X of Euclidean space is contractible. For if $f_1, f_2 : Y \to X$, we define a homotopy by

$$F(y, t) = tf_1(y) + (1 - t)f_2(y) \quad y \in Y, t \in I$$

Call a space *simply connected* if it is pathwise connected and its fundamental group is trivial.

(3.2) *Proposition. A contractible space is simply connected.*

Proof: This is not entirely obvious, because although every loop σ at a point x_0 is homotopic as a map with the constant loop, we do not know they are homotopic *relative to* (0, 1).

(3.3) *Lemma. Given* $F: I \times I \to X$. *Set* $\alpha(t) = F(0, t)$, $\beta(t) = F(1, t)$, $\gamma(s) = F(s, 0)$, $\delta(s) = F(s, 1)$, *so that diagrammatically*

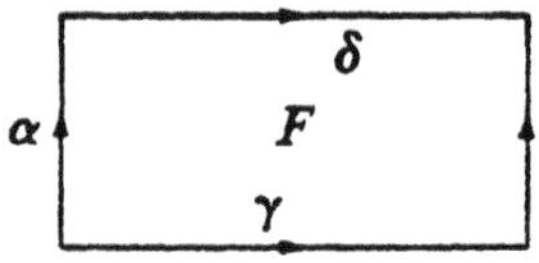

Then $\delta \simeq \alpha^{-1}\gamma\beta$ rel (0, 1).

Proof: The proof is by juxtaposing 3 squares

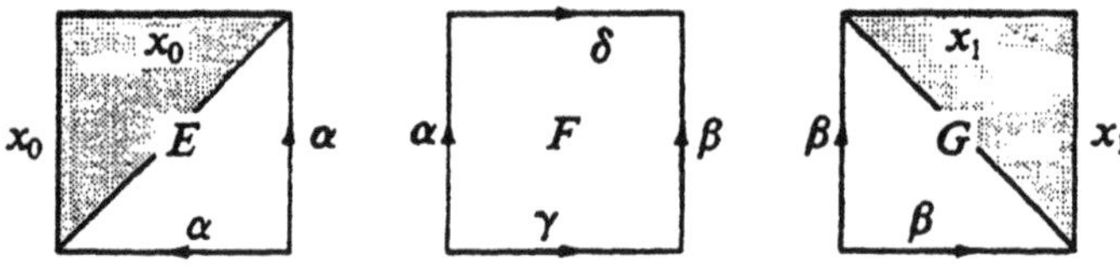

where $x_0 = \delta(0)$, $x_1 = \delta(1)$, and

$$E(s, t) = \begin{cases} x_0 & s \le t \\ \alpha(1 + t - s) & s \ge t \end{cases}$$

$$G(s, t) = \begin{cases} \beta(t + s) & 1 - s \ge t \\ x_1 & 1 - s \le t \end{cases}$$

■

If now X is contractible, we can obtain such an F with $\delta = \sigma$, $\gamma = x_0$, and $\alpha = \beta$ (since σ induces a map of the circle into X which is homotopic to the constant map at x_0), hence σ is homotopically trivial.

(3.4) *Corollary. Let f, g be homotopic maps $Y \to X$ by means of a homotopy $F: Y \times I \to X$. Let $y_0 \in Y$, $x_0 = f(y_0)$, $x_1 = g(y_0)$. Let α be the path from x_0 to x_1 given by*

$$\alpha(t) = F(y_0, t) \quad t \in I$$

Then we have a commutative triangle

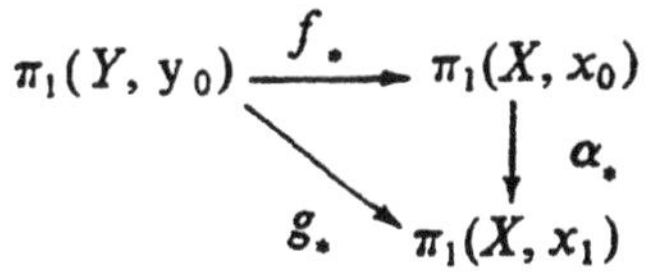

Proof: For any loop σ at y_0, we have

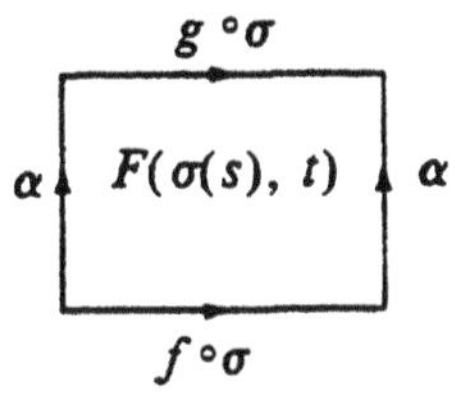

■

(3.5) *Corollary. Under the above conditions, f_* is an isomorphism if and only if g_* is.*

A map $f: Y \to X$ is called a *homotopy equivalence* if there is a map $f': X \to Y$ such that

$$ff' \simeq \text{identity map of } X$$

$$f'f \simeq \text{identity map of } Y$$

If such an f exists we say X and Y are *homotopically equivalent spaces*. For example, X is contractible if and only if it is homotopically equivalent to a point.

(3.6) *Corollary. If f is a homotopy equivalence then f_* is an isomorphism $\pi_1(Y, y_0) \to \pi_1(X, f(y_0))$ for all $y_0 \in Y$.*

For by the previous corollary, $f_* f'_*$ and $f'_* f_*$ are both isomorphisms. ■

Thus the fundamental group of a path-connected space is a *homotopy invariant* (*a fortiori*, a topological invariant). The relation of homotopy equivalence is cruder than topological equivalence. For example, the most elementary homotopy equivalence is a contraction—shrinking a portion of a space to a point. The importance of the idea of homotopy equivalence for algebraic topology lies in the fact that constructions used to attach algebra to spaces usually lead to homotopy invariant structures. Furthermore, the understanding of homotopy types is a base from which to attack more subtle questions involving topological type.

Exercise: Classify the letters of your favorite alphabet according to homotopy type and topological type.

(3.7) *Exercise*. Let X be path connected. Then the following are equivalent:

(1) X is simply connected;

(2) Every map of the unit circle S^1 into X extends to a map of the closed unit disc E^2 into X;

(3) If σ, τ are paths in X with the same initial points and the same terminal points then $\sigma \simeq \tau$ rel $(0, 1)$.

Hints: To show (1) $\Longleftrightarrow$ (2), represent E^2 as a quotient space of $I \times I$ by sending the point $(s, t) \in I \times I$ onto the point $t\,e^{2\pi i s}$. To show (1) $\Rightarrow$ (3), use the transformation of the square described by the diagram

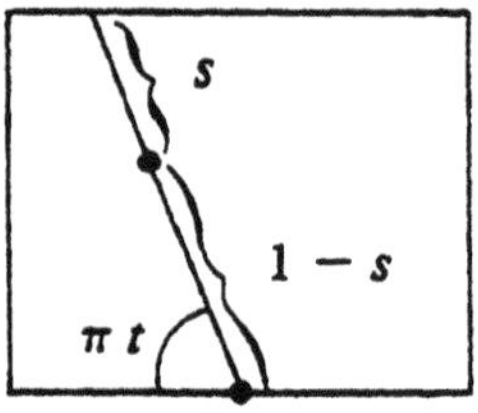

(3.8) *Exercise*. Let $CX = X \times I / X \times \{0\}$, be the *cone* on X. Regard $X \subset CX$ via $x \to (x, 1)$. Generalize (3.7)(2) to show $f : X \to Y$ is homotopically trivial if and only if f extends to $\bar{f} : CX \to Y$.

(3.9) *Exercise*. Suppose Y is contractible to a point y. Show

$$f : X \to X \times Y$$

by $f(x) = (x, y_0)$ and projection $p : X \times Y \to X$ are homotopy equivalences.

(3.10) *Exercise*. Let $f, g : S^n \to S^n$ be maps such that for all $x \in S^n$, $f(x)$ and $g(x)$ are not antipodal. Show $f \simeq g$. If in addition there is $x_0 \in X$ such that $f(x_0) = g(x_0)$, show $f \simeq g$ rel x_0.

(3.11) *Exercise*. Suppose X and Y have the same homotopy type. Show that the arc components of X and Y are in one-to-one correspondence.

(3.12) *Exercise*. Let X be arc connected and suppose every $f: S^1 \to X$ is homotopically trivial but not necessarily by a homotopy leaving the base point fixed. Show $\pi_1(X, x_0) = 0$.

4. Fundamental Group of the Circle

We study the circle S^1 via the line $\mathbf{R}$. It turns out that the homotopy class of a loop is determined by the number of times it "winds around," the number being negative if the "winding" is opposite to the given orientation on S^1.

More precisely, S^1 is the group of complex numbers of absolute value 1. We have a continuous homomorphism $\phi : \mathbf{R} \to S^1$ ($\mathbf{R}$ as additive group) given by

$$\phi(x) = e^{2\pi i x} \qquad x \in \mathbf{R}$$

Moreover, the mapping ϕ is an open mapping, as is easily verified. Hence ϕ maps the open interval $(-\frac{1}{2}, +\frac{1}{2})$ on the line homeomorphically onto $S^1 - \{-1\}$; let ψ be its inverse on that set. We need two key lemmas.

(4.1) *Lifting Lemma. If σ is a path in S^1 with initial point* 1, *there is a unique path σ' in* $\mathbf{R}$ *with initial point* 0 *such that $\phi \circ \sigma' = \sigma$.*

(4.2) *Covering Homotopy Lemma. If also τ is a path in S^1 with the initial point* 1 *such that*

$$F : \sigma \simeq \tau \quad \text{rel } (0, 1)$$

then there is a unique $F' : I \times I \to \mathbf{R}$ such that

$$F' : \sigma' \simeq \tau' \quad \text{rel } (0, 1)$$
$$\phi \circ F' = F$$

Proof: We prove both lemmas at the same time. Let Y be either I or $I \times I$, $f: Y \to S^1$ either σ or F, $0 \in Y$ either 0 or (0, 0). Since Y is compact, f is uniformly continuous, so there exists $\varepsilon > 0$ such that $|y - y'| < \varepsilon \Rightarrow |f(y) - f(y')| < 1$; in particular for such y and y', $f(y) \neq -f(y')$, so $\psi(f(y)/f(y'))$ is defined. We can find N so large that $|y| < N\varepsilon$ for all $y \in Y$. Set

$$f'(y) = \psi\left(f(y)/f\left(\frac{N-1}{N}y\right)\right)$$

$$+ \psi\left(f\left(\frac{N-1}{N}y\right)\Big/ f\left(\frac{N-2}{N}y\right)\right)$$

$$+ \circ\circ\circ + \psi\left(f\left(\frac{1}{N}y\right)/f(0)\right)$$

Then f' is continuous $Y \to \mathbf{R}$, $f'(0) = 0$, and $\phi \circ f' = f$.

If we had $f'': Y \to \mathbf{R}$, $f''(0) = 0$, and $\phi \circ f'' = f$, then $f' - f''$ would be a (continuous) map of Y into the kernel of ϕ, i.e., into $\mathbf{Z}$. Since Y is connected, $f' - f''$ is constant, hence $f' = f''$.

In the case $Y = I \times I$, $f = F$, $f' = F'$, we see $F': \sigma \simeq \tau$. In fact, the homotopy is relative to (0, 1), for on $0 \times I$, $\phi \circ F' = F = 1$, hence $F'(0 \times I) \subset \mathbf{Z}$, so by connectedness again, $F'(0 \times I) = 0$. Similarly $F'(1 \times I)$ is constant. (For another proof, see 5.1–5.3.) ■

(4.3) *Corollary. The end point of σ' depends only on the homotopy class of σ.*

Define a map $\chi: \pi_1(S^1, 1) \to \mathbf{Z}$ by $\chi[\sigma] = \sigma'(1)$. We have just shown χ is well defined. It is a homomorphism: Given $[\sigma]$, $[\tau] \in \pi_1(S^1, 1)$. Let $m = \sigma'(1)$, $n = \tau'(1)$. Let τ'' be the path from m to $m + n$ in $\mathbf{R}$ given by $\tau''(s) = \tau'(s) + m$. Then $\phi \circ \tau'' = \tau$ also, so $\sigma'\tau''$ is the lifting of $\sigma\tau$ with initial point 0; its end point is $m + n$. Hence $\chi([\sigma][\tau]) = \chi[\sigma] + \chi[\tau]$.

χ is onto: Given n, define $\sigma'(s) = ns$. If $\sigma = \phi \circ \sigma'$, $\chi[\sigma] = n$.

χ is a monomorphism: Suppose $\chi[\sigma] = 0$, so σ' is a loop in $\mathbf{R}$ at 0. $\mathbf{R}$ being contractible, $\sigma' \simeq 0$ rel (0, 1), whence applying ϕ, $\sigma \simeq 1$ rel (0, 1), $[\sigma] = 1$. This proves the following theorem.

(4.4) *Theorem.*

$$\pi_1(S^1) \cong \mathbf{Z}$$

Remark: The only property of S^1 used in this proof is that it is a topological group, quotient of **R** by **Z**. The only property of **R** used in this proof is that it is a simply connected topological group. The only property of **Z** used is that it is a discrete subgroup of **R**. Thus exactly the same argument gives a more general result.

(4.5) *Theorem. If G is a simply connected topological group, H a discrete normal subgroup, then*

$$\pi_1(G/H, 1) \cong H$$

There is one detail to check: We must find an open neighborhood V of 1 in G which is mapped homeomorphically onto an open neighborhood of 1 in G/H by $\phi : G \to G/H$, so that we can use ψ as before. Since H is discrete, there is an open neighborhood U of 1 such that $U \cap H = \{1\}$. By continuity of the map $(g_1, g_2) \to g_1 g_2^{-1}$, there is another open neighborhood $V \subset U$ of 1 such that $g_1, g_2 \in V \Rightarrow g_1 g_2^{-1} \in U$. This is the V we need. ■

(4.6) *Exercise.* A discrete normal subgroup of a connected topological group is central. Hence $\pi_1(G/H)$ is commutative.

(4.7) *Corollary. The fundamental group of a torus is* $\mathbf{Z} \times \mathbf{Z}$.

For the torus is topologically $S^1 \times S^1$, hence is a topological group isomorphic to $(\mathbf{R} \times \mathbf{R})/(\mathbf{Z} \times \mathbf{Z})$. ■

The last result could have been derived in another way

(4.8) *Proposition. Given spaces X, Y, $x_0 \in X$, $y_0 \in Y$, we have* $\pi_1(X \times Y, (x_0, y_0)) \cong \pi_1(X, x_0) \times \pi_1(Y, y_0)$.

Proof: The isomorphism is obtained as follows. Let

$$\begin{array}{ccc} & X \times Y & \\ {\scriptstyle p_X} \swarrow & & \searrow {\scriptstyle p_Y} \\ X & & Y \end{array}$$

be the projection maps. They induce homomorphisms

$$\begin{array}{ccc} & \pi_1(X \times Y, (x_0, y_0)) & \\ {\scriptstyle (p_X)_*} \swarrow & & \searrow {\scriptstyle (p_Y)_*} \\ \pi_1(X, x_0) & & \pi_1(Y, y_0) \end{array}$$

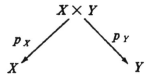

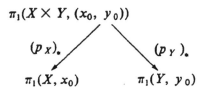

hence a homomorphism $((p_X)_*, (p_Y)_*)$ of $\pi_1(X \times Y, (x_0, y_0))$ into $\pi_1(X, x_0) \times \pi_1(Y, y_0)$. This homomorphism is an isomorphism, because it has the following inverse: Given loops σ at x_0, τ at y_0, assign to the pair $([\sigma], [\tau])$ the class of the loop (σ, τ) at (x_0, y_0) defined by

$$(\sigma, \tau)(t) = (\sigma(t), \tau(t)) \quad \text{for all} \quad t \in I$$

We leave the details as an exercise, as well as the verification that the isomorphism is *functorial* in (X, Y). ■

As an application our theorem and as an illustration of the method of algebraic topology, we prove the following theorem.

(4.9) *Theorem. The circle is not a retract of the closed unit disc.*

This means there is no map f of E^2 onto S^1 whose restriction to S^1 is the identity. Suppose we had such an f. Let $i: S^1 \to E^2$ be the inclusion map, so $fi =$ identity. Applying the fundamental group functor we get

$$\underbrace{\pi_1(S^1, 1) \xrightarrow{i_*} \pi_1(E^2, (1, 0)) \xrightarrow{f_*} \pi_1(S^1, 1)}_{\text{identity}}$$

But this means $\mathbf{Z} \to 0 \to \mathbf{Z}$ is the identity, which is impossible. ■

Note: In fact, there is no map $f: E^2 \to S^1$ such that $fi \simeq$ identity, for if there were, $f_* i_*$ would be an isomorphism (by 3.5), which is impossible.

(4.10) *Exercise*. Prove the circle is a strong deformation retract of the closed disc minus the origin. (A subspace A of a space X is a *strong deformation retract* of X if there is a homotopy $F_t: X \to X$ such that $F_0 =$ identity of X, $F_t | A =$ identity of A for all t, F_1 maps X into A.)

(4.11) *Corollary. Any continuous map of the closed disc into itself has a fixed point.*

This is the case $n = 2$ of the *Brouwer Fixed Point Theorem*, to be proved later for all n. (Exercise: Do the case $n = 1$.) Suppose $f: E^2 \to E^2$ has no fixed point. For any $x \in E^2$, join x to $f(x)$ by a line; move along this line in the direction from $f(x)$ to x until the point $r(x)$ on S^1 is reached. Then r is a retraction of E^2 on S^1, contradiction. ■

(4.12) *Exercise*. Suppose the space X is the union of two open sets U and V, such that $U \cap V$ is nonempty and pathwise connected, and U, V are each simply connected. Then X is simply connected. (This is a special case of Van

Kampen's Theorem, which says that $\pi_1(X)$ is "the amalgamated sum" of $\pi_1(U)$ and $\pi_1(V)$; cf. Crowell and Fox [16] or Massey [67].

On occasion, we shall use the following special cases of the Seifert-Van Kampen theorem for $X = U \cup V$ where U, V are open and $U \cap V$ is not empty and is arc connected.

a) If $U \cap V$ is simply connected, then $\pi_1(X)$ is the free product of $\pi_1(U)$, $\pi_1(V)$.

b) If U is simply connected, then $\pi_1(X)$ is the quotient of $\pi_1(V)$ by the smallest normal subgroup containing the image of $\pi_1(U \cap V)$ under inclusion $U \cap V \to V$.

(4.13) *Exercise*. The n-sphere S^n is simply connected for $n \geq 2$.

(4.14) *Exercise*. Show that every 3×3 matrix with positive real entries has an eigenvector with positive eigenvalue. Suggestion: Consider the triangle $T = \{x + y + z = 1; x, y, z \geq 0\}$ and the self-map of T obtained by composing the associated linear transformation with central projection onto T. Apply (4.11).

Remark. Other results from linear algebra have topological proofs. Material in chapter 16 can be used to show that every invertible linear transformation of R^{2n+1} has an invariant one dimensional subspace.

(4.15) *Exercise*. Suppose $A \subset X$ is a retract and $\pi_1(A)$ is normal in $\pi_1(X)$. Show $\pi_1(X) \cong \pi_1(A) \times \pi_1(X)/\pi_1(A)$. Referring to the figure in exercise (12.12), show that the circle represented by arc b is not a retract of the Klein bottle. Suggestion: use (4.12)(b) to calculate the fundamental group of the Klein bottle.

5. Covering Spaces

We have seen that the technique for determining $\pi_1(S^1)$ generalizes to arbitrary topological groups which can be represented as a quotient of a simply connected group by a discrete subgroup. We now try to generalize to a space X without a group structure, by representing X as a quotient space of a simply connected space $\widetilde{X}$ with the fibres of $\widetilde{X} \to X$ discrete.

Definition: $E \xrightarrow{p} X$ is a *covering space* of X if every $x \in X$ has an open neighborhood U such that $p^{-1}(U)$ is a disjoint union of open sets S_i in E, each of which is mapped homeomorphically onto U by p. Such U are said to be *evenly covered*, and the S_i are called *sheets* over U.

As immediate consequences we see:

(1) the fibre $p^{-1}(x)$ over any point is discrete;

(2) p is a local homeomorphism;

(3) p maps E onto X and X has the quotient topology from E.

Condition (2) tells us that X and E have the same properties *locally*, e.g., X is locally connected iff E is, etc.

We show that this definition captures the essentials of the situation $\mathbf{R} \to S^1$ (except that for greater generality we do not assume E to be simply connected) by proving the analogues of (4.1) and (4.2).

(5.1) *Unique Lifting Theorem. Let* $(E, e_0) \xrightarrow{p} (X, x_0)$ *be a covering space with basepoints,* $(Y, y_0) \xrightarrow{f} (X, x_0)$ *any map. Assume* Y *connected. If there is a map* $(Y, y_0) \xrightarrow{f'} (E, e_0)$ *such that* $pf' = f$, *it is unique.*

Proof: Suppose $f'' : (Y, y_0) \to (E, e_0)$, $pf'' = f$. Let

$$A = \{ y \in Y \mid f'(y) = f''(y) \}$$

$$D = \{ y \in Y | \; f'(y) \neq f''(y) \}$$

Then Y is the disjoint union of A and D, and $y_0 \in A$. We will show both sets are open, so by the connectedness of Y, D is empty.

Given $y_1 \in Y$. Let U be a neighborhood of $f(y_1)$ evenly covered by p. If $y_1 \in A$, $f'(y_1) = f''(y_1)$ lies on some sheet S of $p^{-1}(U)$. Hence $f'^{-1}(S) \cap f''^{-1}(S)$ is an open neighborhood of y_1 contained in A. If $y_1 \in D$, $f'(y_1)$ lies on some sheet S_1, $f''(y_1)$ on some other sheet S_2. Hence $f'^{-1}(S_1) \cap f''^{-1}(S_2)$ is an open neighborhood of y_1 contained in D. ■

(5.2) *Path Lifting Theorem. For $(E, e_0) \xrightarrow{p} (X, x_0)$ as above, if σ is a path in X with initial point x_0, there is a unique path σ'_{e_0} in E with inital point e_0 such that $p\,\sigma'_{e_0} = \sigma$.*

Proof: The uniqueness of σ'_{e_0} follows from the previous theorem.

Case 1: The whole space X is evenly covered. If e_0 lies on the sheet S, and ψ is the homeomorphism $X \to S$ inverse to $p | S$, then $\sigma'_{e_0} = \psi \circ \sigma$ is the desired lifting.

General case: By definition of covering space and compactness of I, we can partition I by $0 = t_0 < t_1 \circ\circ\circ < t_n = 1$ in such a way that σ maps the entire closed interval $[t_i, t_{i+1}]$ into an evenly covered neighborhood of $\sigma(t_i)$, for all i. By Case 1, we can lift $\sigma | [0, t_1]$ to a map $\sigma_1 : [0, t_1] \to E$ such that $\sigma_1(0) = e_0$. Assume by induction we can lift $\sigma | [0, t_i]$ to a map $\sigma_i : [0, t_i] \to E$ such that $\sigma_i(0) = e_0$. Then by Case 1 we can lift $\sigma | [t_i, t_{i+1}]$ to a map taking t_i into $\sigma_i(t_i)$; combining this map with σ_i gives σ_{i+1}. Then $\sigma_n = \sigma'_{e_0}$. ■

(5.3) *Covering Homotopy Theorem. $(E, e_0) \xrightarrow{p} (X, x_0)$ as above. Let (Y, y_0) be arbitrary and $f: (Y, y_0) \to (X, x_0)$ a map which has a lifting $f': (Y, y_0) \to (E, e_0)$.* Then *any homotopy $F: Y \times I \to X$ with $F(y, 0) = f(y)$ for all $y \in Y$ can be lifted to a homotopy $F': Y \times I \to E$ with $F'(y, 0) = f'(y)$ all $y \in Y$.*

(5.4) *Corollary. If σ, τ are paths in X with initial point x_0, and $\sigma \simeq \tau$* rel (0, 1), *then*

$$\sigma'_{e_0} \cong \tau'_{e_0} \text{ rel } (0, 1)$$

In particular σ'_{e_0} and τ'_{e_0} have the same end point.

Proof: The theorem is proved in several steps.

1. If all of X is evenly covered, it is obvious.
2. By definition of covering space and compactness of I, we can find for

each $y \in Y$ an open neighborhood N_y and a partition $0 = t_0 < t_1 \circ\circ\circ < t_n = 1$ (depending on y) such that F maps $N_{y} \times [t_i, t_{i+1}]$ into an evenly covered neighborhood of $F(y, t_i)$. By step 1 and the same inductive argument as before, we can lift F on $N_y \times I$ to a map $F' : N_y \times I \to E$ such that $F'(y', 0) = f'(y')$ for all $y' \in N_y$.

3. The liftings in step 2 on $N_y \times I$ and $N_{y'} \times I$ agree on $(N_y \cap N_{y'}) \times I$, hence we can piece them together to obtain the desired lifting F' on $Y \times I$: F or let $y_1 \in N_y \cap N_{y'}$. Then we have two lifitings of $F | y_1 \times I$ which agree at the point $(y_1, 0)$. By the unique lifting theorem ($y_1 \times I$ is connected!), these two liftings coincide. ■

(5.5) *Corollary.* $p_* : \pi_1(E, e_0) \to \pi_1(X, x_0)$ *is a monomorphism.*

If σ' is a loop at e_0 such that $p \circ \sigma' \simeq x_0$ rel $(0, 1)$ then lifting this homotopy gives $\sigma' \simeq e_0$ rel $(0, 1)$ (since $\sigma' = (p \circ \sigma')'_{e_0}$). ■

Warning: If σ is a loop at x_0, its lifting σ'_{e_0} with initial point e_0 need not be a loop at e_0 (look at the example $X = S^1$). However, its end point will be some point in the fibre $p^{-1}(x_0)$, and since that point depends only on the homotopy class of σ, we can define an operation of $\pi_1(X, x_0)$ on the fibre $p^{-1}(x_0)$ by

$$e \cdot [\sigma] = \sigma_e'(1)$$

for all $e \in p^{-1}(x_0)$, $[\sigma] \in \pi_1(X, x_0)$. (A group G operates on a set S if there is given a map $(s, g) \to s \cdot g$ of $S \times G$ into S satisfying

$$s \cdot 1 = s \qquad \text{all} \quad s \in S$$

$$s \cdot (gg') = (s \cdot g) \cdot g' \qquad \text{all} \quad s \in S, g, g' \in G)$$

Moreover, the stabilizer of a point $e_0 \in p^{-1}(x_0)$ is the subgroup $p_*\pi_1(E, e_0)$ of $\pi_1(X, x_0)$, since σ lifts to a loop at e_0 iff σ is the image by p of a loop at e_0. (In general, the *stabilizer* of $s_0 \in S$ is the subgroup $G_{s_0} = \{g \in G \mid s_0 \cdot g = s_0\}$ of G). Now if E is pathwise connected, $\pi_1(X, x_0)$ operates *transitively* (which in the general case means for all $s, s' \in S$, there exists $g \in G$ such that $s \cdot g = s'$), for a path from e to e' can be written σ_e', σ being its projection. Hence the different subgroups $p_*\pi_1(E, e)$ as e runs through $p^{-1}(x_0)$ are all *conjugate* (if $s_0 \cdot g = s_1$, $G_{s_0} = g\, G_{s_1}\, g^{-1}$).

(5.6) *Corollary. If E is pathwise connected, the map $[\sigma] \to e_0 \cdot [\sigma]$ induces a bijection of the set of all cosets $p_*\pi_1(E, e_0)[\sigma]$ onto the fibre. In particular,* *if $p^{-1}(x_0)$ is finite, the number of points in the fibre is equal to the index of the subgroup $p_*\pi_1(E, e_0)$.*

(5.7). *Exercise.* If E is pathwise connected, all the fibres have the same cardinality.

For any covering space $E \xrightarrow{p} X$, the group G of *covering transformations* is the group of all homeomorphisms of E which preserve the fibres:

$$\begin{array}{ccc} E & \xrightarrow{\phi} & E \\ & {\scriptstyle p}\searrow \quad \swarrow{\scriptstyle p} & \\ & X & \end{array} \qquad p\phi = p$$

(5.8) *Theorem. Given a covering space $(E, e_0) \xrightarrow{p} (X, x_0)$ with group of covering transformations G. If E is simply connected and locally pathwise connected, G is canonically isomorphic to $\pi_1(x, x_0)$.*

This achieves our objective of describing the fundamental group in terms of a simply connected covering space.

Proof: Given $\phi \in G$. Since E is simply connected, all paths from e_0 to $\phi(e_0)$ are homotopic rel (0, 1) (by 3.7), so if σ' is such a path, the element $[p \circ \sigma']$ of $\pi_1(X_1 x_0)$ depends only on e_0 and $\phi(e_0)$; we denote it by $\chi(\phi)$. Clearly χ is a homomorphism $G \to \pi_1(X, x_0)$. Now $\phi(e_0) = e_0 \cdot \chi(\phi)$. Hence $\chi(\phi) = 1$ implies ϕ leaves e_0 fixed. But for a connected covering space, a covering transformation is uniquely determined by its effect on one point (unique lifiting theorem—ϕ lifts p !); thus the only covering transformation with a fixed point is the identity, and χ is a monomorphism.

We use the assumption of local pathwise connectedness to prove χ is onto: Given $[\sigma] \in \pi_1(X, x_0)$. For any $e \in E$, let τ' be a path from e_0 to e, $\tau = p \circ \tau'$. Then $\tau^{-1}\sigma\tau$ is a loop at $x = p(e)$, and we define

$$\phi(e) = e \cdot [\tau^{-1}\sigma\,\tau]$$

Since E is simply connected, ϕ depends only on $[\sigma]$. Taking $e = e_0$ we see that $\chi(\phi) = [\sigma]$, provided we show ϕ is continuous. Note that for any point $e_1 \in E$, if τ' is a path from e_1 to e, $\tau = p \circ \tau'$, then $\phi(e)$ is the end point of the path $\tau'_{\phi(e_1)}$ (lifting of τ through $\phi(e_1)$). Now there is an open pathwise connected neighborhood U of $x_1 = p(e_1)$ which is evenly covered. e_1 is on some sheet S_1 over U and $\phi(e_1)$ on some sheet S'_1. For $e \in S_1$, we can join e_1 to e by a path τ' *in* S_1; then $\tau'_{\phi(e_1)}$ will be a path in S'_1. Hence its end point $\phi(e)$ is in S'_1. Since $\phi(e_1)$ has arbitrarily small neighborhoods of type S'_1, ϕ is continuous. ■

(5.9) *Exercise.* Assume only E is connected and locally pathwise connected. Let N be the normalizer of $p_*\pi_1(E, e_0)$ in $\pi_1(X, x_0)$. Modify the above argument to obtain a homomorphism of N onto G with kernel $p_*\pi_1(E, e_0)$. If $N = \pi_1(X, x_0)$, i.e., $p_*\pi_1(E, e_0)$ is a normal subgroup, the covering space is called *normal*; a necessary and sufficient condition is that G operate transitively on the fibre $p^{-1}(x_0)$.

(5.10) *Exercise.* Given a space E connected and locally pathwise connected. Let G be a group of homeomorphisms of E which operates *properly discontinuously* (i.e., for any $e \in E$, there is an open neighborhood V such that $V \cap gV = \phi$ for all $g \neq 1$ in G). Let $X = E/G$ *be the space of orbits,* $p : E \to X$ the map sending any e onto its orbit Ge. Then $E \xrightarrow{p} X$ is a covering space, G is its group of covering transformations, and $p_*\pi_1(E, e_0)$ is a normal subgroup of $\pi_1(X, x_0)$ for all $e_0 \in E$. (Note that any finite group operating without fixed points on a Hausdorff space operates properly discontinuously.)

This exercise tells us that if we know a simply connected covering space E of X and its group G of covering transformations, then not only do we know $\pi_1(X) \cong G$, but we also can recover X (up to homeomorphism) as E/G.

(5.11) *Example*: Projective n-space $\mathbf{P}^n$ is defined as the quotient space of S^n obtained by identifying antipodal points. The group of covering transformations of $S^n \to \mathbf{P}^n$ consists of the identity and the antipodal mapping only (because S^n is connected, $n > 0$), and since S^n is simply connected for $n \geq 2$, we have

$$\pi_1(\mathbf{P}^n) \cong \mathbf{Z}/2 \qquad n \geq 2$$

(Exercise. Show that $\mathbf{P}^1 \approx S^1$ (homeomorphic).)

(5.12) *Exercise.* Show that $\pi_1(\mathbf{P}^n)$, $n \geq 2$, is generated by the composition pg where $g : I \to S^n$ is any continuous map satisfying $g(0) = -g(1)$ and $p : S^n \to \mathbf{P}^n$ is defined in (5.11). Suggestion: Use (5.1) and (5.8).

A comprehensive introduction to the fundamental group and covering spaces has been written by W.S. Massey [67].

6. A Lifting Criterion

Unless otherwise stated, all spaces considered in this section are assumed to be *connected and locally pathwise connected.*

(6.1) *Theorem. Consider the situation*

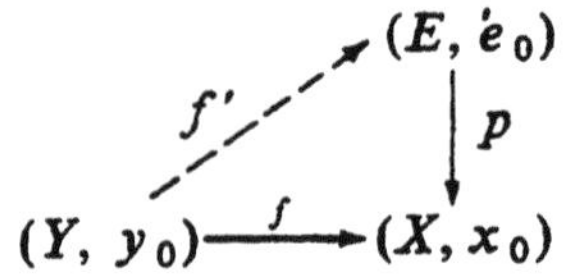

where p *is a covering space map and* f *is arbitrary, There exists a lifting* f' *of* f ($pf' = f$) *if and only if*

$$f_* \pi_1(Y, y_0) \subset p_* \pi_1(E, e_0)$$

Proof: Necessity follows from the functoriality of π_1. Conversely, we define f' as follows. For any $y \in Y$, choose a path σ from y_0 to y. Then $f\sigma$ is a path from x_0 to $x = f(y)$. Set

$$f'(y) = (f\sigma)'_{e_0}(1)$$

The hypothesis shows that this does not depend on the choice of σ. Moreover, we can remove the dependence on y_0: For any $y_1 \in Y$, let $e_1 = f'(y_1)$ and let τ be any path from y_1 to y. Then

$$f'(y) = (f\tau)'_{e_1}(1),$$

for there is a path σ_1 from y_0 to y_1 and

$$(f(\sigma_1 t))'_{e_0} = (f\sigma_1)'_{e_0}(f\tau)'_{e_1}$$

To show f' continuous at y_1, simply choose τ to lie in a suitable neighborhood of y_1 (Y is locally pathwise connected). Compare the proof of 5.8. ■

(6.2) *Exercise.* If, in the situation of (6.1), $f: Y \to X$ is also a covering space, and f' exists, then $f' : Y \to E$ is a covering space. (Show first that if $U \subset X$ is a path-connected open set evenly covered by p and by f, then the sheets of $p^{-1}(U)$ and $f^{-1}(U)$ are the path-connected components.)

(6.3) *Exercise.* Let $p : E \to X$ be a covering space, where X is connected and locally path connected, but E is not connected. Let C be a connected (hence path-connected) component of E. Then $p| C:C \to X$ is a covering space.

(6.4) *Corollary to Theorem. If Y is simply connected, the lifting f' always exists.*

(6.5) *Corollary. If $(E, e_0) \xrightarrow{p} (X, x_0)$, $(E', e'_0) \xrightarrow{p'} (X, x_0)$ are both simply connected covering spaces of X, then there is a unique homeomorphism $\phi:(E', e'_0) \to (E, e_0)$ such that $p\phi = p'$.*

This follows at once from (6.4) and (5.1).

Call two covering spaces of (X, x_0) *equivalent* if there is a homeomorphism ϕ as in (6.5). We have shown that if (X, x_0) has a covering space $(\widetilde{X}, \widetilde{x}_0) \to (X, x_0)$ such that $\widetilde{X}$ is simply connected, then $(\widetilde{X}, \widetilde{x}_0)$ is unique up to equivalence. We call it *the universal covering space* of (X, x_0), since all other covering spaces lie "below it," in the sense of (6.2) and (6.4).

The universal covering space need not exist, in general, for X is locally homeomorphic to $\widetilde{X}$, and so all "small" loops in X can be shrunk to a point. We thus have a necessary condition for the existence of $\widetilde{X}$: For any $x \in X$, there is a neighborhood U such that any loop in U based at x can be shrunk in X to x. (In the process of shrinking the loop, we may have to go outside of U.) A space X with this property is called *semi-locally simply connected.*

(6.6) *Example*: For any $n > 0$, let C_n be the circle with center $(1/n, 0)$ and radius $1/n$, and let

$$X = \bigcup_n C_n$$

Then X has no universal covering space, since the above condition does not hold at the origin.

(6.7) *Theorem. If X is semi-locally simply connected (and of course connected and locally path-connected), then X has a universal covering space.*

(6.8) *Corollary: Every connected manifold has a universal covering space (which is also a manifold).*

Recall that X is a *manifold of dimension n* if every point has an open neighborhood homeomorphic to an open ball in $\mathbf{R}^n$.

Proof (Theorem 6.7): Choose $x_0 \in X$. We consider all paths in X with inital point x_0. Write $\alpha \sim \beta$ if $\alpha(1) = \beta(1)$ and $\alpha \simeq \beta$ rel (0, 1). Let $\langle \alpha \rangle$ be the equivalence class of α. Take $\tilde{X}$ to be the set of all $\langle \alpha \rangle$'s, and set $p\langle \alpha \rangle = \alpha(1)$.

Take as base for a topology on $\tilde{X}$ the sets $\langle \alpha, V \rangle$, where V is an open neighborhood of $p\langle \alpha \rangle$, consisting of all $\langle \alpha\beta \rangle$, β a path in V with initial point $\alpha(1)$. If $\langle \alpha'' \rangle \in \langle \alpha, V \rangle \cap \langle \alpha', V' \rangle$, then $\langle \alpha'', V \rangle = \langle \alpha, V \rangle$, and $\langle \alpha'', V \cap V' \rangle \subset \langle \alpha, V \rangle \cap \langle \alpha', V' \rangle$, so they do form a base. p is continuous and open, since $p\langle \alpha, V \rangle$ is the path component of V containing $p\langle \alpha \rangle$.

Given $x \in X$, take a pathwise connected open neighborhood V such that any loop in V based at x can be shrunk to x in X. Then V is evenly covered: $p^{-1}(V)$ is the disjoint union of the $\langle \alpha, V \rangle$'s such that $p\langle \alpha \rangle \in V$, and clearly $p\langle \alpha, V \rangle = V$. If $p\langle \alpha\beta \rangle = p\langle \alpha\beta' \rangle$, then β and β' have the same end point, whence by choice of V, $\beta \simeq \beta'$ rel (0, 1), so $\langle \alpha\beta \rangle = \langle \alpha\beta' \rangle$.

Let $\tilde{x}_0$ be the class of the constant path c at x_0. We can join any point $\langle \alpha \rangle \in \tilde{X}$ to $\tilde{x}_0$ by a path (hence $\tilde{X}$ is pathwise connected). Let

$$\alpha_s(t) = \alpha(st) \qquad \text{all } s, t \in I$$

One checks easily that $s \to \langle \alpha_s \rangle$ is a path $\tilde{\alpha}$ in $\tilde{X}$ from $\tilde{x}_0$ to $\langle \alpha \rangle$; moreover $\tilde{\alpha}$ lifts α.

Finally, let τ be a loop in $\tilde{X}$ at $\tilde{x}_0$, and let $\alpha = p \circ \tau$. By uniqueness of liftings, $\tau = \tilde{\alpha}$. In particular $\tilde{\alpha}$ is a loop, so $\langle \alpha \rangle = \tilde{\alpha}(1) = \tilde{x}_0 = \langle c \rangle$. Hence $\alpha \sim c$, whence τ is homotopically trivial. Thus $\tilde{X}$ is simply connected. ■

(6.9) *Corollary. Under the same hypothesis, for every subgroup H of $\pi_1(X, x_0)$, there exists a covering space $(E, e_0) \xrightarrow{p} (X, x_0)$, unique up to equivalence, such that $H = p_* \pi_1(E, e_0)$.*

For let $\tilde{X} \to X$ be the universal covering space, G the group of its covering transformations, H' the subgroup of G corresponding to H under the isomorphism $G \cong \pi_1(X, x_0)$. Then we can take $E = \widetilde{X}/H'$ (with appropriate p). (See (5.10).) ∎

(6.10) *Exercise.* When $\widetilde{X}$ exists, the theory of covering spaces of X is entirely analogous to the Galois theory of fields, $\widetilde{X}$ being the analogue of the separable algebraic closure of a field, the group of covering transformations being the analogue of the Galois group. For example, in the situation

$$\begin{array}{ccc} & (E, e_0) & \\ H & \downarrow & \\ & (F, f_0) & G \\ K & \downarrow & \\ & (X, x_0) & \end{array}$$

where G, H, K are the respective groups of covering transformations, if all 3 covering spaces are normal (5.9), then $K \cong G/H$. (For Riemann surfaces, this analogy is not just formal: See Chevalley [13], last chapter.)

Returning to the case of topological groups, we obtain the following result.

(6.11) *Theorem. If X is a topological group, then for any covering space $E \xrightarrow{p} X$ and point e_0 in the fibre of the neutral element x_0 of X, there is a unique structure of topological group on E for which e_0 is the neutral element and p is a homomorphism.*

Proof: Let $m: X \times X \to X$ be the map $m(x_1, x_2) = x_1 x_2^{-1}$. We wish to lift $m \circ (p \times p)$

$$\begin{array}{ccc} (E \times E, (e_0, e_0)) & \xrightarrow{m'} & (E, e_0) \\ p \times p \Big\downarrow & & \Big\downarrow p \\ (X \times X, (x_0, x_0)) & \xrightarrow{m} & (X, x_0) \end{array}$$

so as to be able to define $e_1 e_2^{-1} = m'(e_1, e_2)$. m' is unique by the unique lifting theorem. The criterion for its existence is $m_*(p \times p)_* \pi_1(E \times E, (e_0, e_0)) \subset p_* \pi_1(E, e_0)$. (See (6.1).) This means the following: For any loops σ, τ at x_0, we can define a new loop $\sigma * \tau$ by

$$(\sigma * \tau)(t) = \sigma(t)\tau(t) \qquad \text{all } t \in I$$

the right side being multiplication in the group. We can also define

$$\bar{\tau}(t) = \tau(t)^{-1} \qquad \text{all } t \in I$$

The criterion then says that if $[\sigma], [\tau] \in p_* \pi_1(E, e_0)$ then $[\sigma * \bar{\tau}] \in p_* \pi_1(E, e_0)$. Since $p_* \pi_1(E, e_0)$ is a subgroup of $\pi_1(X, x_0)$, this will follow from a lemma.

Lemma. For any σ, τ *(loops at* x_0*)*

$$\sigma * \tau \simeq \sigma\tau \quad \text{rel}(0, 1)$$

$$\sigma * \tau \simeq \tau\sigma \quad \text{rel}(0, 1)$$

$$\bar{\tau} \simeq \tau^{-1} \quad \text{rel}(0, 1)$$

In particular the fundamental group of a topological group is commutative (as was seen before (4.6) in case the universal covering group exists).

Proof: Consider the homotopies

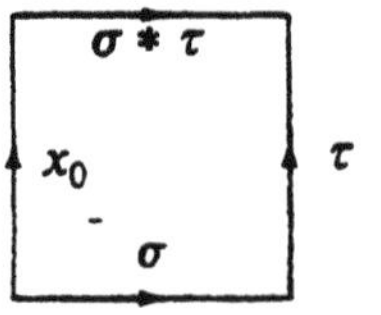

$$F(s, t) = \sigma(s)\tau(st)$$

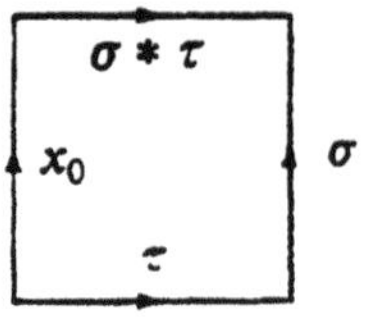

$$G(s, t) = \sigma(st)\tau(s)$$

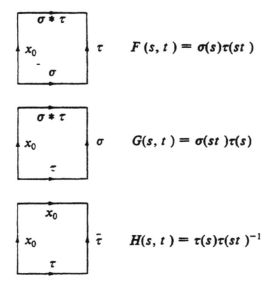

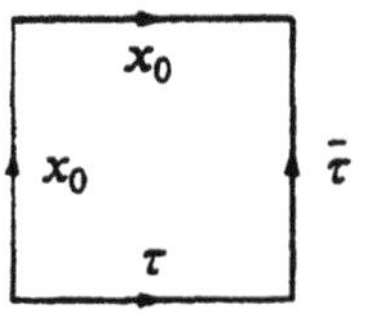

$$H(s, t) = \tau(s)\tau(st)^{-1}$$

and apply Lemma (3.3). ■

(6.12) *Example*: Let $X = \mathbf{SO}(3)$, the group of rotations of $\mathbf{R}^3$. Topologically X is projective 3-space (intuitively, a rotation is determined by an axis of rotation and a number θ, $-\pi \leq \theta \leq \pi$ the angle of rotation; thus X is homeomorphic to the closed ball of radius π in $\mathbf{R}^3$ with antipodal points on the boundary identified). Hence the universal covering group of $\mathbf{SO}(3)$ is S^3 with some structure of topological group. That group turns out to be the group of unit quaternions. For $n > 3$ it is still true that $\pi_1(\mathbf{SO}(n)) \cong \mathbf{Z}/2$, the universal covering group being the group of spinors (cf. Chevalley [14], Chap. 3). Another group of importance in physics is the proper Lorentz group. Topologically it is $\mathbf{P}^3 \times \mathbf{R}^3$, and its universal covering group is SL(2, C), 2 × 2 complex matrices of determinant 1 (cf. Gelfand *et al.* [25]).

(6.13) *Exercise.* Prove $\mathbf{SO}(3)$ is not homotopically equivalent to $S^1 \times S^2$.

(6.14) *Exercise.* We can use (6.13) to give an amusing proof that S^2 supports no continuous nonvanishing tangent vector field (a more sensible argument appears in chapter 16). Suppose there is $t: S^2 \to \mathbf{R}^3 - \{0\}$ such that the inner product $\langle x, t(x) \rangle = 0$ for all $x \in S^2$. We can assume $t(x)$ is a unit vector. Construct $\phi: S^1 \times S^2 \to SO(3)$ as follows. Let $n(x)$ be the cross product of x and $t(x)$. Then $A(x) = (x, t(x), n(x))$ represents an element of $SO(3)$. Define $\phi(\theta, x)$ to be the composition $R_\theta \circ A(x)$ where R_θ is rotation about the line x through an angle θ measured from $t(x)$ to $n(x)$ and $-\pi \leq \theta \leq \pi$. Show ϕ is continuous, one-to-one and onto, hence a homeomorphism, to obtain a contradiction with (6.13).

7. Loop Spaces and Higher Homotopy Groups

Let X^I be the set of all paths in X. If X is a metric space with metric d, we can define a metric d^* on X^I by

$$d^*(\sigma, \tau) = \sup_{t \in I} d(\sigma(t), \tau(t))$$

There is a method of topologizing X^I when X is not necessarily metrizable which gives the d^* topology when X is metric. Consider the sets

$$[K, U] = \{\sigma \mid \sigma(K) \subset U\} \qquad K \text{ compact} \subset I \qquad U \text{ open} \subset X$$

These form a subbase for a topology on X^I called *the compact-open topology*; thus the open sets are arbitrary unions of finite intersections of the $[K, U]$'s.

(7.1) *Exercise*. If X is metric, show this topology on X^I is the same as that defined by d^*.

The main property of this topology is stated here.

(7.2) *Proposition. The evaluation map* $\omega: X^I \times I \to X$ *given by* $\omega(\sigma, t) = \sigma(t)$ *is continuous.*

For the proof of this and the other purely point-set assertions below, see Dugundji [20] or Kelley [34], or better, prove them yourself as exercises. Note that the only property of I needed is local compactness in the strong sense, i.e., for every $t \in I$ and open V containing t, there is an open neighborhood W of t whose closure is compact with $\bar{W} \subset V$.

We will be concerned with the subspace $\Omega_{Xx_0} = \Omega_{x_0}$ of X^I consisting of all *loops* at x_0. This is a closed subspace of X^I if the point x_0 is closed in X, e.g., if X is Hausdorff.

(7.3) *Proposition.* $\sigma, \tau \in \Omega_{x_0}$ *are in the same path-connected components of* Ω_{x_0} *if and only if* $\sigma \simeq \tau$ rel (0, 1).

For a path f from σ to τ corresponds to a homotopy $F{:}\sigma \simeq \tau$ rel (0, 1) by the formula

$$f(s)(t) = F(s, t)$$

The factorization

$$I \times I \xrightarrow{f \times id} \Omega_{x_0} \times I \xrightarrow{\omega} X$$

shows that F is continuous if f is; the converse is an exercise. ■

(7.4) *Corollary.* $\pi_1(X, x_0)$ *is the set of path-connected components of* Ω_{x_0}.

The multiplication of loops defines a map $\Omega_{x_0} \times \Omega_{x_0} \to \Omega_{x_0}$ which is easily seen to be continuous. Let C be the constant loop at x_0; then $CC = C$. Define the maps L_C, $R_C{:}\Omega_{x_0} \to \Omega_{x_0}$ to be left and right multiplication by C, respectively.

(7.5) *Lemma.* L_C *(resp.* R_C*) is homotopic to the identity map of* Ω_{x_0} *relative to* $\{C\}$.

Proof: We know $C\sigma \simeq \sigma$ rel (0, 1) for every σ. Writing out explicitly the homotopy $F_\sigma(s, t)$ used to show this, one sees that it is a continous function of the pair (σ, t), hence a homotopy of L_C with the identity relative to $\{C\}$. ■

The properties continuous multiplication, $CC = C$, and (7.5) are usually expressed by saying that the pair (Ω_{x_0}, C) is an *H-space*. This is weaker than the notion of topological group, but is strong enough to prove the next theorem.

(7.6) *Theorem.* $\pi_1(\Omega_{x_0}, C)$ *is commutative.*

Proof: We modify the proof for topological groups. Let f, g be loops in Ω_{x_0} at C. Define $(f * g)(t) = f(t)g(t)$. We claim

$$fg \simeq f * g \simeq gf \text{ rel } (0, 1)$$

For example, for $fg \simeq f * g$, we use

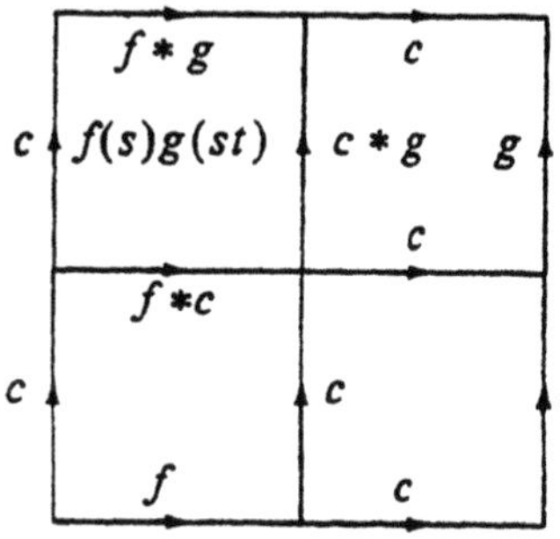

while for $gf \simeq f * g$, we use

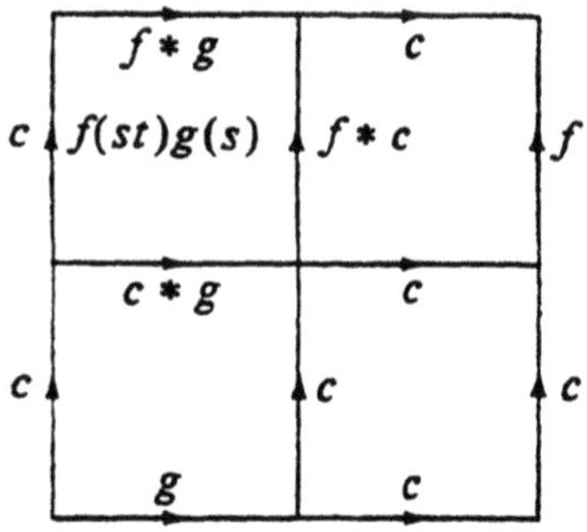

■

We now define the higher homotopy groups of (X, x_0) inductively by

$$\pi_n(X, x_0) = \pi_{n-1}(\Omega_{x_0}, C) \quad n \geq 2$$

(7.7) *Corollary. The higher homotopy groups are all commutative.*

(7.8) *Exercise.* If f is a loop in Ω_{x_0} at C, then defining $\bar{f}$ by $\bar{f}(s, t) = f(s)(t)$ we obtain a map $\bar{f}$ of the square I^2 into X which sends the entire perimeter into the point x_0. Conversely, given such an $\bar{f}$, this equation defines a loop f at C. If g is another loop at C, we see that $f \simeq g$ rel $(0, 1)$ iff $\bar{f} \simeq \bar{g}$ rel ∂I^2, where ∂I^2 is the boundary of the square. This gives a direct interpretation of $\pi_2(X, x_0)$. By induction, 2 can be replaced by n.

From another point of view, if we identify ∂I^n to a point (i.e., define an equivalence relation in I^n by $a \sim b$ iff a and b are both in ∂I^n, and take the quotient space under this relation), we get the n-sphere S^n with a distinguished point s_0. Thus $\pi_n(X, x_0)$ can be interpreted as the homotopy classes of maps $(S^n, s_0) \to (X, x_0)$.

(7.9) *Exercise.* If α is a path from x_0 to x_1, then α induces an isomorphism

$$(\alpha_*)_n : \pi_n(X, x_0) \to \pi_n(X, x_1)$$

for all n.

We wish to make π_n into a *functor*. Given a map $f:(X, x_0) \to (X', x_0')$. Define a map

$$\Omega(f):(\Omega_{x_0}, C) \to (\Omega_{x_0'}, C')$$

by

$$\Omega(f)(\sigma) = f \circ \sigma$$

One can easily show $\Omega(f)$ is continuous; hence, by induction we can define a homomorphism

$$(f_*)_n : \pi_n(X, x_0) \to \pi_n(X', x_0')$$

by

$$(f_*)_n = (\Omega(f)_*)_{n-1} \quad n \geq 2$$

One checks easily by induction on n that

(1) $((\text{identity})_*)_n = \text{identity}$

(2) $((gf)_*)_n = (g_*)_n (f_*)_n$

so that we do obtain a functor. Moreover homotopic maps $f \simeq g$ rel $\{x_0\}$ induce homotopic maps $\Omega(f) \simeq \Omega(g)$ rel $\{C\}$, so that $(f_*)_n = (g_*)_n$ for all n. (Exercise: If $f \simeq g$, but not relative to $\{x_0\}$, what is the relation between $(f_*)_n$ *and* $(g_*)_n$?)

(7.10) *Corollary. If X is contractible then $\pi_n(X, x_0)$ is trivial for all n.*

(7.11) *Exercise.* For all n, there is a canonical isomorphism

$$\pi_n(X \times Y, (x_0, y_0)) \cong \pi_n(X, x_0) \times \pi_n(Y, y_0)$$

(Determine $\Omega_{(x_0, y_0)}$ and apply (4.8).)

(7.12) *Theorem. If $p:(E, e_0) \to (X, x_0)$ is a covering space, then*

$$(p_*)_n : \pi_n(E, e_0) \to \pi_n(X, x_0)$$

is an isomorphism for all $n \geq 2$.

Proof: We use the interpretation of π_n as homotopy classes of maps from (S^n, s_0) into the given pointed space (7.8). To say $(p_*)_n$ is surjective means that any map $f:(S^n, s_0) \to (X, x_0)$ can be lifted to a map $f':(S^n, s_0) \to (E, e_0)$ such that $pf' = f$; since S^n is simply connected for $n \geq 2$ (4.13), this follows from (6.4). To say $(p_*)_n$ is injective means that if f' lifts f, g' lifts g, and $f \simeq g$ rel $\{s_0\}$, then $f' \simeq g'$ rel $\{s_0\}$; this follows from the covering homotopy theorem (5.3). ■

(7.13) *Corollary.* $\pi_n(\mathbf{P}^m) \cong \pi_n(S^m)$ *for* $n \geq 2$, *all* m.

(7.14) *Corollary.* $\pi_n(S^1) = 0$ *for* $n \geq 2$.

Proof: The universal covering space $\mathbf{R}$ of S^1 is contractible. ■

(7.15) *Note.* One might expect that

(a) $\pi_n(S^m) = 0$ for $n < m$.

(b) $\pi_n(S^n) \cong \mathbf{Z}$

(c) $\pi_n(S^m) = 0$ for $n > m$.

Now (a) and (b) are true, (c) is false. One can prove (a) and (b) directly by simplicial approximation (see Dugundji [20]) or one can first calculate the homology groups of S^m (15.4) and then apply a theorem of Hurewicz. (One form of the Hurewicz theorem states that if $n \geq 2$ and $\pi_q(X) = 0$ for all $q < n$, then $\pi_q(X) \cong H_q(X)$ for all $q \leq n$; see Eilenberg [21].) Property (c) is satisfied by the homology groups (15.5). H. Hopf gave the first example of a map $S^3 \to S^2$ not homotopic to a constant (see Hu [33], Chap. 3). Although much has been discovered in recent years (cf. [33], last chapter), the complete determination of the higher homotopy groups of spheres remains one of the major unsolved problems in algebraic topology. A comprehensive introduction to this subject was written by G. W. Whitehead [88].

(7.16) *Note.* Extremely fruitful generalizations of covering spaces are the notions of *fibre space* and *fibre bundle*. There is an infinite exact sequence of homotopy groups attached to any fibre space. However, the homology theory of fibre spaces is more complicated, requiring the use of *spectral sequences*. See Spanier [52], Chaps. 2, 5, 7, and 9.

Part II
SINGULAR HOMOLOGY THEORY

Introduction to Part II

We next develop the theme of *bounding* introduced in the opening passages of Part I. To help fix ideas, the reader might glance at the figure accompanying exercise (12.11) p. 69. Neither curve δ nor γ_1 is deformable to a point on this surface. However δ disconnects the surface, while γ_1 does not. Note also, that the three curves, γ_1, γ_2, γ_3, may be regarded as the complete boundary of a piece of the surface.

The mathematical exploitation of such observations is a delicate matter, because there turn out to be two different approaches. One method leads to "classical" homology theory while the other leads to a theory (bordism) having many features in common with homology theory but differing in fundamental ways. It was not until the 1950's that the true nature of this dichotomy was understood.

In the early stages of algebraic topology, it was regarded as fundamental to replace a space by a rigid combinatorial gadget known as a "simplicial complex" and develop the theory on these objects. Besides making a theory possible, simplical complexes provided a convenient means for discussing manifolds. The Poincaré duality theorem is one of the earliest theorems of the subject.

The simplicial techniques were gradually modified until it was possible (Eilenberg [21]) to introduce singular homology in a topologically invariant manner in a few pages. However, the price paid for this improvement was the loss of an explicit formulation permitting calculation.

A signal triumph was the introduction, by Eilenberg and Steenrod [23], of axioms for homology functors. These gave the subject conceptual coherence and elegance, at the same time providing ready access to calculations. It is along this line that part II is organized.

The basic definitions are given in chapter 9. The fundamental theorems are proved in chapters 11, 14 and 15. These might be regarded as the precise mathematical formulation of the invariance properties expected of homology. The material in chapter 10 is used for the proof of homotopy invariance. Many instructors prefer a more direct proof of homotopy invariance through the construction of "the prism operator". This approach is outlined in exercise (11.7) (consult the earlier *Lectures on Algebraic Topology* for more details).

In Section 12, an important result, due to Poincaré, is proved. Here appears one mathematical connection between the ideas of deformation and bounding.

Special attention is paid to calculation. The ability to calculate is important for applications of algebraic topology. The main tool for our calculations is exact sequences. Their manipulation involves both algebraic and geometric information. When first encountered, they may seem rather formidable. The second author well remembers when the term "diagram chase" was an open invitation to do something else.

We introduce the calculations in a gradual way. Thus the general theory of singular homology is developed to the point where exact sequences first enter. At this stage, some of the standard algebraic manipulations are introduced (five lemma and direct sum lemma) along with geometric illustrations. The key to such calculation is pattern recognition combined with the *formal* process of "chasing around a diagram". In fact, in this text, if an argument involving a diagram is not formal (i.e., requires additional geometric information along the way) we do not call it a diagram chase.

The chief device for introducing geometric information into calculation is the excision theorem in chapter 15. This theorem asserts the invariance of relative homology modules with respect to neglecting certain subspaces. It is analogous to the Noether isomorphism theorem $A/A \cap B \cong A + B/B$ for modules. When combined (in chapter 17) with a pattern recognition device known as the Barratt-Whitehead lemma, a powerful and versatile tool, used both for calculations and theory, emerges: the Mayer-Vietoris sequence. It can be regarded as a generalization of the formula, $\text{card}(A \cup B) = \text{card}(A) + \text{card}(B) - \text{card}(A \cap B)$ for finite sets.

An important theoretical application of the machinery is made in chapter 18 to obtain the Jordan-Brouwer separation properties.

The reader will have noticed that the term "calculation" is being used in two senses. In particular, the above remarks did not address the issue of calculating the values of the homology modules, but rather their exploitation, once known. The key here is to describe spaces in ways that make application of the basic theorems possible. The material in chapters 16 and 19–21 is developed with this idea in mind. Projective spaces offer good examples of how naturally occurring spaces are viewed so as to effect calculations.

8. Affine Preliminaries

Euclidean space stripped naked has neither coordinates, nor addition, nor multiplication by scalars, as does $\mathbf{R}^n$; it has only points, lines, planes, etc., and when thought of this way, without any metric or vector space properties, it is referred to as *affine space*. More precisely, an affine space of dimension n over $\mathbf{R}$ is a set E on which the additive group $\mathbf{R}^n$ operates simply transitively. Thus for each pair of points, P, Q in E there is a unique vector v in $\mathbf{R}^n$ from Q to P:

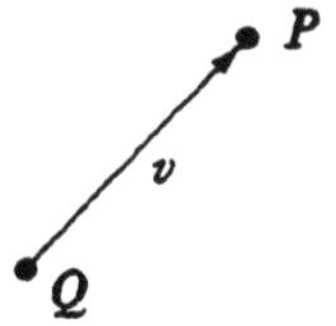

We write $v = P - Q$ and $P = Q + v$. However, the expression $P + Q$ is meaningless in this context. Nevertheless, certain additive expressions do make sense in affine geometry.

Let t be a real number. We define $tP + (1 - t)Q$ to be the unique point S such that $S - Q = t(P - Q)$ (this being a vector equation). If $P \neq Q$, the set of all such points for all $t \in \mathbf{R}$ is the *line* through P and Q (by definition). More generally, given points $P_0, \ldots, P_r$ and real numbers $a_0, \ldots, a_r$ such that $a_0 + \ldots + a_r = 1$, we can define the point

$$\sum_{i=0}^{r} a_i P_i$$

as the unique S such that

$$S - P_0 = \sum_{i=1}^{r} a_i(P_i - P_0).$$

If $P_0, \ldots, P_r$ are *independent* (meaning the vectors $P_1 - P_0, \ldots, P_r - P_0$ are linearly independent), the set of all such S is an affine space of dimension r called the *span* of $P_0, \ldots, P_r$. Each point S in the span has a unique set of coordinates $(a_0, \ldots, a_r)$ called its *barycentric coordinates* relative to $P_0, \ldots, P_r$. These coordinates are arbitrary except for the equation $a_0 + \ldots + a_r = 1$.

Given distinct P, Q, the points $tP + (1 - t)Q$ on the line through P and Q which satisfy $0 \leq t \leq 1$ form the *line segment* through P and Q. A subset of affine space is *convex* if it contains the line segment through any two of its points.

More generally, given independent points $P_0, \ldots, P_r$, the points S in their span having all non-negative barycentric coordinates form the r-dimensional *geometric simplex* spanned by $P_0, \ldots, P_r$. It is equal to the *convex hull* of the set $P_0, \ldots, P_r$ (smallest convex set containing $P_0, \ldots, P_r$). Prove this as an exercise. Intuitively, the point with barycentric coordinates $(a_0, \ldots, a_r)$, all $a_i \geq 0$, is the center of mass when we assign mass a_i to the point P_i. Taking all $a_i = 1/(r + 1)$ gives *the barycenter* of the simplex.

A function f from one affine space E to another E' will be called an *affine map* if

$$f(tP + (1 - t)Q) = tf(P) + (1 - t)f(Q)$$

for all points P, Q and real numbers t. If $f(P) \neq f(Q)$, f sends the line through P, Q into the line through $f(P)$, $f(Q)$, respecting barycentric coordinates. One can show (exercise) that such a map f can always be obtained as follows: Let $\bar{f}$ be a linear transformation from the vector space $\mathbf{R}^n$ to $\mathbf{R}^n$. Choose a point 0 in E and a point 0′ in E'. Then

$$f(P) = 0' + \bar{f}(P - 0) \qquad \text{all } P \in E$$

is an affine map. Conversely, given f, choose 0; for any vector v, set $P = 0 + v$, so that $\bar{f}$ is determined by the equation

$$\bar{f}(v) = f(P) - f(0)$$

Thus if $E = E'$, an affine map is just a linear transformation plus a translation.

Choosing an origin 0 sets up the one-to-one affine correspondence $P \rightarrow 0 + v$ between E and $\mathbf{R}^n$. We topologize E so that this correspondence becomes a homeomorphism. Then clearly affine maps are continuous.

If $P_0, \ldots, P_n$ are independent points which span E, the affine map f is uniquely determined by its effect on these points, since (exercise)

$$f\left(\sum_0^n a_i P_i\right) = \sum_0^n a_i f(P_i)$$

Conversely, given any points $f(P_0), \ldots, f(P_n)$ in E', this equation defines an affine map $f: E \rightarrow E'$.

9. Singular Theory

We take a countably infinite product $\mathbf{R}^\infty$ of copies of $\mathbf{R}$, and consider the vectors

$$E_0 = (0, 0, \ldots, 0, \ldots),$$
$$E_1 = (1, 0, \ldots, 0, \ldots),$$
$$E_2 = (0, 1, \ldots, 0, \ldots), \text{ etc.}$$

We identify $\mathbf{R}^n$ with the subspace having all components after the n-th equal to 0. We let, for any $q \geq 0$, Δ_q denote the q-dimensional geometric simplex spanned by $E_0, \ldots, E_q$, called *the standard (geometric) q-simplex*. Thus Δ_0 is a point, Δ_1 the unit interval, Δ_2 a triangle (including its interior), Δ_3 a tetrahedron, etc.

(9.1) If $P_0, \ldots, P_q$ are points in some affine space E, $(P_0 \ldots P_q)$ will denote the restriction to Δ_q of the unique affine map $\mathbf{R}^q \to E$ taking E_0 into $P_0, \ldots, E_q$ into P_q. Thus $(E_0 \ldots E_q)$ is the identity map of Δ_q, which will be denoted δ_q.

Given a space X, a *singular q-simplex* in X is a map $\Delta_q \to X$. Thus for $q = 0$ it can be identified with a point in X; for $q = 1$ it is a path in X; for $q = 2$ it is a (continuous) map of the standard triangle into X, etc. The affine map $(P_0 \ldots P_q)$ is a singular q-simplex in the affine space E.

We will add and subtract these singular q-simplexes in a purely formal manner. We also give ourselves a commutative unitary ring R and we will multiply by scalars from R (main examples: $R = \mathbf{Z}$, $R = \mathbf{R}$, $R =$ a finite field). Precisely, define $S_q(X)$ to be the free R-module generated by all the singular q-simplexes. The elements of $S_q(X)$ are formal linear combinations

$$\sum_\sigma v_\sigma \sigma$$

where σ runs through singular q-simplexes, and the coefficients v_σ are from

R. (The only way such a sum can vanish is if all the coefficients are 0.) These sums are called *singular q-chains*.

One should take care to recognize the distinction between singular q-simplices (maps $\sigma:\Delta_q \to X$) and singular q-chains (formal linear combinations of singular q-simplices). In particular $-\sigma$ is a singular q-chain. Furthermore, one should not confuse a singular q-simplex with its image. For example $\sigma:\Delta_1 \to X$ and $\sigma':\Delta_1 \to X$ defined by

$$\sigma'(t) = \begin{Bmatrix} \sigma(2t) & 0 \leq t \leq \frac{1}{2} \\ \sigma(1) & \frac{1}{2} \leq t \leq 1 \end{Bmatrix}$$

are different elements of $S_1(X)$.

For $q > 0$, define $F_q^i:\Delta_{q-1} \to \Delta_q$, for $0 \leq i \leq q$, to be the affine map

$$(E_0 \ldots \hat{E}_i \ldots E_q)$$

where $\hat{E}_i$ means "omit E_i"; in other words,

$$F_q^i(E_j) = \begin{cases} E_j & j < i, \\ E_{j+1} & j \geq i. \end{cases}$$

For an arbitrary singular q-simplex σ in a space X define *the i-th face* $\sigma^{(i)}$ of σ to be the singular $(q-1)$-simplex $\sigma \circ F_q^i$. Thus F_q^i is the i-th face of δ_q, and if $\sigma = (P_0 \ldots P_q)$, X affine,

$$\sigma^{(i)} = (P_0 \ldots \hat{P}_i \ldots P_q).$$

Diagramatically, $q = 2$ is shown:

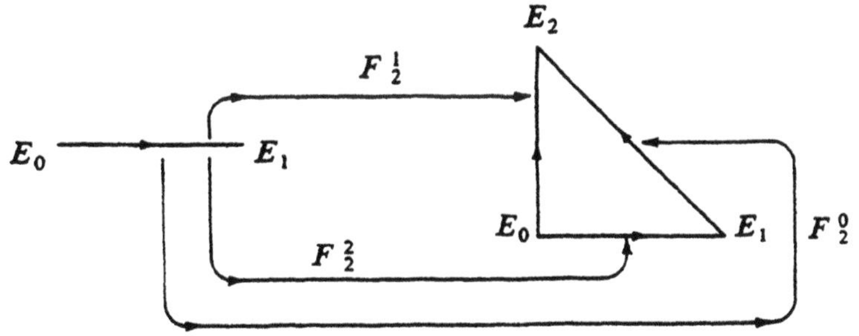

Thus F_q^i maps Δ_{q-1} homeomorphically and affinely onto the set-theoretic face of Δ_q opposite to the vertex E_i.

We now define the *boundary* of a singular q-simplex σ to be the singular $(q-1)$-chain

$$\partial(\sigma) = \sum_{i=0}^{q}(-1)^i \sigma^{(i)}.$$

(In the previous diagram, $\partial\delta_2$ is the sum of the edges of the triangle with signs chosen so that starting with E_0 we travel around in a loop; however, $\partial\delta_2$ is *not* a loop, but a formal sum with $\pm$ signs of 3 paths!) In the special case $\sigma = (P_0 \ldots P_q)$,

$$\partial(P_0 \ldots P_q) = \sum_{i=0}^{q}(-1)^i(P_0 \ldots \hat{P}_i \ldots P_q).$$

We extend ∂ to a module homomorphism $S_q(X) \to S_{q-1}(X)$ by linearity; thus

$$\partial(\Sigma v_\sigma \sigma) = \Sigma v_\sigma \partial(\sigma).$$

For $q = 0$, the boundary of a 0-chain is defined to be 0.

(9.2) *Proposition.* $\partial\partial = 0$.

Proof: It suffices to verify $\partial(\partial c) = 0$ when c is a singular q-simplex σ. (This type of remark, used repeatedly, will usually be omitted.) We derive the following result through calculation:

(9.3) *Lemma.* $F_q^i F_{q-1}^j = F_q^j \, F_{q-1}^{i-1}$ for $j < i$.

Then

$$\begin{aligned}\partial(\partial\sigma) &= \sum_{i=0}^{q}(-1)^i \partial(\sigma^{(i)}) \\ &= \sum_{i=0}^{q}(-1)^i \sum_{j=0}^{q-1}(-1)^j (\sigma \circ F_q^i) \circ F_{q-1}^j \\ &= \sum_{j<i=1}^{q}(-1)^{i+j}\sigma \circ (F_q^j \, F_{q-1}^{i-1}) \\ &\quad + \sum_{0=i\le j}^{q-1}(-1)^{i+j}\sigma \circ (F_q^i \, F_{q-1}^j)\end{aligned}$$

and everything cancels (set $i' = j, j' = i - 1$ in the first sum). ■

A singular q-chain c such that $\partial(c) = 0$ is called a *cycle*; if $c = \partial(c')$ for some $(q+1)$-chain c', c is called a *boundary*. Two q-chains which differ by a boundary are called *homologous*, written $c_1 \sim c_2$. By (9.2) the boundaries form a submodule B_q of the module Z_q of cycles; the quotient module Z_q/B_q is called *the q-th singular homology module* of X, denoted

$$H_q(X; R)$$

or simply $H_q(X)$ when the reference to R is understood.

(9.4) *Example*: X is a single point x. There is a unique singular q-simplex σ_q for each q (constant map on x). We have

$$\partial(\sigma_q) = \begin{cases} \sigma_{q-1} & q \text{ even} > 0, \\ 0 & q \text{ odd}, \end{cases}$$

$$Z_q = B_q = \begin{cases} 0 & q \text{ even} > 0, \\ S_q & q \text{ odd}, \end{cases}$$

so that $H_q = 0$ for all $q > 0$. However $Z_0 = S_0$ while $B_0 = 0$, so that $H_0 \cong R$, the isomorphism being $\nu\sigma_0 \to \nu$.

(9.5) *Proposition. Let* (X_k) *be the family of path connected components of* X. *Then there is a canonical isomorphism*

$$H_q(X) \cong \oplus_k H_q(X_k) \quad \text{all } q \geq 0.$$

(The *direct sum* $\oplus M_k$ of a family of R-modules is defined to be the submodule of the Cartesian product of the M_k's consisting of those families (m_k) such that at most finitely many m_k are different from 0.)

Proof: We have, in fact, an isomorphism

$$S_q(X) \cong \oplus_k S_q(X_k) \quad \text{all } q \geq 0$$

such that the boundary operates component by component. Namely, Δ_q being pathwise connected, a singular q-simplex σ maps Δ_q into some path component X_k. *Thus each* q-chain c decomposes uniquely into a sum

$$c = \sum_k c_k$$

where c_k is a singular q-chain on X_k. ■

(9.6) *Proposition.* *$H_0(X)$ is a free R-module on as many generators as there are path components of X.*

Proof: By (9.5), we may assume X is path connected. Choose a base point x_0 in X. For any $x \in X$, let σ_x be a path from x_0 to x, so that $\partial(\sigma_x) = x - x_0$. Given a 0-chain

$$c = \sum_x v_x x$$

we claim c is a boundary if and only if the sum of its coefficients is 0. If the latter happens,

$$c = \sum_x v_x x - (\sum_x v_x)x_0 = \partial(\sum v_x \sigma_x).$$

The converse is clear. Now every 0-chain is a cycle. The map sending c onto the sum of its coefficients is a homomorphism of S_0 onto R with kernel B_0, hence

$$H_0(X) \cong R.$$

This proposition is the key to all the connectedness theorems in algebraic topology (see sections 18 and 27).

(9.7) *Exercise. The reduced* 0-th *homology module* $H_0^*(X)$ is obtained by defining a different boundary operator on 0-chains:

$$\partial^*(\sum_x v_x x) = \sum_x v_x.$$

Verify that $\partial^*\partial = 0$. $H_0^*(X)$ is the quotient of the kernel of ∂^* by the boundaries of 1-chains. If X is path connected, $H_0^*(X) = 0$, while if X has r path components, $r > 1$, $H_0^*(X)$ is a free R-module on $(r-1)$ generators, so it is called *reduced homology.*

For $q > 0$ we define $H_q^*(X) = H_q(X)$.

Consider now functorial properties. Let $f:X \to X'$. If σ is a singular q-simplex in X, $f \circ \sigma$ is one in X'. We obtain a homomorphism $S_q(f):S_q(X) \to S_q(X')$ by

$$S_q(f)(\sum v_\sigma \sigma) = \sum v_\sigma(f \circ \sigma).$$

Clearly

(i) $S_q(\text{identity}) = \text{identity}$,

(ii) $S_q(gf) = S_q(g)S_q(f)$.

Moreover, we get the following relationship:

(9.8) *Lemma.* $\partial S_q(f) = S_{q-1}(f)\partial$.

Immediate from $(f\circ \sigma)\circ F_q^i = f\circ(\sigma\circ F_q^i)$. ∎

Hence if z is a q-cycle on X, $\bar{z}$ its homology class, we obtain a homomorphism

$$H_q(f):H_q(X) \to H_q(X')$$

by $H_q(f)(\bar{z}) = \overline{S_q(f)(z)}$. Since (i) and (ii) also hold for $H_q(f)$, we see that H_q is a functor from the category of topological spaces to the category of R-modules, for each $q \geq 0$. *Thus the homology modules are topological invariants.*

(9.9) *Exercise.* Let X be a path component of X', f the inclusion map. Then $H_q(f)$ is a monomorphism, and there exists a homomorphism $P_q:H_q(X') \to H_q(X)$ such that $P_qH_q(f) =$ identity.

(9.10) *Maps between singular complexes.* Much of the technical work is the construction of maps between singular complexes. Since $S_q(X)$ is free on the singular q-simplices, a unique homomorphism is defined on $S_q(X)$ by specifying its values on the singular q-simplices. The following device is often used when the source and target are functors of the space X *and* when the properties of the desired homomorphism are essentially independent of spaces.

A singular q-simplex $\sigma \in S_q(X)$ can be written

$$\sigma = S_q(\sigma)(\delta_q)$$

where δ_q is the identity map of Δ_q. Then (for example, to fix ideas) a homomorphism

$$P^X:S_q(X) \to S_q(X)$$

can be defined for all X using a single value $P(\delta_q) \in S_q(\Delta_q)$ by the equation

$$P^X(\sigma) = S_q(\sigma)\,P(\delta_q).$$

The basic feature of this construction is a property called *naturality*; given $f:X \to Y$, then the diagram

(9.11)
$$\begin{array}{ccc} S_q(X) & \xrightarrow{P^X} & S_q(X) \\ {\scriptstyle S_q(f)}\downarrow & & \downarrow{\scriptstyle S_q(f)} \\ S_q(Y) & \xrightarrow{P^Y} & S_q(Y) \end{array}$$

commutes.

Proof. We compute on singular q-simplices $\sigma \in S_q(X)$. Then $P^Y S_q(f)(\sigma) = P^Y(f\sigma) = S_q(f\sigma)\,P(\delta_q) = S_q(f)S_q(\sigma)\,P(\delta_q) = S_q(f)\,P^X(\sigma)$. The general statement follows because $S_q(X)$ is freely generated by all σ. ∎

Conversely, a homomorphism $P^X\colon S_q(X) \to S_q(X)$ defined for all X and satisfying the naturality property (9.11) is characterized by its value $P^{\Delta_q}(\delta_q)$.

The naturality property is the precise meaning of the sense in which P is independent of X. The usual categorical terminology is to call P a *natural transformation* of the functor S_q. We shall usually suppress superscripts X from the notation.

The singular chain complex is discussed in more detail in the next chapter. The material is used in the proof of homotopy invariance, but remains in the background until the introduction of cohomology theory. The reader who prefers to go directly to homotopy invariance may do so after (10.6) and use exercise (11.7) for the proof.

(9.12) *Exercise.* Let A, B be distinct points in an affine space X. Let $\sigma, \tau\colon \Delta_1 \to X$ be the affine maps (A, B), (B, A). Then $\sigma \neq -\tau$. Show σ is homologous to $-\tau$. Suggestion: let $\rho \in S_1(X)$ be the constant map at A. Let $\omega, \rho' \in S_2(X)$ be the affine map (A, B, A) and the constant map at A, respectively. Show $\partial(\omega + \rho') = \tau + \sigma$.

(9.13) *Exercise.* Regard the torus T and the Klein bottle K as quotients of the affine space $I \times I$ obtained by identifications on opposite edges as pictured:

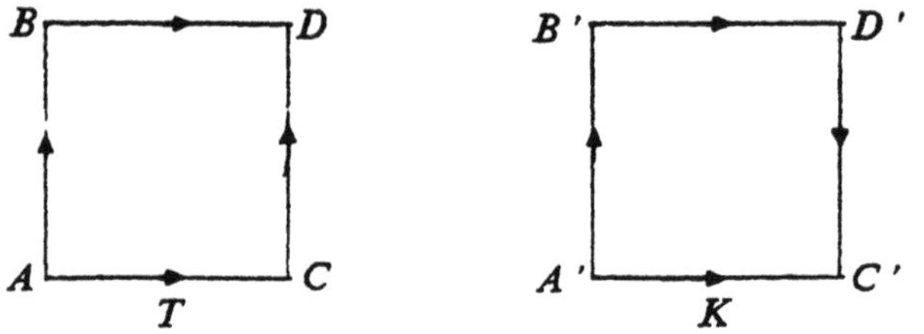

T K

Let $p, p' : I \times I \to T, K$ be the identification maps. Define maps $\sigma_1, \sigma_2 \in S_2(T)$ by $\sigma_1 = p \circ (A, B, D), \sigma_2 = p \circ (A, C, D)$; $\tau_1, \tau_2 \in S_2(K)$ by $\tau_1 = p' \circ (A', B', D')$, $\tau_2 = p' \circ (A', D', C')$; and $\omega \in S_1(K)$ by $\omega = p' \circ (A', B') = p' \circ (D', C')$. Show $\partial(\sigma_1 - \sigma_2) = 0$, $\partial(\tau_1 + \tau_2) = 2\omega$, and $\partial\omega = 0$. Thus if $R = \mathbf{Z}$, we have $\omega \in Z_1(K)$ and $2\omega \in B_1(K)$, whereas if $R = \mathbf{Z}/2\mathbf{Z}$, $\omega \in Z_1(K)$ and $\tau_1 + \tau_2 \in Z_2(K)$.

10. Chain Complexes

This chapter presents algebraic properties of chain complexes with the singular complex a central example. We abstract the central notions of the previous chapter.

(10.1) *Definition. A chain complex over R* is a sequence $C = \{C_q, \partial_q\}$ of free R-modules and homomorphisms $\partial_q : C_q \to C_{q-1}$ such that $\partial_q \partial_{q+1} = 0$,

$$C_{q+1} \xrightarrow{\partial_{q+1}} C_q \xrightarrow{\partial_q} C_{q-1}.$$

In most cases $C_q = 0$ if $q < 0$. An element of C_q has *dimension q*.

The singular complex of a space X is the example where $C_q = S_q(X)$ and ∂_q is the boundary map constructed in (9.1). We often abbreviate $\{S_q(X)\}$ by $S(X)$.

(10.2) *Definition.* A sequence of homomorphism $\{f_q\}$ with $f_q : C_q \to C'_q$ is a *chain map provided* $\partial'_q f_q = f_{q-1}\partial_q$

$$\begin{array}{ccc} C_q & \xrightarrow{f_q} & C'_q \\ {\scriptstyle \partial_q}\downarrow & & \downarrow{\scriptstyle \partial'_q} \\ C_{q-1} & \xrightarrow[f_{q-1}]{} & C'_{q-1} \end{array}.$$

For example, a map of spaces $f:X \to Y$ induces a chain map $S(f):S(X) \to S(Y)$ (9.8).

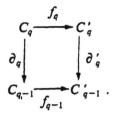

As in the topological case, we introduce $Z_q(C)$, $B_q(C)$, the submodules of C_q defined by

$$Z_q(C) = \text{Ker}\, \partial_q \qquad q\text{-cycles},$$
$$B_q(C) = Im\, \partial_{q+1} \qquad q\text{-boundaries}.$$

(10.3) *Definition.* The q-th *homology* module of C is defined by $H_q(C) = Z_q(C)/B_q(C)$.

By construction, $H_q(C)$ is an R-module. If $z \in Z_q(C)$ we write $\bar{z}$ for the corresponding element in $H_q(C)$.

A chain map $f:C \to C'$ sends cycles to cycles and boundaries to boundaries. Hence f induces a well-defined homomorphism

(10.4)
$$H_q(f):H_q(C) \to H_q(C'),$$
$$H_q(f)(\bar{z}) = \overline{f_q(z)}.$$

Thus H defines a functor from the category of chain complexes over R and chain maps to the category of R-modules and homomorphisms. The verification is an exercise; see the material following (9.8).

The motivation for the next definition will become apparent in the next chapter.

(10.5) *Definition.* A *chain homotopy* between chain maps

$$f = \{f_q:C_q \to C'_q\}$$

and $g = \{g_q:C_q \to C'_q\}$ is a sequence $D = \{D_q:C_q \to C'_{q+1}\}$ of homomorphisms such that $\partial'_{q+1}D_q + D_{q-1}\partial_q = f_q - g_q$.

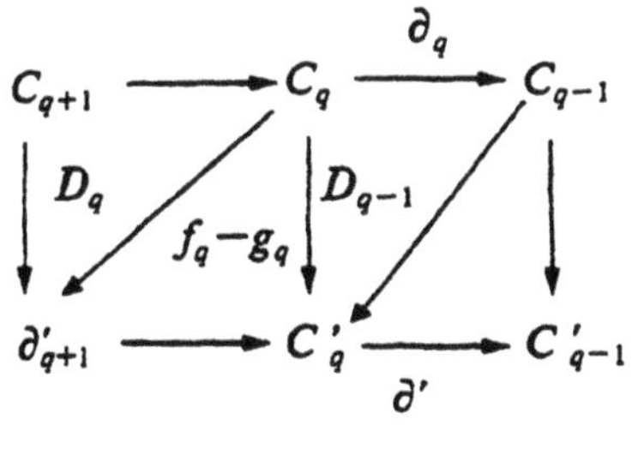

We write $f \simeq g$. (If $C_q = 0$ for $q < 0$, the equation reads $\partial'_1 D_0 = f_0 - g_0$.)

(10.6) *Proposition. Chain homotopic maps induce equal maps in homology.*

Proof. Let $\bar{z} \in H_q(C)$ be a class with representative $z \in Z_q(C)$. Then $f_q(z) - g_q(z) = \partial'_{q+1} D_q(z) \in B(C')$. Hence $H_q(f)(\bar{z}) = H_q(g)(\bar{z})$. ∎

The next sequence of ideas is motivated by properties of $S(P)$ where P is a point. In (9.4) we observed im ∂_{q+1} = ker ∂_q for $q \geq 1$ and $S_0(P)/\text{im}\,\partial_1 \cong R$. We abstract the first of these properties.

(10.7) *Definition.* A chain complex C is *acyclic* if $H_q(C) = 0$ for all q. This means im ∂_{q+1} = ker ∂_q for all q (im $\partial_1 = C_0$ if $C_q = 0$ for $q < 0$). The term *exact* is used synonymously.

Note that $S(P)$ is not acyclic according to (10.7). A systematic device for relating acyclicity to the properties of $S(P)$ is the following

(10.8) *Definition* (for C such that $C_q = 0$ if $q < 0$). An *augmentation over* R for C is an epimorphism $\varepsilon: C_0 \to R$ such that $\varepsilon\partial_1 = 0$,

$$C_1 \xrightarrow{\partial_1} C_0 \xrightarrow{\varepsilon} R.$$

This means im $\partial_1 \subset$ ker ε. Note the isomorphism $C_0/\ker \varepsilon \cong R$.

For example $S(X)$ has an augmentation $\varepsilon: S_0(X) \to R$ given by $\varepsilon = \partial^*$ (9.7). In the case $S(P)$ we have im ∂_1 = ker ε.

Given a chain complex C with augmentation $\varepsilon: C_0 \to R$, we form the *reduced chain complex* $\tilde{C} = \{\tilde{C}_q, \tilde{\partial}_q\}$ by setting $\tilde{C}_q = C_q$ for $q \geq 1$, $\tilde{C}_0 = \ker \varepsilon$, and $\tilde{\partial}_q = \partial_q$. Note $\tilde{\partial}_1: \tilde{C}_1 \to \tilde{C}_0$ since im $\partial_1 \subset$ ker ε. For example, exercise (9.7) asserts that the reduced homology of a space X is the homology of the reduced singular complex $\tilde{S}(X)$.

(10.9) *Definition.* A chain complex with augmentation is *acyclic* if the reduced chain complex is acyclic. An equivalent formulation is im ∂_{q+1} = ker ∂_q and im ∂_1 = ker ε.

(10.10) *Proposition. A complex $C = \{C_q, \partial_q\}$ with augmentation $\varepsilon: C_0 \to R$ is acyclic if and only if $H_q(C) = 0$ for $q > 0$ and $H_0(C) \cong R$.*

Proof. If C is acyclic, then im ∂_{q+1} = ker ∂_q and im ∂_1 = ker ε. Hence $H_q(C) = 0$ for $q \geq 1$ and $H_0(C) = C_0/\text{im}\,\partial_1 = C_0/\ker \varepsilon \cong R$. Conversely, if $H_q(C) = 0$ for $q > 0$, then im ∂_{q+1} = ker ∂_q for $q > 0$. The projection $\varepsilon: C_0 \to H_0(C)$ satisfies ker ε = im ∂_1 and is an augmentation if $H_0(C) \cong R$. ∎

Our next goal is to relate acyclicity with certain chain homotopies. Every complex with augmentation $\varepsilon: C_0 \to R$ has a (nonunique) right inverse $\eta: R \to C_0$ given by $\eta(1) = x$ where x satisfies $\varepsilon(x) = 1$. Then $\varepsilon\eta = 1$. We regard $\eta\varepsilon: C \to C$ as a chain map by defining $\varepsilon|\,C_q = 0$ for $q \geq 1$. For example, in

$S(P)$ we have $\eta(1) = \sigma_0$ and a chain homotopy $1 \simeq \eta\varepsilon$ given by $D_q(\sigma_q) = \sigma_{q+1}$ where $\sigma_q:\Delta_q \to P$ is the constant map. The verification of the equation $\partial_{q+1}D_q + D_{q-1}\partial_q = 1 - \eta\varepsilon$ is an exercise. More generally we have

(10.11) *Proposition. If* $1 \simeq \eta\varepsilon$ *then* C *is acyclic.*

Proof. The equation $\partial_{q+1}D_q + D_{q-1}\partial_q = 1 - \eta\varepsilon$ implies $\operatorname{im}\partial_{q+1} = \ker\partial_q$ for $q \geq 1$ and $\operatorname{im}\partial_1 = \ker\varepsilon$. Then C is acyclic by either (10.9) or (10.10).

We now introduce a geometric condition on a space X which will imply acyclicity (10.9) of $S(X)$.

(10.12) *Definition.* X is *aspherical* if every continuous $f:S^n \to X$ extends to $\bar{f}:E^{n+1} \to X$. Here S^n is the unit sphere in R^{n+1} and E^{n+1} the unit ball. We shall understand the definition to apply to homeomorphic images of the pair (E^{n+1}, S^n) as well.

Remark. Setting $n = 0$ in (10.12) shows X is arc-wise connected.

Remark. In some contexts, the term "aspherical" describes a space X whose universal covering space $\tilde{X}$ is aspherical in our sense.

Example. A convex subset X of euclidean space is aspherical. To see this, pick any point $Q \in X$. Represent points in E^{n+1} by "polar coordinates" (t, x) where $0 \leq t \leq 1, x \in S^n$ and $(0, x) = (0, x')$ is the origin. Points in S^n have coordinates of the form $(1, x)$. Given $f:S^n \to X$, define $\bar{f}:E^{n+1} \to X$ by $\bar{f}(t, x) = (1 - t)Q + tf(x)$.

Example. A contractible space is aspherical. Let $H:X \times I \to X$ be a contracting homotopy such that $H(x, 0) = x$, $H(x, 1) = Q$ for some point $Q \in X$. Define $\bar{f}:E^{n+1} \to X$ by $\bar{f}(t, x) = H(f(x), 1 - t)$.

Exercise. The topologists' sine curve is aspherical. This space is made up of the graph of $y = \sin(1/x)$, $0 < x \leq 1$, the line segment $\{(0, y) | -1 \leq y \leq 1\}$, and an arc from $(0, 0)$ to $(1, \sin 1)$ as pictured. Prove that x is not contractible. Suggestion: use local connectedness to show that the image of a sphere misses the nasty part.

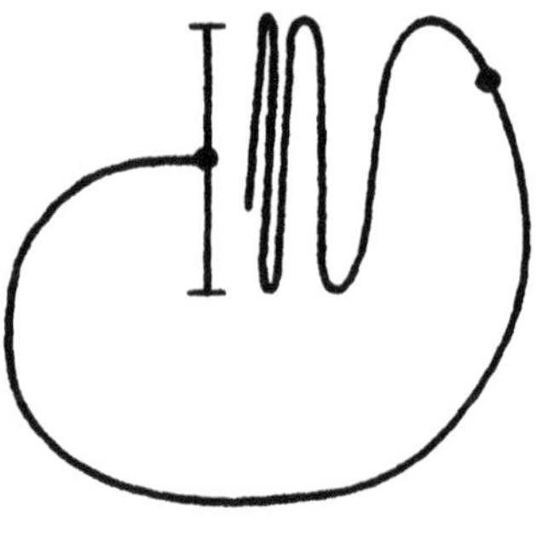

(10.13) *Theorem. If X is aspherical, then $S(X)$ is acyclic.*

Before proving this we recall a method for constructing continuous maps. If $X = A_1 \cup \ldots \cup A_k$ with A_i closed in X and we have continuous maps $f_i: A_i \to Y$ such that $f_i | A_i \cap A_j = f_j | A_i \cap A_j$, then a well-defined, continuous map $g: X \to Y$ is given by $g(x) = f_i(x)$ if $x \in A_i$.

Proof (suggested by A. L. Liulevicius). Let $\varepsilon: S_0(X) \to R$ be the augmentation and $\eta: R \to S_0(X)$ a right inverse. We construct a chain homotopy $1 \simeq \eta\varepsilon$. Let $\dot{\Delta}_{q+1} \subset \Delta_{q+1}$ be the boundary. Then $\dot{\Delta}_{q+1}$ is the union of q-faces $F^i_{q+1}(\Delta_q)$, $0 \leq i \leq q + 1$. The pair $(\Delta_{q+1}, \dot{\Delta}_{q+1})$ is homeomorphic to (E^{q+1}, S^q). Since X is aspherical, any map $\dot{\Delta}_{q+1} \to X$ extends to a map $\Delta_{q+1} \to X$. Let σ be a singular q-simplex of X. We define $D_q(\sigma) \in S_{q+1}(X)$ inductively on q and by specifying its values on each face of $\dot{\Delta}_{q+1}$. The definition is,

$$D_0(\sigma)\, F^0_1 = \sigma; \qquad D_0(\sigma)\, F^1_1 = \eta\varepsilon(\sigma)$$

and for

$$q \geq 1,\ D_q(\sigma)\, F^0_{q+1} = \sigma;\ D_q(\sigma)\, F^i_{q+1} = D_{q-1}(\sigma F^{i-1}_q).$$

A typical intersection of q-faces in $\dot{\Delta}_{q+1}$ has the form $F^i_{q+1}(\Delta_q) \cap F^j_{q+1}(\Delta_q)$ and without loss of generality we can assume $j < i$. The face commutation relations (9.3) yield a commutative diagram

$$\begin{array}{ccc} \Delta_{q-1} & \xrightarrow{F^j_q} & \Delta_q \\ {\scriptstyle F^{i-1}_q}\downarrow & & \downarrow{\scriptstyle F^i_{q+1}} \\ \Delta_q & \xrightarrow[F^j_{q+1}]{} & \dot{\Delta}_{q+1} \end{array}$$

Hence $F^i_{q+1}(\Delta_q) \cap F^j_{q+1}(\Delta_q) = F^i_{q+1}F^j_q(\Delta_{q-1}) = F^j_{q+1}F^{i-1}_q(\Delta_{q-1})$. The verification that $D_q(\sigma)$ is well defined on overlaps is just a calculation:

$$
\begin{aligned}
D_q(\sigma)\, F^i_{q+1}F^j_q &= D_{q-1}(\sigma\, F^{i-1}_q)\, F^j_q \\
&= \begin{cases} \sigma\, F^{i-1}_q & \text{if } j = 0, \\ \eta\varepsilon(\sigma\, F^{i-1}_q) & \text{if } q = 1, j \neq 0, \\ D_{q-2}(\sigma\, F^{i-1}_q F^{j-1}_{q-1}) & \text{otherwise;} \end{cases}
\end{aligned}
$$

and

$$
\begin{aligned}
D_q(\sigma)\, F^j_{q+1}F^{i-1}_q &= \begin{cases} \sigma\, F^{i-1}_q & \text{if } j = 0, \\ D_{q-1}(\sigma\, F^{j-1}_q)\, F^{i-1}_q & \text{otherwise,} \end{cases} \\
&= \begin{cases} \sigma\, F^{i-1}_q & \text{if } j = 0, \\ \eta\varepsilon(\sigma\, F^{j-1}_q) & \text{if } q = 1, j \neq 0, \\ D_{q-2}(\sigma\, F^{j-1}_q F^{i-2}_{q-1}) & \text{otherwise.} \end{cases}
\end{aligned}
$$

These expressions are equal by the definition of $\eta\varepsilon$ and (9.3). Hence we have $D_q(\sigma){:}\Delta_{q+1} \to X$ and $D_q{:}S_q(X) \to S_{q+1}(X)$. We check that D_q is a chain homotopy;

$$
\partial_1 D_0(\sigma) = D_0(\sigma)\, F^0_1 - D_0(\sigma)\, F^1_1 = (1 - \eta\varepsilon)(\sigma),
$$

$$
\begin{aligned}
\partial_{q+1}D_q(\sigma) + D_{q-1}\partial_q(\sigma) &= \sum_{i=0}^{q+1}(-1)^i D_q(\sigma)\, F^i_{q+1} + D_{q-1}\Big(\sum_{i=0}^{q}(-1)^i \sigma F^i_q\Big) \\
&= \sigma + \sum_{i=1}^{q+1}(-1)^i D_{q-1}(\sigma\, F^{i-1}_q) \\
&\quad + \sum_{i=0}^{q}(-1)^i D_{q-1}(\sigma\, F^i_q) = \sigma
\end{aligned}
$$

since the first sum cancels the second. ∎

(10.14) *Exercise.* Let X be a space, $Q \in X$ a point. Let $c{:}X \to X$ be the constant map at Q. What is $S(c){:}S(X) \to S(X)$? Let $c'{:}S(X) \to S(X)$ be defined by: $c'(\sigma) = 0$ if $\sigma \in S_q(X)$, $q \geq 1$, and $c'(\sigma) = \sigma_0$ if $\sigma \in S_0(X)$ where $\sigma_0{:}\Delta_0 \to Q$ and σ is a singular 0-simplex. Show c' is a chain map and is chain homotopic to $S(c)$.

(10.15) *Exercise.* Prove $\simeq$ is an equivalence relation.

(10.16) *Exercise.* Let $f_1, f_2: C \to C'$ and $g_1, g_2: C' \to C''$ be chain maps. If $f_1 \simeq f_2$ and $g_1 \simeq g_2$ prove $g_1 f_1 \simeq g_2 f_2$.

(10.17) *Exercise.* For each q, Δ_q is convex hence aspherical. Theorem (10.13) produces chain homotopies $D_q^{\Delta_q}: S_q(\Delta_q) \to S_{q+1}(\Delta_q)$. Following the principles in (9.10) we can use the values $D_q^{\Delta_q}(\delta_q)$ to define natural maps $P_q: S_q(X) \to S_{q+1}(X)$ by $P_q(\sigma) = S_{q+1}(\sigma) D_q^{\Delta_q}(\delta_q)$. Why doesn't this construction prove the acyclicity of $S(X)$ for all X?

11. Homotopy Invariance of Homology

A fundamental feature of the homology functors is their homotopy invariance; homotopic maps induce the same map in homology. Spaces of the same homotopy type have isomorphic homologies.

(11.1) *Theorem. If f, g are homotopic maps $X \to Y$, then $S(f)$ and $S(g)$ are chain homotopic maps $S(X) \to S(Y)$.*

Since chain homotopic maps induce equal maps in homology we have

(11.2) *Theorem. If f, g are homotopic maps $X \to Y$, then for every $q \geq 0$, the induced homomorphisms $H_q(f)$ and $H_q(g)$ on the homology modules are equal.*

(11.3) *Theorem. If $f: X \to Y$ is a homotopy equivalence, then for every $q \geq 0$, $H_q(f)$ is an isomorphism $H_q(X) \to H_q(Y)$.*

Proof. Let $g: Y \to X$ be a homotopy inverse to f. Then $fg \simeq id_Y$ and $gf \simeq id_X$. By (9.8), $H_q(f)H_q(g) = 1$ and $H_q(g)H_q(f) = 1$. Hence $H_q(f)$ and $H_q(g)$ are inverse isomorphisms. ■

To prove (11.1) we first reduce it to a theorem about X. Write $H: X \times I \to Y$ for the homotopy such that $H(\ ,0) = f$ and $H(\ ,1) = g$. Let $i_0, i_1: X \to X \times I$ be the top and bottom inclusions such that $f = Hi_0$, $g = Hi_1$. By functoriality it is enough to prove

(11.4) *Theorem. $S(i_0)$ and $S(i_1)$* are chain homotopic maps $S(X) \to S(X \times I)$.

Proof. We start by proving the result in dimension 0. Identify Δ_1 with $\Delta_0 \times I$ so that E_i in Δ_1 corresponds with $E_0 \times \{i\}, i = 0, 1$ by the unique affine map. Define $D_0: S_0(X) \to S_1(X \times I)$ on singular 0-simplices σ by $D_0(\sigma) = \sigma \times id$

$$\Delta_1 = \Delta_0 \times I \xrightarrow{\sigma \times id} X \times I.$$

Extend D_0 to all of $S_0(X)$ by linearity. Then $\partial_1 D_0 = S_0(i_1) - S_0(i_0)$ since this holds for each σ. Note that D_0 is natural; if $h: X \to X'$ then

$$\begin{array}{ccc} S_0(X) & \xrightarrow{D_0} & S_1(X \times I) \\ {\scriptstyle S_0(h)} \downarrow & & \downarrow {\scriptstyle S_1(h \times id)} \\ S_0(X') & \xrightarrow{D_0} & S_1(X' \times I) \end{array}$$

To check this, it is enough to check on singular 0-simplices σ. Then computing: $S_1(h \times id)D_0(\sigma) = (h \times id) \circ (\sigma \times id) = h\sigma \times id = D_0(h\sigma) = D_0 S_0(h)(\sigma)$.

We construct D_q by induction on q. The inductive hypothesis is the existence of chain homotopies (one for each space)

$$D_j: S_j(X) \to S_{j+1}(X \times I), \qquad j \leq q - 1$$

satisfying

(a) $\partial_{j+1} D_j + D_{j-1} \partial_j = S_j(i_1) - S_j(i_0)$ and

(b) (naturality) for $h: X \to X'$, $D_j S_j(h) = S_{j+1}(h \times id) D_j$. Computing: $\partial_q(S_q(i_1) - S_q(i_0)) = (S_{q-1}(i_1) - S_{q-1}(i_0))\partial_q = \partial_q D_{q-1} \partial_q$ by (a) with $j = q - 1$. Rewriting, we have

(c) $\partial_q(S_q(i_1) - S_q(i_0) - D_{q-1}\partial_q) = 0$ for all spaces X. To construct D_q we use (10.13). Applying (c) to the element $\delta_q \in S_q(\Delta_q)$ yields

$$(S_q(i_1) - S_q(i_0) - D_{q-1}\partial_q)(\delta_q)$$

is a cycle. By (10.13), $S(\Delta_q \times I)$ is acyclic, so there is $\omega \in S_{q+1}(\Delta_q \times I)$ such that $\partial_{q+1}(\omega) = (S_q(i_1) - S_q(i_0) - D_{q-1}\partial_q)(\delta_q)$. We set

$$D_q(\delta_q) = \omega$$

and have

$$\partial_{q+1} D_q(\delta_q) + D_{q-1}\partial_q(\delta_q) = (S_q(i_1) - S_q(i_0))(\delta_q) \quad .$$

Then for a singular q-simplex σ in $S_q(X)$ define

$$D_q(\sigma) = S_{q+1}(\sigma \times id)D_q(\delta_q).$$

Extend D_q to map $S_q(X) \to S_{q+1}(X \times I)$ by linearity. The naturality property (b) follows automatically from (9.11):

$$\begin{aligned} D_q S_q(h)(\sigma) &= D_q(h\sigma) = S_{q+1}(h\sigma \times id)D_q(\delta_q) \\ &= S_{q+1}(h \times id)S_{q+1}(\sigma \times id)D_q(\delta_q) = S_{q+1}(h \times id)D_q(\sigma). \end{aligned}$$

To verify condition (a) we calculate:

$$\partial_{q+1}D_q(\sigma) = \partial_{q+1}S_{q+1}(\sigma \times id)(D_q(\delta_q)) = S_q(\sigma \times id)\partial_{q+1}D_q(\delta_q).$$

Writing $\sigma = S_q(\sigma)(\delta_q)$ and applying ∂_q gives

$$\partial_q \sigma = S_{q-1}(\sigma)\partial_q(\delta_q).$$

Then

$$\begin{aligned} D_{q-1}\partial_q\sigma &= D_{q-1}S_{q-1}(\sigma)\partial_q(\delta_q) \\ &= S_q(\sigma \times id)D_{q-1}\partial_q(\delta_q) \quad \text{by (b)}. \end{aligned}$$

Putting these together, we have

$$\begin{aligned} (\partial_{q+1}D_q + D_{q-1}\partial_q)(\sigma) &= S_q(\sigma \times id)(\partial_{q+1}D_q + D_{q-1}\partial_q)(\delta_q) \\ &= S_q(\sigma \times id)(S_q(i_1) - S_q(i_0))(\delta_q) \\ &= (S_q(i_1) - S_q(i_0))S_q(\sigma)(\delta_q) \\ &\qquad \text{since } (\sigma \times id)i_j = i_j\sigma,\ j = 0, 1 \\ &= (S_q(i_1) - S_q(i_0))(\sigma). \end{aligned}$$

This completes the inductive step. ■

Remark. We have proved more than was stated: $S(i_0)$ and $S(i_1)$ are chain homotopic by a natural chain homotopy.

Here are some exercises to aid in digesting the ideas used in our proof.

(11.5) *Exercise.* To obtain D_0 for all spaces it is enough to construct $D_0{:}S_0(\Delta_0) \to S_1(\Delta_0 \times I)$ such that $\partial_1 D_0 = S_0(i_1) - S_0(i_0)$.

(11.6) *Exercise*. The augmentation $\varepsilon_X: S_0(X) \to R$ defined in (9.7) is natural, $\varepsilon_X = \varepsilon_Y S_0(f)$ for $f: X \to Y$. Prove that a ***natural*** right inverse η to ε implies $S(X)$ is acyclic for all X. Assume for all X, ε_X has a right inverse $\eta_X: R \to S_0(X)$ and $S_0(f)\eta_X = \eta_Y$. Define $D_0: S_0(\Delta_0) \to S_1(\Delta_0)$ by $D_0(\sigma_0) = \sigma_1$ (9.4). Then $\partial_1 D_0 = 1 - \eta_{\Delta_0}\varepsilon_{\Delta_0}$. Regard R as a chain complex concentrated in dimension 0 and ε, η as chain maps. Define $D_0: S_0(X) \to S_1(X)$ by $D_0(\sigma) = S_1(\sigma)D_0(\sigma_0)$. Use naturality to obtain $\partial_1 D_0 = 1 - \eta_X \varepsilon_X$. Complete the construction $D_q: S_q(X) \to S_{q+1}(X)$ such that $\partial D + D\partial = 1 - \eta\varepsilon$ by the same sequence of steps as in the proof of (11.4).

(11.7) *Exercise*. A direct proof of (11.4) can be given without reference to acyclicity, naturality or the equation $\sigma = S(\sigma)(\delta_q)$. In $\Delta_q \times I$ denote $\Delta_q \times \{0\}$ by $(E_0, \ldots, E_q)$ and $\Delta_q \times \{1\}$ by $(E'_0, \ldots, E'_q)$. For each $q \geq 0$, $0 \leq i \leq q$, define $U^i_q: \Delta_{q+1} \to \Delta_q \times I$ to be the affine map $(E_0, \ldots, E_i, E'_i, \ldots, E'_q)$.

Given a singular q-simplex $\sigma: \Delta_q \to X$ define $D_q(\sigma) = S_q(\sigma \times id)(\sum_{i=0}^{q} (-1)^i U^i_q) \in S_{q+1}(X \times I)$. Show by direct calculation that $(\partial_{q+1}D_q + D_{q-1}\partial_q)(\sigma) = (S_q(i_1) - S_q(i_0))(\sigma)$. An intermediate position is to verify the formula for $D_q(\delta_q)$ and avoid reference to acyclicity. This D_q is called the *prism* operator.

(11.8) *Remark*. Another explicit chain homotopy will be available in (15.26). The construction of D_q is a special case of the construction in the proof of the Theorem on Acyclic Models (29.25).

12. Relation Between π_1 and H_1

Throughout this section we take our coefficient ring R to be the integers and write $H_1(X)$ for $H_1(X; \mathbf{Z})$.

(12.1) *Theorem. There is a natural homomorphism* $\chi : \pi_1(X, x_0) \to H_1(X: \mathbf{Z})$ *which sends the homotopy class of a loop* γ *into the homology class of the singular 1-simplex* γ. *If* X *is path connected,* χ *is surjective, and its kernel is the commutator subgroup.*

(12.2) *Corollary. If* X *is path connected, then* χ *is an isomorphism if and only if the fundamental group of* X *is commutative.*

Proof of theorem : Suppose γ, γ' are loops at x_0 and

$$F : \gamma \simeq \gamma' \text{ rel } (0, 1)$$

Define a singular 2-simplex σ in X as follows: $\sigma(E_0) = x_0$. For any point Q of Δ_2 other than E_0, the line through E_0 and Q meets the edge opposite E_0 in a point Q'. We write

$$Q' = tE_2 + (1-t)E_1, \qquad \text{and} \qquad Q = sQ' + (1-s)E_0$$

Define $\sigma(Q) = F(s, t)$. σ is continuous because Δ_2 is the quotient space of I^2 under the map $(s, t) \to Q$, $s \neq 0$, $(0, t) \to E_0$.

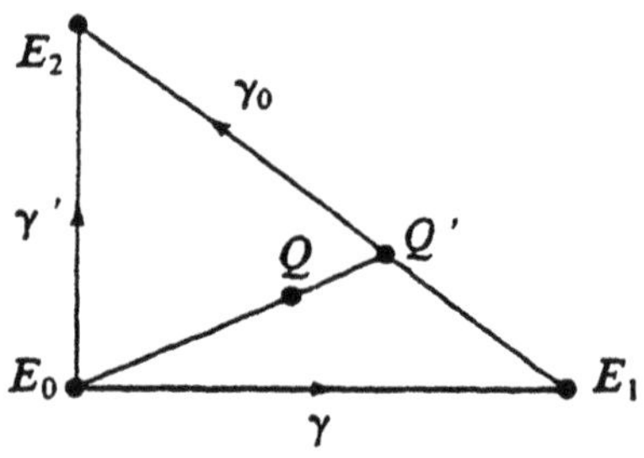

If γ_0 is the constant loop at x_0, we have

$$\partial(\sigma) = \gamma - \gamma' + \gamma_0.$$

Since γ_0 is the boundary of the trivial 2-simplex at x_0, we have $\gamma \sim \gamma'$, and χ is well defined.

To see that it is a homomorphism, use the diagram

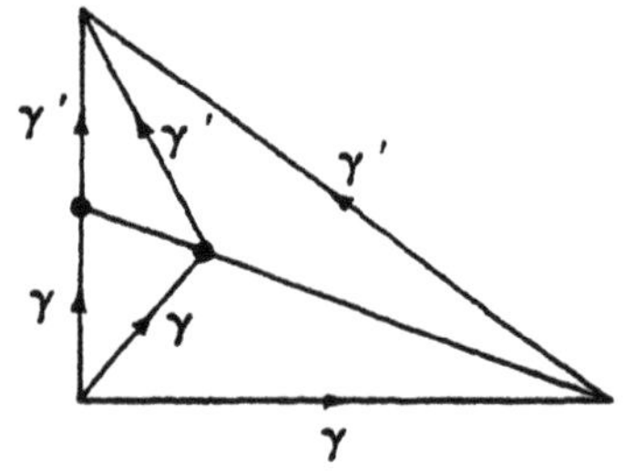

to define a singular 2-simplex σ such that

$$\partial(\sigma) = \gamma + \gamma' - \gamma\gamma'.$$

(This formula applies more generally for paths γ, γ' such that $\gamma\gamma'$ is defined.)

Suppose now that X is path connected.

Let $z = \Sigma\nu_i\alpha_i$ be a 1-cycle, so that

$$0 = \Sigma\nu_i(\alpha_i(1) - \alpha_i(0)) = \partial z.$$

This means that after collecting terms, all coefficients are 0. Choose a path η_{i0} from x_0 to $\alpha_i(0)$, a path η_{i1} from x_0 to $\alpha_i(1)$, and choose the paths to depend only on the vertices, not on the indexing. If the initial or terminal point of α_i is x_0, then the path must be the constant path. This remark holds throughout the proof. Then, collecting terms,

$$0 = \Sigma\nu_i(\eta_{i1} - \eta_{i0}).$$

Setting $\beta_i = \eta_{i0} + \alpha_i - \eta_{i1}$, we get

$$z = \sum \nu_i \beta_i.$$

Now if γ_i is the loop $\eta_{i0}\alpha_i\eta_{i1}^{-1}$, we see that

$$\chi[\prod \gamma_i^{\nu_i}] = \bar{z}$$

and χ is surjective.

Next we identify Ker χ. The main difficulty is recognizing commutators when they are expressed as products of paths. Suppose a loop γ is expressed

$$\gamma = \prod_i \alpha_i^{\varepsilon_i}$$

where α_i are paths which are not necessarily distinct and $\varepsilon_i = \pm 1$. Write

$$\exp(\alpha_i) = \sum_j \varepsilon_j$$

where the sum is over all j such that $\alpha_j = \alpha_i$.

(12.3) *Lemma. If* $\exp(\alpha_i) = 0$ *for each distinct factor* α_i *of* γ, *then* $[\gamma]$ *is in the commutator subgroup.*

Proof. Choose paths η_{i0} and η_{i1} from x_0 to the initial and terminal points of α_i and choose the paths to depend only on the endpoints, not the indexing. Then

$$\gamma = \prod_i \alpha_i^{\varepsilon_i} \simeq \prod_i (\eta_{i0}\alpha_i\eta_{i1}^{-1})^{\varepsilon_i} \text{ rel } (0, 1).$$

Let π_1' be the quotient of $\pi_1(X, x_0)$ by its commutator subgroup and let $\bar{\gamma}$ be the coset of $[\gamma]$. Then writing $\beta_i = \eta_{i0}\alpha_i\eta_{i1}^{-1}$ we have

$$\bar{\gamma} = \sum_i \varepsilon_i \bar{\beta}_i = \sum \exp(\alpha_i)\bar{\beta}_i = 0$$

where the last sum is over distinct path classes. ∎

We continue with the proof of the theorem.
Let γ be a loop homologous to 0. Thus

$$\gamma = \partial(\sum \nu_i \sigma_i).$$

Write $\partial(\sigma_i) = \alpha_{i0} - \alpha_{i1} + \alpha_{i2}$, so that after collecting terms in the sum

$$\text{(12.4)} \qquad \sum \nu_i(\alpha_{i0} - \alpha_{i1} + \alpha_{i2})$$

γ occurs with coefficient 1, all other paths with coefficient 0. We again choose paths η_{ij}, $j = 0, 1, 2$, from x_0 to $\alpha_{i2}(0)$, $\alpha_{i0}(0)$, $\alpha_{i1}(1)$ so as to depend only on the vertices, not the indexing, and we choose the constant path to the vertex x_0.

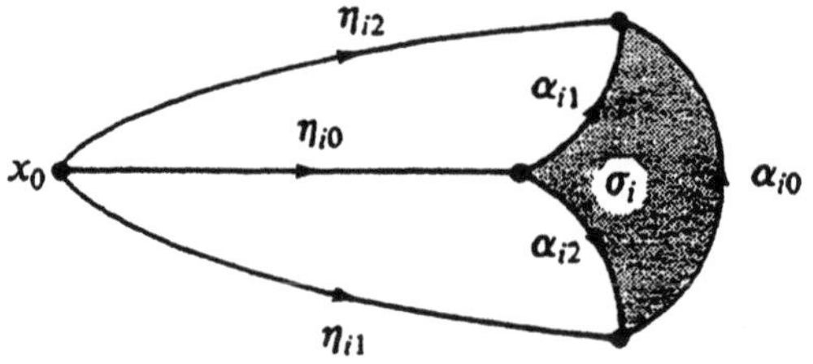

Consider next the loops at x_0

$$\beta_{i0} = \eta_{i1}\alpha_{i0}\eta_{i2}^{-1},$$
$$\beta_{i1} = \eta_{i0}\alpha_{i1}\eta_{i2}^{-1},$$
$$\beta_{i2} = \eta_{i0}\alpha_{i2}\eta_{i1}^{-1}.$$

Then

$$\beta_i = \beta_{i0}\beta_{i1}^{-1}\beta_{i2} \simeq \eta_{i1}\alpha_{i0}\alpha_{i1}^{-1}\alpha_{i2}\eta_{i1}^{-1} \simeq x_0 \text{ rel}(0, 1).$$

Hence

$$\prod_i [\beta_i]^{\nu_i} = 1.$$

By the observation on coefficients in (12.4), the composite of paths

$$\gamma(\prod_i \beta_i^{\nu_i})^{-1}$$

satisfies (12.3). Thus $[\gamma]$ belongs to the commutator subgroup.

Since $H_1(X)$ is commutative, the kernel of χ must contain the commutator subgroup, and we are done. ■

(12.5) *Example*: If X is the figure 8, its fundamental group is the free group on 2 generators (by Van Kampen's theorem; cf. Crowell and Fox [16]). By our theorem, $H_1(X)$ is the free Abelian group on 2 generators, i.e., $\mathbf{Z} \times \mathbf{Z}$.

(12.6) *Exercise.* Show that any map $(I^q, \partial I^q) \to (X, x_0)$ (here ∂I^q is in the sense of (7.8)) induces by passage to the quotient a singular q-simplex each of whose faces is the constant map on x_0. Deduce from this a homomorphism $\chi: \pi_q(X, x_0) \to H_q(X; \mathbf{Z})$ for all $q \geq 1$. Show that these homomorphisms are functorial in the sense that a map $f:(X, x_0) \to (Y, y_0)$ induces a commutative diagram

$$\begin{array}{ccc} \pi_q(X, x_0) & \xrightarrow{\chi} & H_q(X) \\ {\scriptstyle (f_*)_q}\downarrow & & \downarrow{\scriptstyle H_q(f)} \\ \pi_q(Y, y_0) & \xrightarrow[\chi]{} & H_q(Y) \end{array}$$

(12.7) *Remark.* A more geometric description of the relation $\chi[\gamma] = 0$ is possible. Recall that the closed orientable surface M_g of genus g can be obtained by identifying the edges of a $4g$-polygon. The case $g = 2$ is indicated (ignore the interior loop).

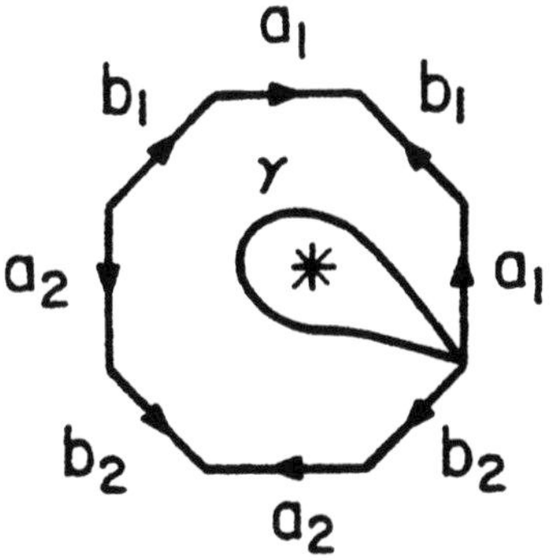

If we remove the interior of a small disc, we obtain W, an orientable surface with boundary γ, and W has the homotopy type of a bouquet of $2g$ circles. Hence $\pi_1 W$ is free on $\{\alpha_i, \beta_i; 1 \leq i \leq g\}$ where $\alpha_i = [a_i]$, $\beta_i = [b_i]$. Since $[\gamma] = \prod_{i=1}^{g} [\alpha_i, \beta_i]$, it follows that γ is homologous to 0 in $H_1(W)$. More generally we have

(12.8) *Proposition. Let γ be a loop in X regarded as a map $f: S^1 \to X$. For $\chi[\gamma] = 0$ it is necessary and sufficient that f extend to $\bar{f}: W \to X$ where W is a compact orientable surface with boundary S^1.*

One should contrast (12.8) with the observation that f is homotopically trivial if and only if f extends to a map of a disc $\bar{f}: E^2 \to X$.

We sketch the proof of (12.8) leaving details for an exercise. On the one hand, if f has an extension, then regarded as a loop in W, $\chi[\gamma] = 0$. On the other hand, suppose $\chi[\gamma] = 0$. Then there is a homotopy $H: S^1 \times I \to X$ with $H(\ ,0) = f$ and by (12.1) $H(\ ,1)$ is a product of commutators. Make suitable identifications on $S^1 \times \{1\}$ to obtain W and $\bar{f}$.

(12.9) *Remark.* There are nice results concerning the representation of integral homology classes by embedded circles. For the torus T, $H_1(T) = \mathbf{Z} \oplus \mathbf{Z}$. Denote the generators by $(1,0)$ and $(0,1)$. Then (a,b) is represented by an embedded circle if and only if $a = b = 0$ or $gcd(a,b) = 1$ ($gcd(a,0) = a$ in this statement).

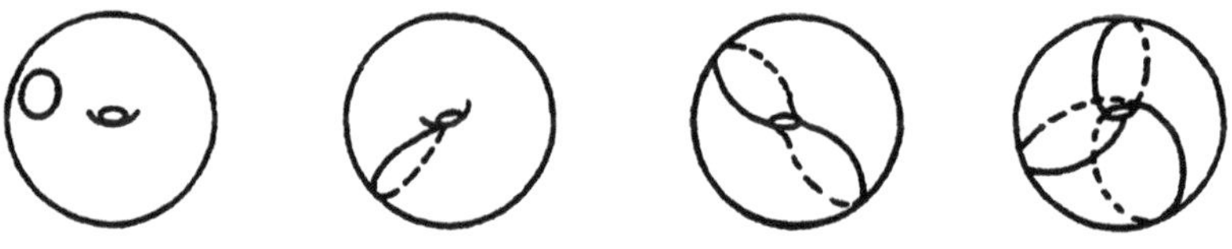

(0,0) (1,0) (2,1) (3,2)

A proof is given in Rolfsen [85], p. 19. The analogous results for the other surfaces appear in Schafer, Canadian Math. Bull. *19* (1976); Meyerson, Proc. Amer. Math. Soc. *61* (1976); Meeks and Patrusky, Illinois J. Math. *22* (1978).

The representation of $H_2(M^4)$ by embedded surfaces is an important open problem, where M^4 is a closed 4-dimensional manifold.

(12.10) *Remark.* Let K be a knot in $\mathbf{R}^3$. A method, the Wirtinger presentation, for calculating $\pi_1(\mathbf{R}^3 - K)$ is described in Crowell and Fox [16] or Rolfsen [85]. Applying (12.1) results in $H_1(\mathbf{R}^3 - K) \cong \mathbf{Z}$ (this will also appear in chapter 18). If J and K are nonintersecting circles in $\mathbf{R}^3$, possibly knotted, then $\chi[J] \in H_1(\mathbf{R}^3 - K)$ defines an integer called the *linking number* of J and K. The precise definition depends on orientations. This topological invariant was formulated by Gauss. It has been used by W. R. Bauer, F. H. C. Crick, and J. H. White, *Supercoiled DNA, Scientific American 243* July (1980), 118–133, in the study of DNA.

(12.11) *Exercise.* Consider the curves γ_i, $i = 1,2,3$ and δ on the surface of genus 2 as pictured with orientations on γ_i.

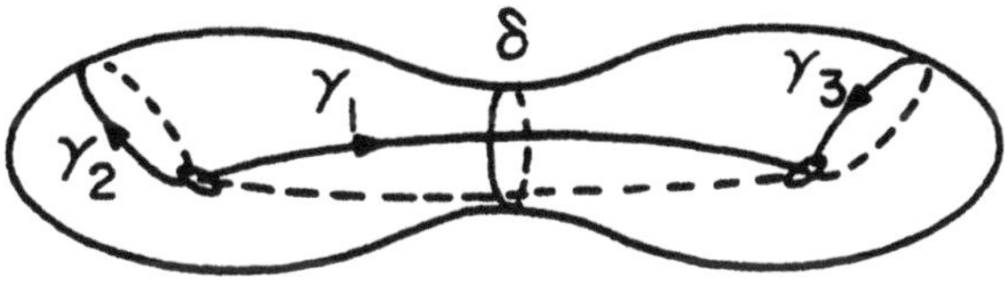

Show $\delta = 0$ and $\gamma_1 = \gamma_2 + \gamma_3$ in H_1.

(12.12) *Exercise.* Regard the Klein bottle K as the indentification space pictured (the shaded part will enter later).

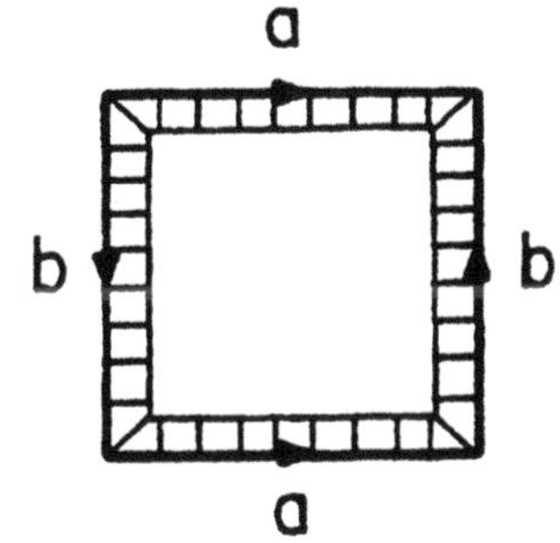

Let $\alpha = [a]$, $\beta = [b]$ in $\pi_1(K)$. Show $[\alpha, \beta] \neq 1$ in $\pi_1 K$. Find a map of a punctured torus to K representing the commutator $[\alpha, \beta]$. Suggestion: visualize the punctured torus

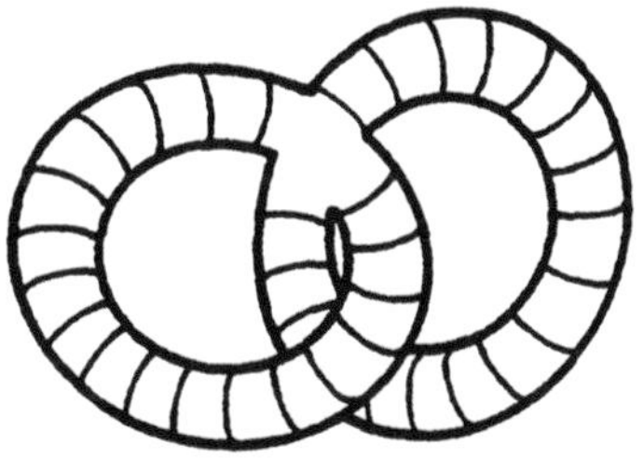

What is the subspace of K obtained by identifying the outer edges of the shaded region in the picture for K? The suggested map is not an embedding.

(12.13) *Exercise.* Show that the boundary of the Möbius strip is not a retract. You can use the relation between π_1 and H_1 to identify generators. Calculations like (9.13) can be made to compute the map $H_1(\partial M) \to H_1(M)$ induced by inclusion. Exhibit a deformation retraction of the Möbius strip onto its median circle.

13. Relative Homology

Let A be a subspace of X. Then for every $q \geq 0$, $S_q(A)$ is the submodule of $S_q(X)$ consisting of linear combinations of singular q-simplexes $\Delta_q \to X$ which actually map into A. We can then form the quotient module, and since the boundary operator sends $S_q(A)$ into $S_{q-1}(A)$, it induces a homomorphism $\bar{\partial}$ which makes the diagram

$$\text{(13.1)} \qquad \begin{array}{ccc} S_q(X) & \longrightarrow & S_q(X)/S_q(A) \\ \partial \big\downarrow & & \big\downarrow \bar{\partial} \\ S_{q-1}(X) & \longrightarrow & S_{q-1}(X)/S_{q-1}(A) \end{array}$$

commutative [i.e., if $c \in S_q(X)$, we define $\bar{\partial}$ (coset of c mod $S_q(A)$) = coset of ∂c mod $S_{q-1}(A)$]. Clearly $\bar{\partial}\bar{\partial} = 0$. We can then consider, as before, the modules

(a) Kernel $(S_q(X)/S_q(A) \xrightarrow{\bar{\partial}} S_{q-1}(X)/S_{q-1}(A))$

(b) Image $(S_{q+1}(X)/S_{q+1}(A) \xrightarrow{\bar{\partial}} S_q(X)/S_q(A))$

Since (b) is a submodule of (a), we can again form the quotient module, which is denoted $H_q(X, A)$ [or $H_q(X, A; R)$ if we want to make explicit the coefficient ring] and called the qth *relative homology module of X mod A*.

We can obtain this module directly from $S_q(X)$ if we like. Start with $c \in S_q(X)$. Suppose that in going around the square (13.1) to $S_{q-1}(X)/S_{q-1}(A)$, c is sent to 0; clearly this means that $\partial c \in S_{q-1}(A)$. The set of all such c's forms a submodule $Z_q(X, A)$ of $S_q(X)$ whose elements are called *relative* q-cycles *on* X *mod* A.

(13.2) *Example*: If σ is a path in X, it is a relative 1-cycle mod A iff its end points lie in the subspace A. More generally, a singular q-simplex is a relative q-cycle iff its faces are in A.

The module $Z_q(X, A)$ is just the pre-image by the quotient homomorphism of the module (a) above. What is the pre-image of (b)? Clearly it is the submodule $B_q(X, A)$ of $S_q(X)$ consisting of chains homologous to chains in $S_q(A)$; they are called *relative q-boundaries on* X *mod* A (write $c \sim c'$ mod A if $c - c'$ is a relative q-boundary).

(13.3) *Lemma.* $H_q(X, A) \cong Z_q(X, A)/B_q(X, A)$

This is a consequence of the isomorphism theorem $(M/P)/(N/P) \cong M/N$ in algebra. ■

(13.4) *Example*: If X is the cylinder $I \times S^1$, A the subspace $1 \times S^1$, then any horizontal loop $s \to (t, e^{2\pi is})$ is a relative 1-boundary, since it is homologous to the loop $s \to (1, e^{2\pi is})$ in A (prove this).

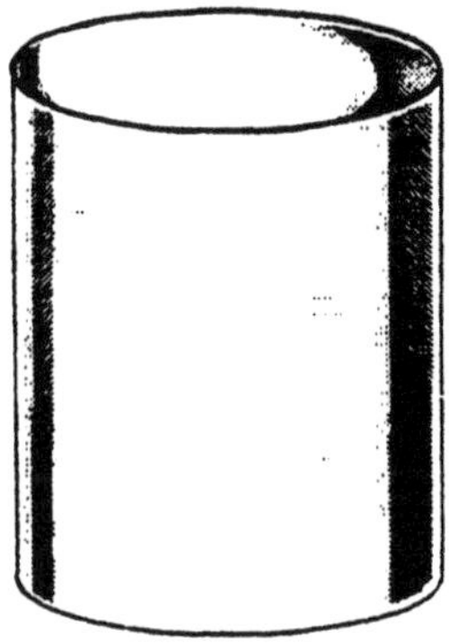

(13.5) *Note.* If A is empty, $S_q(A) = 0$ for all q, by definition, hence $H_q(X, \phi) = H_q(X)$. Thus any discussion of relative homology modules includes the absolute ones as a special case.

(13.6) The relative homology modules are functorial in the pair (X, A). Thus given another pair (X', A') (the word *pair* will henceforth mean two topological spaces such that the second is a subspace of the first) and a map $f:(X, A) \to (X', A')$ [meaning f is a map $X \to X'$ such that $f(A) \subset A'$], the induced chain homomorphism

$$S_q(f):S_q(X) \to S_q(X')$$

sends $S_q(A)$ into $S_q(A')$, hence $Z_q(X, A)$ into $Z_q(X', A')$ and $B_q(X, A)$ into $B_q(X', A')$, hence induces by passage to the quotient a homorphism

$$H_q(f):H_q(X, A) \to H_q(X', A')$$

As usual

(i) $H_q(\text{identity}) = \text{identity}$

(ii) $H_q(gf) = H_q(g)H_q(f)$

(13.7) *Example*: We have always a map $j:(X, \phi) \to (X, A)$ which is the identity map on X, hence an induced homomorphism.

$$H_q(j):H_q(X) \to H_q(X, A)$$

On the other hand, the inclusion map $i:A \to X$ induces a homomorphism

$$H_q(i):H_q(A) \to H_q(X)$$

What is the composite homomorphism

$$H_q(ji):H_q(A) \to H_q(X, A)\,?$$

Since $Z_q(A) \subset B_q(X, A)$, it is the zero homomorphism.

(13.8) *Exercise*. Show that both rectangles in the diagram

$$\begin{array}{ccccc} H_q(A) & \to & H_q(X) & \to & H_q(X, A) \\ \downarrow & & \downarrow & & \downarrow \\ H_q(A') & \to & H_q(X') & \to & H_q(X', A') \end{array}$$

are commutative (the vertical arrows are homomorphisms induced by a map $f:(X, A) \to (X', A')$, the horizontal ones from 13.7).

(13.9) *Proposition. Let (X_k) be the family of path components of X, and put $A_k = X_k \cap A$. Then there is a canonical isomorphism for all $q \geq 0$*

$$H_q(X, A) \cong \bigoplus_k H_q(X_k, A_k).$$

Proof. Just modify the proof of (9.5). Of course the subspaces A_k need not be path connected. ■

(13.10) *Proposition. If A is nonempty and X is path connected then* $H_0(X, A) = 0$.

Proof. Choose $x_0 \in A$. Given a 0-chain on X, $c = \Sigma v_x x$, choose a path α_x from x_0 to x. Then $\partial(\Sigma v_x \alpha_x) = c - (\Sigma v_x)x_0$, so that c is homologous to a 0-chain on A. ■

(13.11) *Corollary. If (X_k) are the path components of X, then $H_0(X, A)$ is a free module with as many generators as indices k such that X_k does not meet A.*

For $H_0(X_k, A_k) \simeq R$ whenever A_k is empty (9.6). ■

(13.12) *Exercise.* If A is a single point x_0 in X, then $H_q(X) \to H_q(X, x_0)$ is an isomorphism for all $q > 0$.

(13.13) Two maps $f, g{:}(X, A) \to (Y, B)$ are *homotopic* if there is a homotopy $H{:}X \times I \to Y$ between f and g such that H maps $A \times I \to B$. If $f|A = g|A = H|A \times I$ then we write $f \simeq g$ rel A. The top and bottom inclusions $i_0, i_1{:}(X, A) \to (X \times I, A \times I)$ are maps of pairs. Theorem (11.4) provides natural chain homotopies fitting in the diagram

$$\begin{array}{ccccccc} 0 \to S_q(A) & \longrightarrow & S_q(X) & \longrightarrow & S(X)/S_q(A) & \longrightarrow & 0 \\ \downarrow D_q & & \downarrow D_q & & \downarrow \bar{D}_q & & \\ 0 \to S_{q+1}(A \times I) & \to & S_{q+1}(X \times I) & \to & S_{q+1}(X \times I)/S_{q+1}(A \times I) & \to & 0. \end{array}$$

Hence a map $\bar{D}_q$ is induced giving a natural chain homotopy of $S(i_0)$ with $S(i_1)$ on the relative singular complexes. We then have

(13.14) *Proposition. Homotopic maps of pairs $f, g{:}(X, A) \to (Y, B)$ induce equal maps on homology.*

We leave the proof for an exercise. One reworks the argument of (11.2).

(13.15) *Example.* Let $X = X'$ be the closed disc E^2, A the circle S^1, and $A' \supset A$ the closed annulus defined by $\frac{1}{2} \leq |z| \leq 1$. We claim that the inclusion map $f{:}(X, A) \to (X, A')$ is a homotopy equivalence, hence

$$H_q(f):H_q(X, A) \to H_q(X, A')$$

is an isomorphism. Define $g:(X, A') \to (X, A)$ by

$$g(z) = \begin{cases} 2z & |z| \leq \frac{1}{2}, \\ e^{i\theta} & \text{if } z = re^{i\theta} \text{ with } \frac{1}{2} \leq r \leq 1. \end{cases}$$

We must show fg and gf homotopic to the identity maps.

Define a homotopy $F_t: fg \simeq$ identity by

$$F_t(z) = \begin{cases} (1+t)z & |z| \leq 1/1+t, \\ e^{i\theta} & \text{if} z = re^{i\theta} \text{with} (1/1+t) \leq r \leq 1. \end{cases}$$

Define a homotopy $G_t: gf \simeq$ identity by

$$G_t(z) = \begin{cases} (1+t)z & |z| \leq \frac{1}{2}, \\ (1+s+t-st/2)e^{i\theta} & \text{if } z = ((1+s)/2)e^{i\theta}, 0 \leq s \leq 1. \end{cases}$$

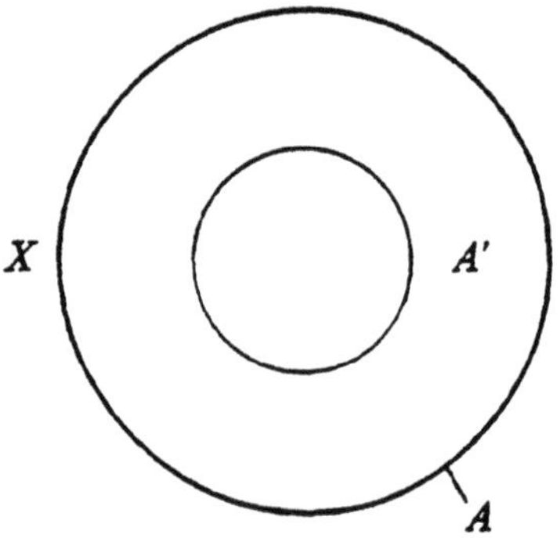

(13.16) *Exercise*. A homotopy equivalence of pairs induces a pair of homotopy equivalences, but the converse is false. For example, in (13.15) replace A' by the punctured disc A'' defined by $0 < |z| \leq 1$. Then the inclusion $(X, A) \to (X, A'')$ is not a homotopy equivalence. Suggestion: $X = \bar{A}''$ and a homotopy inverse to the inclusion yields a retraction of X to S^1.

14. The Exact Homology Sequence

The most important property of the relative homology modules $H_q(X, A)$ is the existence of a *connecting homomorphism*

$$H_q(X, A) \to H_{q-1}(A)$$

by means of which we obtain an infinite sequence of homomorphisms

$$\ldots \to H_q(A) \to H_q(X) \to H_q(X, A) \to H_{q-1}(A) \to \ldots$$

called *the homology sequence* of the pair (X, A) (see 13.7).

We define this homomorphism as follows: Given a relative q-cycle z representing a relative homology class $\bar{z}$. By definition, ∂z is a $(q-1)$-chain on A, but since $\partial\partial = 0$, ∂z is actually a $(q-1)$-cycle on A, and we can consider its homology class $\overline{\partial z} \in H_{q-1}(A)$. This class depends only on $\bar{z}$: If $z \sim z' \bmod A$, then $z = z' + w + \partial z''$, where w is a q-chain on A, z'' a $(q+1)$-chain on X; then $\partial z = \partial z' + \partial w$, i.e., ∂z and $\partial z'$ are homologous on A. We therefore define the connecting homomorphism, denoted also ∂ by $\partial \bar{z} = \overline{\partial z}$.

(14.1) *Theorem. The homology sequence of* (X, A) *is exact.*

This means (a) the composite of any two homomorphisms in the sequence is zero, and (b) the image of one homomorphism equals the kernel of the next one.

Proof: We will verify exactness at the stage $H_q(X, A)$ and leave the exactness verfication at stages $H_q(A)$, $H_q(X)$ as an exercise. Let z be a q-cycle on X, i.e., $\partial z = 0$. Then $\partial H_q(j)(\bar{z}) = \overline{\partial z} = 0$, so that the composite $\partial H_q(j)$ is zero. Let z be a relative q-cycle such that the image $\partial \bar{z}$ of $\bar{z}$ under the connecting homomorphism is zero. This means that $\partial z = \partial w$, where w is a q-chain on A. Therefore, $z - w$ is a cycle on X. Moreover, the relative homology class of $z - w$ is the same as that of z; thus $H_q(j)(\overline{z-w}) = \bar{z}$. This shows that the image of $H_q(j)$ equals the kernel of ∂. ■

(14.2) *Remark*. The homology sequence ends on the right with

$$\rightarrow H_0(X) \xrightarrow{H_0(j)} H_0(X, A) \rightarrow 0$$

Exactness at $H_0(X, A)$ means the image of $H_0(j)$ equals the kernel of the zero homomorphism, i.e., $H_0(j)$ is surjective. This has been proved above.

(14.3) *Example*: Exercise (13.12) can now be carried out as an easy consequence of (14.1). For if A is a single point x_0, $H_q(A) = 0$ for all $q > 0$. Thus for all $q \geq 2$,

$$0 \rightarrow H_q(X) \xrightarrow{H_q(j)} H_q(X, x_0) \rightarrow 0$$

is exact, which says $H_q(j)$ is an isomorphism. For $q = 1$ the right hand 0 must be replaced by $H_0(x_0) \simeq R$. But clearly $\partial: Z_1(X, x_0) \rightarrow H_0(x_0)$ is the zero homomorphism, so $H_1(j)$ is also an isomorphism.

(14.4) *Example*: Let X be the n-ball $E^n = \{v \in \mathbf{R}^n; |v| \leq 1\}$, $A = S^{n-1}$ its frontier. Since E^n is contractible, $H_q(E^n) = 0$ for all $q \geq 1$, so by the exact homology sequence

$$\partial: H_q(E^n, S^{n-1}) \rightarrow H_{q-1}(S^{n-1})$$

is an isomorphism for all $q \geq 2$. For $q = 1$ we get

$$0 \rightarrow H_1(E^n, S^{n-1}) \rightarrow H_0(S^{n-1}) \rightarrow H_0(E^n) \rightarrow 0$$

For $n > 1$, S^{n-1} is path connected and we see $H_1(E^n, S^{n-1}) = 0$. For $n = 1$, we get $H_1(E^1, S^0) \cong R \cong$ Kernel $(H_0(S^0) \rightarrow H_0(E^1))$.

In the next section we will develop a technique of calculating the relative homology modules in certain cases; this combined with (14.1) will enable us to determine certain absolute homology modules, such as $H_q(S^n)$.

(14.5) *Proposition. The homology sequence is functorial in the pair* (X, A).

Proof. This means that a map $f:(X, A) \rightarrow (X', A')$ induces an infinite sequence of commutative squares

$$\begin{array}{ccccccccc}
\to & H_q(A) & \to & H_q(X) & \to & H_q(X, A) & \to & H_{q-1}(A) & \to \\
 & \downarrow & & \downarrow & & \downarrow & & \downarrow & \\
\to & H_q(A') & \to & H_q(X') & \to & H_q(X', A') & \to & H_{q-1}(A') & \to
\end{array}$$

the vertical homomorphisms being induced by f. By 13.8 we need only check that

$$\begin{array}{ccc}
H_q(X, A) & \xrightarrow{\partial} & H_{q-1}(A) \\
\downarrow & & \downarrow \\
H_q(X', A') & \xrightarrow{\partial} & H_{q-1}(A')
\end{array}$$

is commutative; this is immediate from the fact that the chain homomorphisms $S_q(f)$ commute with the boundary operator (9.8). ■

The term "natural" is used interchangeably with "functorial" in the context of (14.5) and elsewhere.

(14.6) *Exercise.* Suppose $A \subset X \subset X'$. Generalize the previous to obtain an exact homology sequence

$$\ldots \to H_q(X, A) \to H_q(X', A) \to H_q(X', X) \to H_{q-1}(X, A) \to \ldots$$

which is functorial in the *triple* (X', X, A). *Suggestion*: One can establish a short exact sequence of chain complexes

$$0 \to S(X)/S(A) \to S(X')/S(A) \to S(X')/S(X) \to 0$$

or use the diagram

$$\begin{array}{ccccccccc}
H_q(X) & \to & H_q(X') & \searrow & & & & & \\
\downarrow & & \downarrow & & & & & & \\
H_q(X, A) & \to & H_q(X', A) & \to & H_q(X', X) & & & & \\
& \searrow & \downarrow & & \downarrow & & & & \\
& & H_{q-1}(A) & \to & H_{q-1}(X) & \to & H_{q-1}(X') & & \\
& & & & \downarrow & & \downarrow & & \\
& & & & H_{q-1}(X, A) & \to & H_{q-1}(X', A). & &
\end{array}$$

Ladders of the sort occurring in (14.5) appear frequently in the subject. We shall develop a number of algebraic lemmas to handle them. The most important is known as the

(14.7) *Five Lemma. Given a diagram of R-modules and homomorphisms with all rectangles commutative*

$$\begin{array}{ccccccccc} A_1 & \xrightarrow{f_1} & A_2 & \xrightarrow{f_2} & A_3 & \xrightarrow{f_3} & A_4 & \xrightarrow{f_4} & A_5 \\ \alpha\downarrow & & \beta\downarrow & & \gamma\downarrow & & \delta\downarrow & & \varepsilon\downarrow \\ B_1 & \xrightarrow[g_1]{} & B_2 & \xrightarrow[g_2]{} & B_3 & \xrightarrow[g_3]{} & B_4 & \xrightarrow[g_4]{} & B_5 \end{array}$$

such that the rows are exact (at joints 2, 3, 4) and the four outer homomorphisms α, β, δ, ε are isomorphisms, then γ is an isomorphism.

Proof. Show γ is monic. If $a \in A_3$ and $\gamma(a) = 0$ then $\delta f_3(a) = 0$. Since δ is monic, $f_3(a) = 0$. By exactness at A_3, $a = f_2(a')$. Then $g_2\beta(a') = \gamma f_2(a') = 0$. By exactness at B_2, $\beta(a') = g_1(b)$. Since α is epic, $b = \alpha(a'')$. Then $\beta(a' - f_1(a'')) = 0$. Since β is monic, $a' = f_1(a'')$. Then $a = f_2 f_1(a'') = 0$. The proof that γ is epic is left as an exercise. Start with an element $b \in B_3$ and do the only obvious thing: apply g_3. ■

Remark. Arguments of this short are called "diagram chases". Only perseverence is required. At each stage there is one obvious new thing to do.

Remark. The hypotheses on α and ε are excessive. The argument requires only α epic and ε monic. In the example

$$\begin{array}{ccccccccc} Z & \xrightarrow{0} & Z & = & Z & \xrightarrow{0} & Z & = & Z \\ 0\downarrow & & \| & & \downarrow 0 & & \| & & \downarrow 0 \\ Z & = & Z & \xrightarrow{0} & Z & = & Z & \xrightarrow{0} & Z \end{array}$$

the need for this requirement is obvious.

Example (13.16) continued. The inclusion $(E^2, S^1) \to (E^2, E^2 - \{0\})$ is not a homotopy equivalence; however by the five lemma, the inclusion induces an isomorphism in homology.

(14.8) *Application to retracts.* By definition, a subspace $A \subset X$ is a *retract* if there is $r: X \to A$ such that $ri = id_A$, where $i: A \to X$ is the inclusion. The long exact sequence of the pair (X, A) where A is a retract has special properties. Conversely, the impossibility of A being a retract can sometimes be proved by showing the incompatibility of these properties with the values of the homology modules. We first treat the relevant algebra.

(14.9) *Definition.* An exact sequence of R-modules of the form $0 \to A \xrightarrow{i} B \xrightarrow{j} C \to 0$ is called a *short exact sequence.* In particular, i is monic, $\operatorname{im} i = \ker j$, and j is epic. The Noether isomorphism theorem then says $C = \operatorname{im} j \cong B/\operatorname{im} i$ with $A \cong \operatorname{im} i$. So, up to isomorphism, C is B/A. Among the short exact sequences, certain special cases are singled out.

(14.10) *Definition.* A short exact sequence is *split* if either (a) there is a $k{:}B \to A$ such that $ki = id_A$; or (b) there is $l{:}C \to B$ such that $jl = id_C$. For example $0 \to \mathbf{Z} \xrightarrow{i} \mathbf{Z} \oplus \mathbf{Z} \xrightarrow{j} \mathbf{Z} \to 0$, $i(1) = (2, 3)$, $j(1, 0) = 3$, $j(0, 1) = -2$ is split. Take $l{:}\mathbf{Z} \to \mathbf{Z} \oplus \mathbf{Z}$ by $l(1) = (1, 1)$. But $0 \to \mathbf{Z} \xrightarrow{\times 2} \mathbf{Z} \to \mathbf{Z}/2\mathbf{Z}$ is not split because $\mathbf{Z}$ accepts no nontrivial homomorphisms from $\mathbf{Z}/2\mathbf{Z}$.

(14.11) *Proposition. Conditions* (a) *and* (b) *in* (14.10) *are equivalent.*

Proof. Show (a) implies (b). Given $c \in C$ let $b \in B$ be any element such that $j(b) = c$. Define $l(c) = b - ik(b)$. We check l is well defined. If b' satisfies $j(b) = j(b')$ then $b - b' = i(a)$. Then the difference

$$(b - ik(b)) - (b' - ik(b')) = b - b' - iki(a) = b - b' - i(a) = 0.$$

By construction l is a homorphism. Then $jl(c) = j(b - ik(b)) = j(b) = c$ so l satisfies condition (b).

Show (b) implies (a). Given $b \in B$, we have $j(b - lj(b)) = 0$. Since i is monic, there is a unique $a \in A$ such that $i(a) = b - lj(b)$. Define $k(b) = a$. We check k is a homomorphism. If $k(b) = a$, $k(b') = a'$, then $i(a) = b - lj(b)$ and similarly for a'. Hence $i(a + a') = b + b' - lj(b + b')$. By definition $k(b + b') = a + a' = k(b) + k(b')$. To check condition (a) we have $i(a) = i(a) - lji(a)$. Hence $ki(a) = a$. ■

(14.12) *Remark.* The existence of $k{:}B \to A$ such that $ki = id_A$ implies that i is monic and k is epic. Similarly $l{:}C \to B$ such that $jl = id_C$ implies j is epic and l is monic. Furthermore $\ker k = \operatorname{im} l$ if one map is used to construct the other as in the above proof. (Prove this.)

In a split short exact sequence, the structure of the module B is determined by A and C. Precisely we have

(14.13) *Direct sum lemma. Consider the diagram of R-modules with commutative triangles*

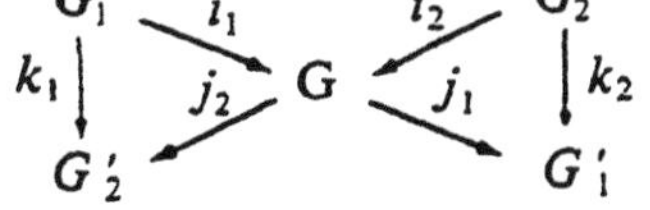

with im i_s = ker j_s, *and k_s an isomorphism for $s = 1, 2$. Then the compositions*

$$G_1 \oplus G_2 \xrightarrow{i_1 \oplus i_2} G \oplus G \xrightarrow{\nabla '} G$$

$$G \xrightarrow{\Delta} G \oplus G \xrightarrow{j_1 \oplus j_2} G_1' \oplus G_2'$$

are isomorphisms where $\nabla '(g, g') = g + g'$, $\Delta(g) = (g, g)$.

Proof. Since both k_s are isomorphisms, then both i_s are monic and both j_s are epic. To check $\nabla ' \circ (i_1 \oplus i_2)$ is monic, we note $\nabla ' \circ (i_1 \oplus i_2)(g, g') = 0$ implies $i_1(g) + i_2(g') = 0$. Applying j_1 yields $j_1 i_2(g') = k_2(g') = 0$. Hence $g' = 0$. Similary $g = 0$. To check the map is epic, take $g \in G$. Then $j_1(g) = k_2(g')$. Hence $j_1(g - i_2(g')) = 0$. So $g - i_2(g') = i_1(g'')$. Then

$$g = \nabla ' \circ (i_1 \oplus i_2)(g'', g').$$

We leave the proof that $(j_1 \oplus j_2) \circ \Delta$ is an isomorphism for an exercise. ∎

Remark. A split short exact sequence is an example

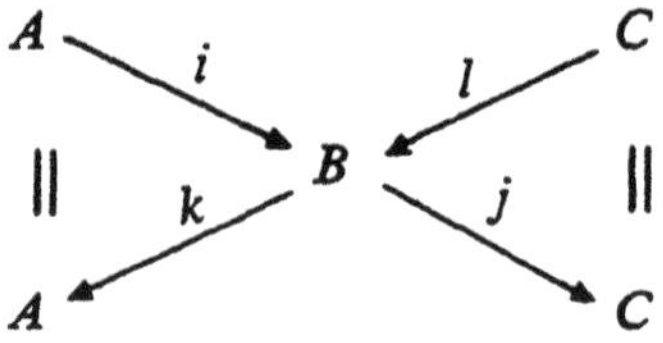

Returning to spaces we have

(14.14) *Proposition. If $A \subset X$ is a retract then the long exact homology sequence of the pair (X, A) breaks into split short exact sequences*

$$0 \to H_q(A) \underset{H_q(r)}{\overset{H_q(i)}{\rightleftarrows}} H_q(X) \to H_q(X, A) \to 0$$

for all $q \geq 0$. For $q = 0$ either ordinary or reduced homology may be used. In particular $H_q(X)$ is isomorphic to the direct sum $H_q(A) \oplus H_q(X, A)$.

Proof. Since $ri = id_A$ we have $1 = H_q(r)H_q(i)$. Then (14.12) implies the long exact sequence breaks into short exact sequences which are split by (14.10). ∎

(14.15) *Exercise.* Let $X = X_1 \cup X_2$ and $A = X_1 \cap X_2$. Using the exact sequences of triples, show that if the inclusion $(X_1, A) \hookrightarrow (X, X_2)$ induces an isomorphism in homology then the same holds for the inclusion $(X_2, A) \hookrightarrow (X, X_1)$.

(14.16) *Exercise.* Verify that the sequence of reduced homology modules (9.7) is also exact, where we define $H_0^*(X, A)$ to be $H_0(X, A)$ if A is nonempty, $H_0^*(X, \phi) = H_0^*(X)$

(14.17) *Note.* If $x_0 \in A \subset X$, one can also define $\pi_q(X, A, x_0)$ for all $q \geq 1$; however, for $q = 1$ it is only a set with a distinguished element, not a group. For $q > 1$ it is a group, and for $q > 2$ it is commutative. Moreover, there is an exact homotopy sequence completely analogous to the exact homology sequence (cf. Hu [33], Chapter 4), or [88].

(14.18) *Note.* The proof of Theorem 14.1 can easily be generalized to prove *the fundamental lemma of homological algebra*, which states that any short exact sequence of chain complexes has an associated connecting homomorphism which gives a long exact homology sequence. See Lang [35], Chapter IV.

(14.19) *Note.* The relative homology modules $H_q(X, A)$ are not determined by the complement $X - A$ of A in X. A relative homeomorphism f: $(X, A) \to (Z, Y)$ (i.e., a map which induces a homeomorphism $X - A \approx Z - Y$) need not induce an isomorphism in homology without further hypotheses as in (19.14); consider the inclusion $([0,1), \{0\} \to (I, \{0, 1\})$. There is another homology theory H' for *locally compact* spaces (due to Steenrod) in which there is an exact homology sequence $\ldots \to H'_q(A) \to H'_q(X) \to H'_q(X - A) \to H'_{q-1}(A) \to \ldots$ whenever A is *closed* (cf. W. S. Massey, *Homology and Cohomology Theory*, Dekker, N.Y., 1978).

15. The Excision Theorem

This states that certain subspaces $U \subset A$ may be cut out or *excised* from the space without affecting the relative homology modules. More precisely, the inclusion map $(X - U, A - U) \to (X, A)$ is called *an excision* if it induces an isomorphism

$$H_q(X - U, A - U) \to H_q(X, A)$$

for all q.

(15.1) *Theorem. If the closure of U is contained in the interior of A, then U can be excised.* (Proof later.)

(15.2) *Theorem. Suppose $V \subset U \subset A$ and*

(i) *V can be excised*
(ii) *$(X - U. A - U)$ is deformation retract of $(X - V, A - V)$. Then U can be excised.*

Proof of (15.2). Condition (ii) means that the identity map of

$$(X - V, A - V)$$

is homotopic to ir, where r is a retraction on $(X - U, A - U)$, and i is the inclusion map $(X - U, A - U) \to (X - V, A - V)$. By the homotopy theorem (13.14), $H_q(i)$ is an isomorphism for all q. Since homology is a functor and V may be excised, $(X - U, A - U) \to (X, A)$ is an excision. ■

Before proving (15.1), let's give some applications.

(15.3) *Theorem. Let E_n^+, E_n^- be the closed northern and southern hemi-*

spheres of the n-sphere S^n, $n \geq 1$, *(so that* $E_n^+ \cap E_n^-$ *is the equator* S^{n-1}*). Then*

$$(E_n^+, S^{n-1}) \to (S^n, E_n^-)$$

is an excision.

Proof. We are excising the open southern hemisphere

$$U = \{x \in S^n \mid x_{n+1} < 0\}.$$

We can't apply (15.1) directly since its hypothesis is not satisfied. Thus we proceed in 2 steps. Let

$$V = \{x \in S^n \mid x_{n+1} < -\tfrac{1}{2}\}.$$

By (15.1) V can be excised. But (E_n^+, S^{n-1}) is a deformation retract of $(S^n - V, E_n^- - V)$ (move up along great circles), hence (15.2) applies. ■

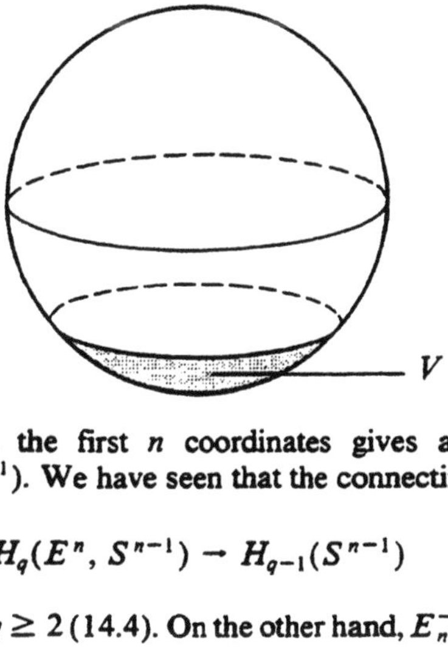

Now projecting on the first n coordinates gives a homeomorphism $(E_n^+, S^{n-1}) \to (E^n, S^{n-1})$. We have seen that the connecting homomorphism

$$H_q(E^n, S^{n-1}) \to H_{q-1}(S^{n-1})$$

is an isomorphism for $q \geq 2$ (14.4). On the other hand, E_n^- being contractible implies that

$$H_q(S^n) \to H_q(S^n, E_n^-)$$

is an isomorphism for $q \geq 2$. Combining these facts with (15.3) gives an isomorphism

$$H_q(S^n) \to H_{q-1}(S^{n-1})$$

for all $q \geq 2$, $n \geq 1$.

It remains to settle the case $q = 1$. By (14.4),

$$H_1(E^n, S^{n-1}) = \begin{cases} 0 & n > 1, \\ R & n = 1. \end{cases}$$

We also have the exact sequence for $n \geq 1$

$$0 \to H_1(S^n) \xrightarrow{a} H_1(S^n, E_n^-) \xrightarrow{b} H_0(E_n^-) \xrightarrow{c} H_0(S^n) \to 0$$

Since c is an isomorphism, we again get a to be an isomorphism ($b = 0$).

(15.4) *Exercise.* The above can be expressed in a neater way by using the reduced homology modules (9.7): The homomorphisms

$$H_q^*(S^n) \to H_q^*(S^n, E_n^-) \leftarrow H_q^*(E^n, S^{n-1}) \to H_{q-1}^*(S^{n-1})$$

are isomorphisms for all $q \geq 0$, $n \geq 1$.

Thus we get

(15.5) *Corollary. For $q \geq 1$ and $n \geq 1$*

$$H_q(S^n) \cong \begin{cases} R & q = n, \\ 0 & q \neq n, \end{cases}$$

$$H_q(E^n, S^{n-1}) \cong \begin{cases} R & q = n, \\ 0 & q \neq n. \end{cases}$$

(15.6) *Corollary. S^{n-1} is not a retract of E^n.*

The argument is the same as in (4.9), using homology functors instead of homotopy: If we had a map $f: E^n \to S^{n-1}$ whose restriction to S^{n-1} is the identity, then the induced homomorphisms give

$$\underbrace{H_{n-1}(S^{n-1}) \to H_{n-1}(E^n) \to H_{n-1}(S^{n-1})}_{\text{identity}}.$$

For $n \geq 2$ this is

$$\underbrace{R \to 0 \to R}_{\text{identity}}$$

which is ridiculous. (Exercise: do $n = 1$.) ■

(15.7) *Brouwer's Fixed Point Theorem. Any continuous map $E^n \to E^n$ has a fixed point.*

Same proof as (4.11). ■

(15.8) *Remark.* If we take $R = \mathbf{Z}$, and pick a generator α of $H_n(S^n)$ corresponding to the integer 1, then for any map $f:S^n \to S^n$, $H_n(f)(\alpha)$ corresponds to some integer called the *degree* of the map f. This can be described explicitly (see Dugundji [20], Chapter XVI).

The proof of 15.1 involves some completely new ideas.

First, let $\mathscr{V} = (V_i)$ be a family of open sets which cover X. Call a singular q-simplex *small of order* $\mathscr{V}$ if it maps Δ_q into one of the V_i. A key fact we will need follows:

(15.9) *Theorem. Every homology class in $H_q(X, A)$ can be represented by a relative cycle which is a linear combination of simplexes small of order $\mathscr{V}$.*

We will apply this to the covering consisting of the two open sets $X - \bar{U}$ and $\mathring{A}$ (interior of A) and deduce (15.1).

Secondly, to obtain these small simplexes we will use the technique of *barycentric subdivision*, which for $q = 2$ is illustrated by

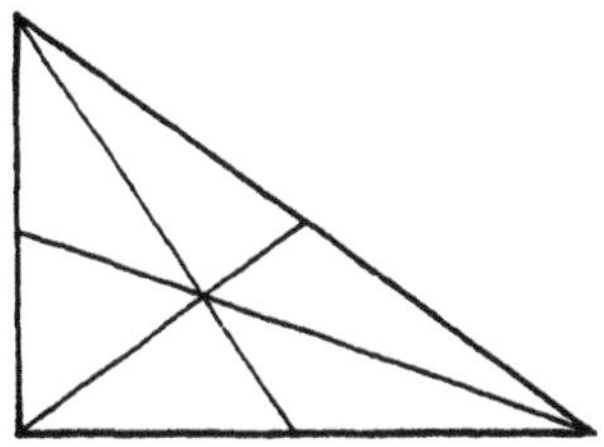

More precisely, we will construct subdivision operators

$$Sd: S_q(X) \to S_q(X)$$

which commute with the boundary operators. To be able to compare c and Sdc, we will also need operators

$$T: S_q(X) \to S_{q+1}(X)$$

(which for $q = 2$ corresponds to constructing a degenerate 3-dimensional polyhedron on a given subdivided triangle as base). For an excellent intuitive discussion of these constructions, see Wallace [59], pp. 133–137 (where the notation is somewhat different).

(15.10) Let B be a point, $\sigma = (P_0 \dots P_q)$ an affine singular q-simplex in some affine space. We define *the join* $B\sigma$ to be the affine singular $(q + 1)$-simplex

$$B\sigma = (BP_0 \ldots P_q)$$

For example, if $q = 1$, its image is

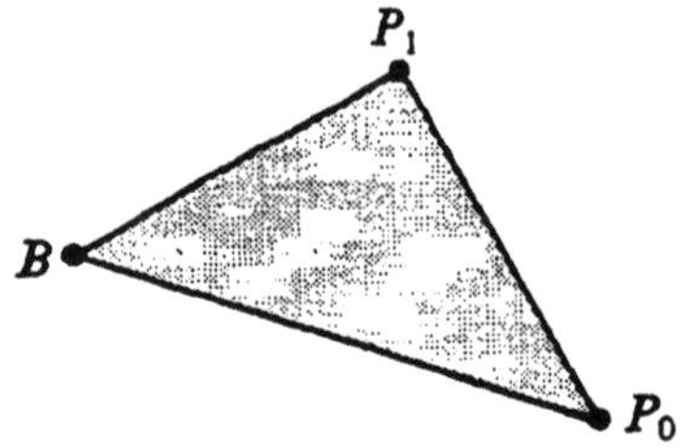

We extend this operation by linearity to linear combinations c of affine σ's (we do not define it for singular simplexes which are not affine):

$$Bc = B(\sum_i v_i\sigma_i) = \sum_i v_i B\sigma_i$$

Then we have the formulas

$$\partial Bc = c - B\partial c \qquad q > 0$$

$$\partial Bc = c - (\sum_i v_i)B \qquad q = 0$$

as is easily checked.

The operators Sd, T will be defined so as to be functorial: If $f: X \to X'$, then we will have commutative diagrams

$$\begin{array}{ccc} S_q(X) & \xrightarrow{Sd} & S_q(X) \\ {\scriptstyle S_q(f)}\downarrow & & \downarrow{\scriptstyle S_q(f)} \\ S_q(X') & \xrightarrow{Sd} & S_q(X') \end{array}$$

$$\begin{array}{ccc} S_q(X) & \xrightarrow{T} & S_{q+1}(X) \\ {\scriptstyle S_q(f)}\downarrow & & \downarrow{\scriptstyle S_{q+1}(f)} \\ S_q(X') & \xrightarrow{T} & S_{q+1}(X') \end{array}$$

Thus it suffices to define these operators for the space $X = \Delta_q$ and the singular q-simplex δ_q (see 9.10). For $q = 0$, set $Sd\delta_\circ = \delta_\circ$, $T\delta_\circ = 0$. Assuming these operators defined in dimension $<q$, we use the join operation to define them in dimension q:

$$Sd\ \delta_q = B_q Sd\ \partial\delta_q$$

$$T\ \delta_q = B_q(\delta_q - Sd\ \delta_q - T\ \partial\delta_q)$$

where B_q is the barycenter of Δ_q:

$$\sum_{i=0}^{q} \frac{1}{q+1} E_i = B_q$$

(15.11) *Lemma. We have the operator equations*

$$\partial Sd = Sd\partial$$

$$\partial T = Id - Sd - T\partial$$

where Id is the identity operator.

Proof: Induction on q, $q = 0$ being obvious. For $q > 0$, it suffices by functoriality to evaluate both sides of the equations on δ_q. Using (15.10) (and dropping the subscripts q for simplicity) we get

$$\partial Sd\ \delta = \partial B\ Sd\ \partial\delta = Sd\ \partial\delta - B\partial Sd\ \partial\delta$$

but by inductive hypothesis

$$B\partial Sd\ \partial\delta = B\ Sd\ \partial^2\delta = 0$$

Similarly

$$\begin{aligned}\partial T\ \delta &= \partial B(\delta - Sd\ \delta - T\ \partial\delta)\\ &= \delta - Sd\ \delta - T\ \partial\delta - B(\partial\delta - \partial Sd\ \delta - \partial T\ \partial\delta)\\ &= \delta - Sd\delta - T\partial\delta - B(\partial\delta - Sd\partial\delta - \partial\delta + Sd\partial\delta + T\partial^2\delta)\\ &= \delta - Sd\ \delta - T\ \partial\delta\end{aligned}$$ ∎

Now let $\sigma = (P_0 \ldots P_q)$ be an affine q-simplex in some affine space. The image $\sigma(\Delta_q)$ is a compact set, and we can consider its diameter $d(\sigma)$.

(Exercise: $d(\sigma) =$ maximum of the lengths of the edges.)

(15.12) *Lemma. Each affine singular simplex appearing in the q-chain $Sd\sigma$ has diameter at most*

$$\frac{q\, d(\sigma)}{q+1}$$

Proof: Exercise. ■

(15.13) *Proposition. Let σ be a singular simplex in X, $\mathscr{V}$ an open covering of X. Then there is an $r > 0$ such that $Sd^r\, \sigma$ is a linear combination of singular simplexes small of order $\mathscr{V}$.*

Proof: Since Δ_q is compact, there is an $\varepsilon > 0$ such that σ maps the ε-neighborhood of any point in Δ_q into one of the sets in $\mathscr{V}$ (exercise). By (15.12), since

$$\lim_{r\to\infty} \frac{q^r}{(q+1)^r} = 0$$

there is an $r > 0$ such that $Sd^r\, \delta_q$ is a linear combination of affine singular simplexes of diameter $< \varepsilon$. But $Sd^r\sigma = S_q(\sigma)Sd^r\, \delta_q$. ■

(15.14) *Note*. In (15.13), we can replace σ by an arbitrary q-chain (exercise).

We can now prove Theorem (15.9): Let z be any relative q-cycle. Then (15.11) gives

$$z - Sdz = \partial Tz + T\partial z$$

Since ∂z is a chain on A, so is $T\partial z$, so we have

$$z \sim Sdz \qquad \text{mod } A$$

Hence by induction, for all r

$$z \sim Sd^r z \qquad \text{mod } A$$

whence by the above note we are done. ■

Finally we prove the excision theorem (15.1).

Proof: Given a homology class in $H_q(X, A)$, represent it by a relative cycle

$$z = \sum v_i \sigma_i$$

such that each σ_i is small of order $(X - \bar{U}, \dot{A})$. Now any σ_i which does not map entirely into $X - U \supset X - \bar{U}$ must map into $\dot{A} \subset A$, hence can be dropped from this expression without changing the homology class of z mod A. Having omitted such σ_i we see that z can be regarded as a relative cycle on $X - U$ mod $A - U$. Thus

$$H_q(X - U, A - U) \to H_q(X, A)$$

is onto.

Suppose z is a relative cycle on $X - U$ mod $A - U$ such that $z \sim 0$ on X mod A.

Thus

$$z = z' + \partial w$$

where z' is a q-chain on A and w a $(q + 1)$-chain on X. Subdivide r times (which does not change the homology class $\bar{z} \varepsilon H_q(X - U, A - U)$)

$$Sd^r z = Sd^r z' + \partial Sd^r w$$

where r is chosen so that $Sd^r w$ is a linear combination of simplexes small of order $(X - \bar{U}, \dot{A})$. Hence we can write

$$Sd^r w = w_1 + w_2$$

where all simplexes occurring in w_1 map in $X - U$, and simplexes occurring in w_2 map into A. We get

$$Sd^r z - \partial w_1 = Sd^r z' + \partial w_2$$

Since the left side is a chain on $X - U$, the right side a chain on A, we see both sides are chains on $A - U$, and solving for $Sd^r z$ again shows

$$Sd^r z \sim 0 \quad \text{on} \quad X - U \bmod A - U$$

Thus $H_q(X - U, A - U) \to H_q(X, A)$ is a monomorphism. ■

Here is a picture of what is happening in the proof of (15.9). In figures 1 and 2, the space X is the rectangle, A is the triangle below the diagonal, and U is a disc inside A.

In figure 1 we have the image of a relative singular 2-simplex. In figure 2 we have subdivided to get a singular 2-chain. Throwing away the pieces in A does not change the relative singular 2-chain. The result (shaded) is a relative singular 2-chain of $(X - U, A - U)$. Regarded as an element of $S_2(X)/S_2(A)$, the new chain is homologous to the singular 2-simplex in figure 1.

(15.15) *Remark.* The excision theorem holds for reduced homology if we assume $U \neq A$, for in that case

$$H_0^*(X - U, A - U) = H_0(X - U, A - U),$$

$H_0^*(X, A) = H_0(X, A)$ (we assume of course $A \neq \phi$). $U = A$ can occur only under the hypothesis of (15.1) when A is both open and closed.

(15.16) *Note.* A stronger form of Theorem 15.9 will be needed later: *Let $S(\mathcal{V})$ be the subcomplex of $S(X)$ generated by all the simplexes small of order $\mathcal{V}$. Then the inclusion homomorphism $S(\mathcal{V}) \to S(X)$ is an equivalence of chain complexes.*

We shall establish (15.16) by the general method of *algebraic mapping cones*. In order that submodules of free R-modules be free, we assume R is a p.i.d. (see Lang [35]).

(15.17) *Definition.* Let $f: C \to C'$ be a chain map of chain complexes. The *mapping cone* Cf is the complex

$$(Cf)_q = C'_q \oplus C_{q-1},$$

$$\partial_q^f(x, y) = (\partial'_q x + f_{q-1} y, -\partial_{q-1} y).$$

The verification $\partial_{q-1}^f \partial_q^f = 0$ is straightforward,

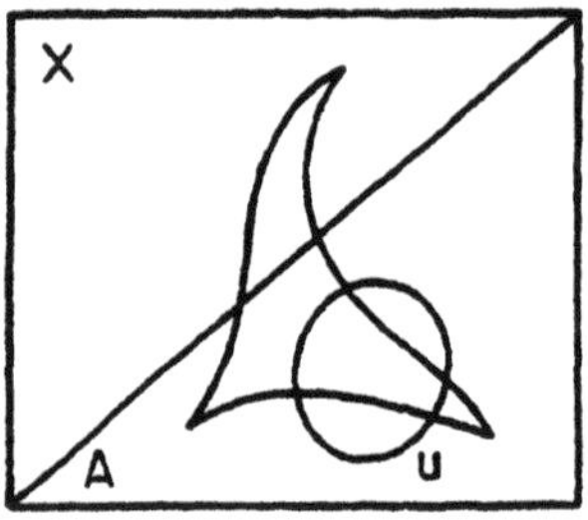

Figure 1.

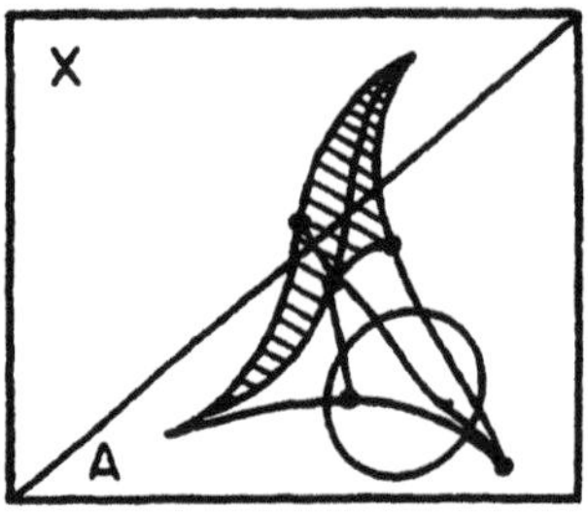

Figure 2.

$$\partial^f_{q-1}(\partial'_q x + f_{q-1}y, -\partial_{q-1}y) = (\partial'_{q-1}\partial'_q x + \partial'_{q-1}f_{q-1}y - f_{q-2}\partial_{q-1}y, \partial_{q-2}\partial_{q-1}y) = 0.$$

We form the exact sequence of chain complexes

$$0 \to C' \xrightarrow{i} Cf \xrightarrow{j} C^+ \to 0$$

where $(C^-)_{q+1} = C_q$, $\partial^+_{q+1} = -\partial_q$, $i(x) = (x, 0)$, $j(x, y) = y$. We define a connecting homomorphism

$$\partial: H_q(C^+) \to H_{q-1}(C')$$

as in the topological case. Given $y \in C^+$ such that $\partial y = 0$ then $\partial^f(x, y) = (\partial' x + fy, 0)$. Thus $\partial \bar{y} = \overline{fy}$ is a well-defined homomorphism (verify the remaining details). Just as in the topological case, we obtain a long exact sequence

$$\ldots \to H_q(C') \xrightarrow{H(i)} H_q(Cf) \xrightarrow{H(j)} H_{q-1}(C) \xrightarrow{H(f)} H_{q-1}(C') \to \ldots$$

where we have used $H_q(C^-) = H_{q-1}(C)$, and identified ∂ with $H(f)$. Hence

(15.18) *If $H(f)$ is an isomorphism, then $H(Cf) = 0$, i.e., Cf is acyclic in the sense of* (10.7). ∎

(15.19) *Lemma. If $H(C) = 0$ then $id \simeq 0$.*

Proof. By hypothesis $Z_q C = B_q C$. Since R is assumed to be a p.i.d., $B_q C$ is a free submodule of C_q for all q. Hence the chain complex C splits

$$0 \to B_q C \to C_q \underset{\partial}{\overset{k}{\rightleftarrows}} B_{q-1} C \to 0$$

$\partial k = id$. By (14.13), C_q can be represented as a direct sum $C_q \cong B_q C \oplus B_{q-1} C$ and $\partial_q: C_q \to C_{q-1}$ has the form $\partial_q(x, y) = (y, 0)$. Using this representation, define $D_q: C_q \to C_{q+1}$ by $D_q(x, y) = (0, x)$. Then $(\partial_{q+1} D_q + D_{q-1}\partial_q)(x, y) = \partial_{q+1}(0, x) + D_{q-1}(y, 0) = (x, 0) + (0, y) = (x, y)$. ∎

(15.20) *Definition*. A chain map $f: C \to C'$ is a *chain homotopy equivalence* provided there is a chain map $g: C' \to C$ such that $fg \simeq id_{C'}$, $gf \simeq id_C$.

(15.21) *Lemma. If $f:C \to C'$ satisfies $H(Cf) = 0$, then f is a chain homotopy equivalence.*

Proof. By (15.19) there is a chain homotopy $D:Cf \to Cf$ such that $\partial^f D + D\partial^f = id$. Then $D_q:C'_q \oplus C_{q-1} \to C'_{q+1} \oplus C_q$ defines four maps $S_q:C'_q \to C'_{q+1}, g_q:C'_q \to C_q, E_{q-1}:C_{q-1} \to C'_{q+1}, T_{q-1}:C_{q-1} \to C_q$ such that

$$D_q(x, y) = (S_q x + E_{q-1}y, g_q x + T_{q-1}y).$$

Calculating: $\partial^f D(x, y) = (\partial' Sx + \partial' Ey + fgx + fTy, -\partial gx - \partial Ty)$.

$$\begin{aligned} D\partial^f(x, y) &= D(\partial' x + fy, -\partial y) \\ &= (S\partial' x + Sfy - E\partial y, g\partial' x + gfy - T\partial y). \end{aligned}$$

Substituting $(x, 0)$ and adding, we obtain

$$(x, 0) = (\partial' Sx + fgx + S\partial' x, -\partial gx + g\partial' x).$$

Hence g is a chain map and $id - fg = \partial' S + S\partial'$. Substituting $(0, y)$ and taking the second component, we obtain

$$y = -\partial Ty + gfy - T\partial y$$

or

$$id - gf = -\partial T - T\partial. \qquad \blacksquare$$

(15.22) *Theorem. If the chain map $f:C \to C'$ induces an isomorphism $H(f)$, then f is a chain homotopy equivalence.*

Proof. By (15.18), $H(Cf) = 0$. hence by (15.21), f is a chain homotopy equivalence. ■

(15.23) The proof of (15.16) is now at hand. Theorem (15.9) yields that the inclusion $S(\mathcal{V}) \to S(X)$ induces an isomorphism in homology. Thus (15.16) is implied by (15.22). Similarly, for an excision $(X - U, A - U) \subset (X, A)$ we obtain a chain equivalence $S(X - U, A - U) \to S(X, A)$. ■

Remark. We cannot obtain the chain equivalence of (15.23) by the method of proof used for homotopy invariance, since these equivalences depend on the spaces, e.g. U and its placement in A. An explicit construction is given in Eilenberg and Steenrod [23], p. 207.

(15.24) *Exercise.* Generalize the argument of (15.3) to prove the *suspension isomorphism* $H_q^*(X) \cong H_{q-1}^*(\Sigma X)$. Here ΣX is the quotient of $X \times I$ obtained by identifying $X \times \{0\}$ and $X \times \{1\}$ to points P and Q. We can regard X as the subspace $X \times \{\frac{1}{2}\}$ in ΣX.

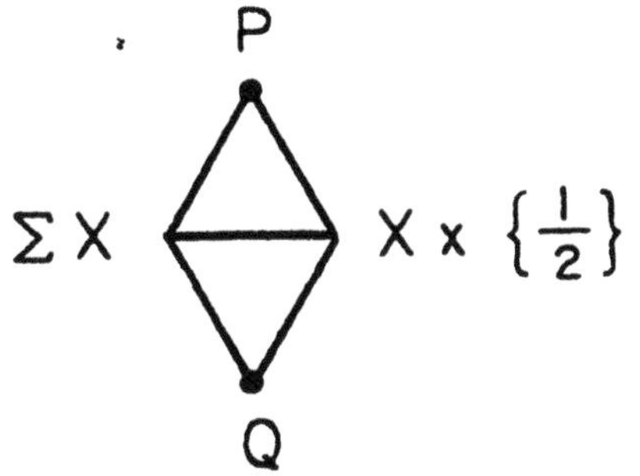

(15.25) *Exercise.* Concoct an example where excision fails for reduced homology, cf. (15.15).

(15.26) *Exercise.* Let $M_q \in \Delta_q \times I$ be the point $(B_q, \frac{1}{2})$, cf. (15.10). Use the join with M_q to prove (11.4). Suggestion: Consider $T\delta_q = M_q(i_1\delta_q - i_0\delta_q - T\partial\delta_q)$. Draw some pictures comparing the chain homotopy T with the prism operator (11.7).

(15.27) *Exercise.* Let I be a homeomorph of [0, 1] in S^2 with x a point in I. Show the inclusion $(S^2 - x, I - x) \to (S^2, I)$ is not an excision.

(15.28) *Exercise.* Assume C and $H(C)$ are free. If the chain map $f: C \to C'$ induces $H(f) = 0$, then $f \cong 0$.

16. Further Applications to Spheres

(16.1) *Proposition. Let $r: S^n \to S^n$ be the reflection $r(x_0, \ldots, x_n) = (-x_0, x_1, \ldots, x_n)$. Then the induced homomorphism*

$$H_n(r): H_n(S^n) \to H_n(S^n)$$

is multiplication by -1 for all $n \geq 1$.

Proof. This assertion is valid for $n = 0$ provided we use the reduced homology module $H_0^*(S^0)$ (exercise). But then we can do an induction on n, since we have the commutative diagram

$$\begin{array}{ccc} H_n(S^n) & \xrightarrow{\sim} & H_{n-1}^*(S^{n-1}) \\ {\scriptstyle H_n(r)}\downarrow & & \downarrow{\scriptstyle H_{n-1}(r)} \\ H_n(S^n) & \xrightarrow{\sim} & H_{n-1}^*(S^{n-1}) \end{array}$$

(see 15.4). ■

(16.2) *Proposition. Any rotation of S^n is homotopic to the identity map of S^n.*

Proof. If $A \in SO(n+1)$, there exists B such that BAB^{-1} has the form $[\frac{n-1}{2}]$ 2×2 matrices of the form

$$\begin{pmatrix} \cos\theta & \sin\theta \\ -\sin\theta & \cos\theta \end{pmatrix}$$

along the diagonal, 1 in the last diagonal place if n is even, and 0 elsewhere (Halmos [28], p. 164). Replacing θ by $t\theta$ gives a homotopy H_t such that $H_0 = id$ and $H_1 = BAB^{-1}$. The desired homotopy is $B^{-1}H_tB$. ∎

(16.3) *Proposition. Let $g:S^n \to S^n$ be the restriction of an orthogonal transformation of $\mathbf{R}^{n+1}$. Then the induced homorphism*

$$H_n(g):H_n(S^n) \to H_n(S^n)$$

is multiplication by the determinant of g (which is ± 1).

Proof. By (16.2) we may assume $\det g = -1$. But then rg and gr are rotations (r as in (16.1)), hence by (16.2) again (and the homotopy theorem) $H_n(g) = H_n(r)^{-1}$ = mult. by -1. (Exercise: If $\det g = -1$ then in fact $g \simeq r$.) ∎

(16.4) *Corollary. Let $a:S^n \to S^n$ be the antipodal map $a(x) = -x$. Then the induced homomorphism $H_n(a)$ is multiplication by $(-1)^{n+1}$.*

Because $\det a = (-1)^{n+1}$. ∎

We say that a has *degree* $(-1)^{n+1}$, (15.8).

This result leads to a classical theorem about vector fields on spheres: A *vector field* on S^n is a continuous map $v:S^n \to \mathbf{R}^{n+1}$ such that x is perpendicular to $v(x)$ for all $x \in S^n$ (which makes sense because x and $v(x)$ are both vectors in $(n + 1)$-space; if $v(x)$ is placed with its initial point at the end point of x, it will be tangent to the sphere).

(16.5) *Theorem. S^n has a nowhere vanishing vector field if and only if n is odd.*

Proof. For $n = 2m + 1$, define

$$v(x_0, x_1, \ldots, x_{2m+1}) = (-x_1, x_0, -x_3, x_2, \ldots, -x_{2m+1}, x_{2m}).$$

Conversely, given v, with $v(x) \neq 0$ for all $x \in S^n$. Then

$$w(x) = v(x)/|v(x)|$$

is a map $S^n \to S^n$, with $x \perp w(x)$ for all $x \in S^n$.

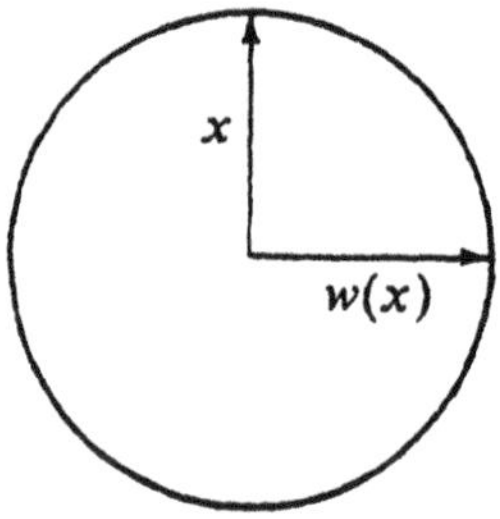

Now, as the diagram suggests, we can deform $w(x)$ back to x, or in the other direction to $-x$. To be precise,

$$F(x, t) = x \cos t\pi + w(x) \sin t\pi$$

defines a homotopy such that

$$F(x, 0) = x$$

$$F(x, \tfrac{1}{2}) = w(x)$$

$$F(x, 1) = -x$$

Thus

$$Id \simeq w \simeq a$$

but by (16.4) and the homotopy theorem, the antipodal map is not homotopic to the identity map for n even.

Note. If n is odd, we have shown in the course of the proof that any field of unit vectors w on S^n is, when considered as a map $S^n \to S^n$, homotopic to the identity map.

(16.6) *Exercise.* Let f, g be maps $S^n \to S^n$ such that $f(x) \neq g(x)$ for all $x \in S^n$. Then $f \simeq ag$, where a is the antipodal map, hence

$$H_n(f) = (-1)^{n+1} H_n(g)$$

In particular, any map $S^n \to S^n$ without fixed points is homotopic to the antipodal map.

(16.7) *Exercise.* Let $f:S^n \to S^n$ be homotopic to a constant, e.g., any map which is not onto. Then f has a fixed point and also a point x at which $f(x) = -x$ (use (16.7)).

(16.8) *Exercise.* Any $f:S^{2n} \to S^{2n}$ either has a fixed point or sends some point into its antipode.

(16.9) *Exercise.* Adapt the construction in (16.5) to produce 3 linearly independent nonvanishing tangent vector fields on S^{4n+3}.

(16.10) *Vector fields on spheres Theorem.* Write $n+1 = (\text{odd})(2^{4a+b})$, $0 \le b \le 3$. Then S^n admits exactly $2^b + 8a - 1$ linearly independent nonvanishing tangent vector fields. The construction is due to Hurwitz-Radon and Eckmann. The proof that no more exist is a hard theorem due to J. F. Adams [1].

Along the lines of (16.9), construct 7 independent tangent vector fields on S^7 and 8 on S^{15}. Can you see a general pattern for producing the Hurwitz-Radon number of independent vector fields on S^n of the form $(\pm x_{\sigma(0)}, \ldots, \pm x_{\sigma(n)})$ where σ is permutation of $(0, \ldots, n)$?

(16.11) *Exercise.* Let G be a group of homeomorphisms acting freely on S^{2n}, i.e., for all $g \in G$, $gx = x$ for some $x \Leftrightarrow g = 1$. Prove $|G| \le 2$.

(16.12) *Exercise.* Every $f:\mathbf{P}^{2n} \to \mathbf{P}^{2n}$ has a fixed point. Suggestion: Use (16.9) and the lifting theorem (6.1). Construct $f:\mathbf{P}^{2n+1} \to \mathbf{P}^{2n+1}$ without fixed points.

(16.13) *Exercise.* Prove that the covering projection $S^n \to \mathbf{P}^n$ is not null-homotopic. Suggestion: The lifting theorem provides a set equivalence $[X, S^n] \to [X, \mathbf{P}^n]$ for all simply connected X.

(16.14) *Exercise.* Refine (15.4) to observe that replacing the pair (S^n, E_n^-) by (S^n, E_n^+) has the effect of multiplying $H_n(S^n)$ by -1.

(16.15) *Exercise.* We define $f, g:S^n \to S^n$ to be *orthogonal* at a point $x \in S^n$, if the inner product (in $\mathbf{R}^{n+1}$) $f(x) \cdot g(x) = 0$. Prove; if $|\deg f| \ne |\deg g|$, then f, g are orthogonal at some x.

17. Mayer-Vietoris Sequence

We now consider *triads* (X, X_1, X_2), ordered triples of spaces such that X_1 and X_2 are subspaces of X. We have inclusion maps

$$k_1:(X_2, X_1 \cap X_2) \to (X_1 \cup X_2, X_1),$$

$$k_2:(X_1, X_1 \cap X_2) \to (X_1 \cup X_2, X_2)$$

obtained by excising $X_1 - X_1 \cap X_2$, $X_2 - X_1 \cap X_2$ from $X_1 \cup X_2$. If both k_1 and k_2 are excisions,[†] the triad is called *exact* (or "proper" in some books): Thus

$$H_q(k_i):H_q(X_{i'}, X_1 \cap X_2) \longrightarrow H_q(X_1 \cup X_2, X_i)$$

is an isomorphism for all q, $(i, i') = (1, 2)$ or $(2, 1)$. These isomorphisms are analogous to the first isomorphism theorem in group theory. Exactness depends only on $X_1 \cup X_2$, not X.

(17.1) *Example.* If X_1, X_2 are both open sets, then (X, X_1, X_2) is exact. For we can assume $X = X_1 \cup X_2$. For $A = X_1$, $U = X_1 - X_1 \cap X_2$, then $X - U = X_2$, so U is a closed subset of X contained in the open A, and the excision theorem applies.

(17.2) *Example.* (S^n, E_n^+, E_n^-) is an exact triad, by (15.3).

Assume until further notice $X = X_1 \cup X_2$. let $A = X_1 \cap X_2$.

The inclusion $(X_1, A) \to (X, X_2)$ induces the ladder

[†] If one is then so is the other (cf. Dold [64], p. 47)

(17.3)

$$\begin{array}{ccccccccccc}
\to H_q(A) & \to & H_q(X_1) & \to & H_q(X_1, A) & \to & H_{q-1}(A) & \to & H_{q-1}(X_1) & \to \cdots \\
\downarrow & & \downarrow & & \downarrow & & \downarrow & & \downarrow & \\
\to H_q(X_2) & \to & H_q(X) & \to & H_q(X, X_2) & \to & H_{q-1}(X_2) & \to & H_{q-1}(X) & \to \cdots
\end{array}$$

in which all rectangles are commutative. A useful lemma for dealing with this set-up is

(17.4) *Barratt-Whitehead Lemma. Given a diagram of R-modules and homomorphisms in which all rectangles commute and rows are exact*

$$\begin{array}{ccccccccccccc}
\longrightarrow & C_{i+1} & \longrightarrow & A_i & \xrightarrow{f_i} & B_i & \xrightarrow{g_i} & C_i & \xrightarrow{h_i} & A_{i-1} & \longrightarrow & B_{i-1} & \longrightarrow \\
 & \downarrow{\scriptstyle \gamma_{i+1}} & & \downarrow{\scriptstyle \alpha_i} & & \downarrow{\scriptstyle \beta_i} & & \downarrow{\scriptstyle \gamma_i} & & \downarrow{\scriptstyle \alpha_{i-1}} & & \downarrow{\scriptstyle \beta_{i-1}} & \\
\longrightarrow & C'_{i+1} & \longrightarrow & A'_i & \xrightarrow{f'_i} & B'_i & \xrightarrow{g'_i} & C'_i & \xrightarrow{h'_i} & A'_{i-1} & \longrightarrow & B'_{i-1} & \longrightarrow \quad ;
\end{array}$$

if the γ_i are isomorphisms, then there is a long exact sequence

(17.5) $$\to A_i \xrightarrow{\Phi_i} A'_i \oplus B_i \xrightarrow{\Psi_i} B'_i \xrightarrow{\Gamma_i} A_{i-1} \to \cdots$$

where $\Phi_i = (\alpha_i \oplus f_i)\Delta$, $\Psi_i = \nabla'(-f'_i \oplus \beta_i)$, $\Gamma_i = h_i\gamma_i^{-1}g'_i$. Recall $\Delta(a) = (a, a)$, $\nabla'(x, y) = (x + y)$ (14.13).

Proof. The proof is a diagram chase. We prove exactness at B'_i. The composite $\Gamma_i\Psi_i(a, b) = \Gamma_i(-f'_i(a) + \beta_i(b)) = h_i\gamma_i^{-1}g'_i\beta_i(b)$ since $g'_if'_i = 0$. But $\gamma_i^{-1}g'_i\beta_i = g_i$ and $h_ig_i = 0$. Hence $\Gamma_i\Psi_i = 0$. If $b \in B'_i$ satisfies $\Gamma_i(b) = 0$, then there is $b_1 \in B_i$ such that $g_i(b_1) = \gamma_i^{-1}g'_i(b)$. Since $g'_i(b - \beta_ib_1) = 0$, there is $a \in A'_i$ such that $f'_i(a) = b - \beta_ib_1$. Then $\Psi_i(-a, b_1) = b$. We leave the remaining steps for the reader. ■

We shall refer to (17.5) as the *Barratt-Whitehead sequence* of the ladder.

(17.6) Functoriality of Barratt-Whitehead sequences is an immediate consequence of the construction. A map of ladders induces a map of Barratt-Whitehead sequences.

(17.7) *Definition.* If (X, X_1, X_2) is an exact triad, then (17.3) is a ladder with $H_q(X_1, A) \to H_q(X, X_2)$ an isomorphism. The associated Barratt-Whitehead sequence is the *Mayer-Vietoris* sequence of the triad,

$$\to H_{q+1}(X) \xrightarrow{\Gamma_{q+1}} H_q(A) \xrightarrow{\Phi_q} H_q(X_1) \oplus H_q(X_2) \xrightarrow{\Psi_q} H_q(X) \to \dots .$$

Functoriality is a consequence of (17.6).

(17.8) *Corollary. Suppose that for some* q, $H_{q+1}(X) = 0$. *Then necessary and sufficient conditions that an element* $a \in H_q(A)$ *be zero are that* $H_q(m_1)(a) = 0 = H_q(m_2)(a)$, where $m_i{:}A \to X_i$ are inclusions.

Proof. The conditions are certainly necessary. Conversely, these conditions imply $\Phi(a) = 0$, by definition of Φ. By exactness, a is in the image of Γ, hence by hypothesis is zero. ■

This little corollary will be very useful in the next section.

(17.9) *Remark.* If $A \neq \emptyset$, the Mayer-Vietoris sequence can be terminated in

$$\to H_1(X) \to H_0^\#(A) \to H_0^\#(X_1) \oplus H_0^\#(X_2) \to H_0^\#(X) \to 0,$$

by remark (15.15).

(17.10) *Relative Mayer-Vietoris sequence.* Suppose (X, X_1, X_2) is an exact triad but not necessarily $X = X_1 \cup X_2$. Let $Y = X_1 \cup X_2$ and $A = X_1 \cap X_2$. Then there is a *relative Mayer-Vietoris sequence* which is exact:

$$\dots \to H_q(X, A) \to H_q(X, X_1) \oplus H_q(X, X_2) \to H_q(X, Y)$$
$$\to H_{q-1}(X, A) \dots$$

and functorial for maps of exact triads. The proof is to apply (17.4) to the ladder obtained from the homology long exact sequence of triples (14.6) and the diagram of inclusion maps

$$\begin{array}{ccccc} (X_1, A) & \to & (X, A) & \to & (X, X_1) \\ \downarrow & & \downarrow & & \downarrow \\ (Y, X_2) & \to & (X, X_2) & \to & (X, Y) \end{array}$$

in which the left vertical map is an excision.

(17.11) For certain calculations (notably (26.6)) we want to evaluate Γ on the chain level. Consider

$$\begin{array}{ccccc} S(A) & \longrightarrow & S(X_1) & \xrightarrow{k_1} & S(X_1, A) \\ \downarrow i & & \downarrow i & & \downarrow i \; \uparrow j \\ S(X_2) & \longrightarrow & S(X) & \xrightarrow[k]{} & S(X, X_2) \end{array}$$

where i is the chain map induced by inclusion and j is a chain inverse, $ij - id = \partial D + D\partial$, given by (15.22). First note, if $z \in Z_{q+1}(X)$ and $w \in S_{q+1}(X_1)$ satisfy

$$ki(w) \sim kz \quad \text{in } S(X, X_2),$$
$$k_1(\partial w) = 0 \quad \text{so } \partial w \text{ lies in } S_q(A),$$

then $\overline{\partial w} = \Gamma\bar{z}$. Furthermore a suitable w is one satisfying $k_1 w = jkz$, since $kiw = ik_1 w = ijkz = kz + \partial Dkz$; and $k_1(\partial w) = \partial k_1 w = jk\partial z = 0$.

(17.12) *Application*. A *graph* is a space which is a union of finitely many closed arcs, any 2 of which have at most end points in common (i.e., no crossings); a *closed arc* is a homeomorph of the closed unit interval or of S^1. We wish to compute the homology of a graph. By direct sum decomposition (9.5), it suffices to consider connected graphs. Now a connected graph is homotopically equivalent to an r-leaved rose G_r

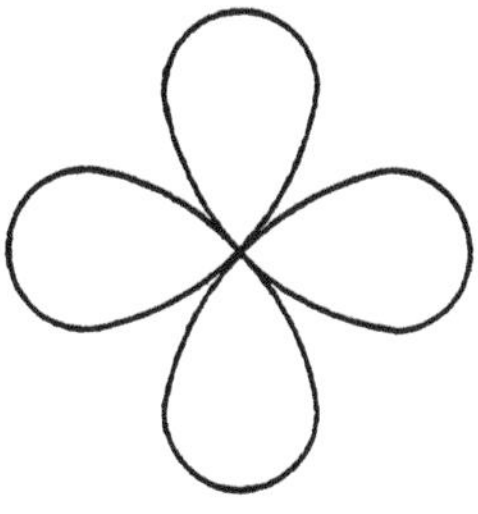

G_4

the union of r topological circles having one point P in common (see Artin [3], p. 49). So it suffices to compute the homology of G_r.

Now $G_1 = S^1$ and we know the answer. For $r \geq 2$, $G_r = S^1 \cup G_{r-1}$, with $S^1 \cap G_{r-1} = \{P\}$. Now (G_r, G_{r-1}, S^1) is an exact triad. The maps

$$k_1 : (G_{r-1}, P) \to (G_r, S^1),$$
$$k_2 : (S^1, P) \to (G_r, G_{r-1})$$

are obtained by excising $S^1 - P$, $G_{r-1} - P$, respectively. By excising slightly smaller opens, we can apply the excision theorem; then a deformation retract argument shows k_1 and k_2 are excisions. Thus we can use the Mayer-Vietoris sequence: For $q > 0$,

$$0 = H_q(P) \xrightarrow{\Phi} H_q(S^1) \oplus H_q(G_{r-1}) \xrightarrow{\Psi} H_q(G_r) \xrightarrow{\Gamma} H_{q-1}(P) = 0$$

which gives

$$H_q(G_r) \cong H_q(S^1) \oplus H_q(G_{r-1})$$

for all $q > 1$. The same result is also valid for $q = 1$ as can be seen by directly showing that $\Gamma: H_1(G_r) \to H_0(P)$ is the zero map or by remarking that in this case the Mayer-Vietoris sequence holds for reduced homology (17.9).

Thus by induction we get

$$H_q(G_r) = 0 \qquad q > 1,$$
$$H_1(G_r) \cong \underbrace{R \oplus R \oplus \cdots \oplus R}_{r \text{ copies}}.$$

(17.13) *Example.* We compute the homology of the torus T by two methods.

(a). The torus can be regarded as two annuli A identified along their common boundary. (By abuse of notation both annuli are denoted A.) The boundary is the disjoint union $C_1 + C_2$ of two circles C_1, C_2 with the inclusions $C_i \to A$ homotopy equivalences. Using the excision theorem (15.1) and (15.2) we observe that $(A, \partial A) \subset (T, A)$ induces an isomorphism in homology. Hence using (17.7) we get an exact Mayer-Vietoris sequence

$$0 \to H_2(T) \to H_1(C_1 + C_2) \xrightarrow{\Phi_1} H_1(A) \oplus H_1(A)$$
$$\to H_1(T) \to H_0^*(C_1 + C_2) \to 0.$$

By (9.5) and (15.5), $H_1(C_1 + C_2) \cong R \oplus R$, and by (9.7) $H_0^*(C_1 + C_2) \cong R$. Using the homotopy equivalences to identify $H_1(A) \oplus H_1(A) = R \oplus R$, the matrix of Φ_1 with respect to the natural basis is $\left(\begin{smallmatrix}1&1\\1&1\end{smallmatrix}\right)$. It follows that ker $\Phi_1 \cong R$ generated by $(1, -1)$ and im $\Phi_1 \cong R$ generated by $(1, 1)$. Then $H_2T \cong R$ and $H_1(T) \cong R \oplus R$.

(b) The torus can be regarded as $S^1 \times S^1$. Let $x_0 \in S^1$ be a point and write $S^1 \vee S^1$ for the subspace $S^1 \times \{x_0\} \cup \{x_0\} \times S^1$. We use the long exact sequence for the pair $(S^1 \times S^1, S^1 \vee S^1)$. Let $p_i: S^1 \times S^1 \to S^1$ be projections, $j_i: S^1 \to S^1 \vee S^1$ inclusions $j_1(x) = (x, x_0)$, $j_2(x) = (x_0, x)$. From (17.12) we have a direct sum decomposition (14.13)

$$\begin{array}{ccccc} H_q(S^1) & & & & H_q(S^1) \\ & \searrow^{H_q(j_1)} & & {}^{H_q(j_2)}\swarrow & \\ id\downarrow & & H_q(S^1 \vee S^1) & & \downarrow id \\ & \swarrow_{H_q(p_1)} & & {}_{H_q(p_2)}\searrow & \\ H_q(S^1) & & & & H_q(S^1) \end{array}$$

Let $V = S^1 \times U \cup U \times S^1$. Since each j_i factors through $S^1 \times S^1$, the map induced in homology by the inclusion $S^1 \vee S^1 \to S^1 \times S^1$ has a left inverse $\phi: H(S^1 \times S^1) \to H_1(S^1) \oplus H_1(S^1)$ given by $\phi(x) = (H(p_1)(x), H(p_2)(x))$. Hence $H_q(S^1 \vee S^1) \to H_q(S^1 \times S^1)$ is a monomorphism by (14.12). (In (21.17) we shall see that $S^1 \vee S^1$ is not a retract of $S^1 \times S^1$.) To calculate $H_q(S^1 \times S^1, S^1 \vee S^1)$ we first thicken $S^1 \vee S^1$ to $U = S^1 \times J \cup J \times S^1$, where J is a small interval with center x_0, figure 3.

Then by using a homotopy similar to (15.3) and (13.15),

$$H_q(S^1 \times S^1, S^1 \vee S^1) \cong H_q(T, V).$$

The inclusion $(T - S^1 \vee S^1, V - S^1 \vee S^1) \subset (T, V)$ satisfies (15.1) so is an excision. Let D be the disc in T pictured in figure 3. The inclusion $(D, \partial D) \subset (T - S^1 \vee S^1, V - S^1 \vee S^1)$ is a homotopy equivalence. Combining, we have $H_q(S^1 \times S^1, S^1 \vee S^1) \cong H_q(D, \partial D)$ which has the same values as the homology modules of the 2-sphere, (15.5). Substituting these results in the long exact homology sequence of the pair $(T, S^1 \vee S^1)$ we obtain $H_2(T) \cong H_2(D, \partial D)$ and $H_1(T) \cong H_1(S^1 \vee S^1) \cong R \oplus R$. The reader might contemplate the geometric significance (if any) of the facts $H_2(T) \neq 0$ and $\pi_2 T = 0$.

(17.14) *Exercise.* Let D_k be the surface obtained by removing k small disjoint open 2-discs from the unit disc E^2. Show that $D_k \simeq G_k$, the k-leaved

rose. Let M_k be the surface obtained by identifying two copies of D_k along

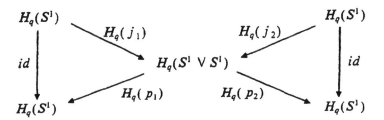

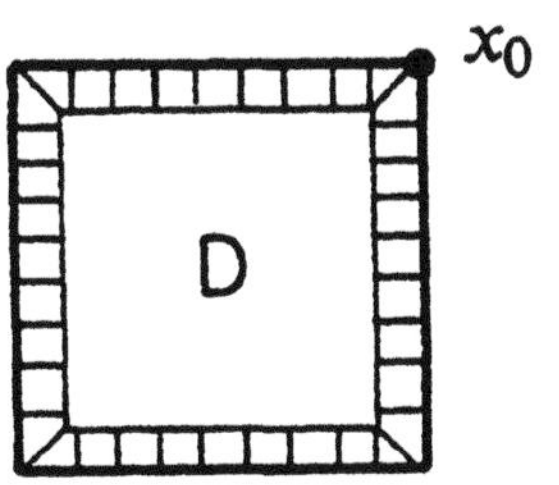

Figure 3. T is an identification space of $I \times I$.

their boundaries. Use the method of (17.13)(a) to calculate $H_2(M_k) \cong R$ and $H_1(M_k) \cong R \oplus \ldots \oplus R$ ($2k$ copies).

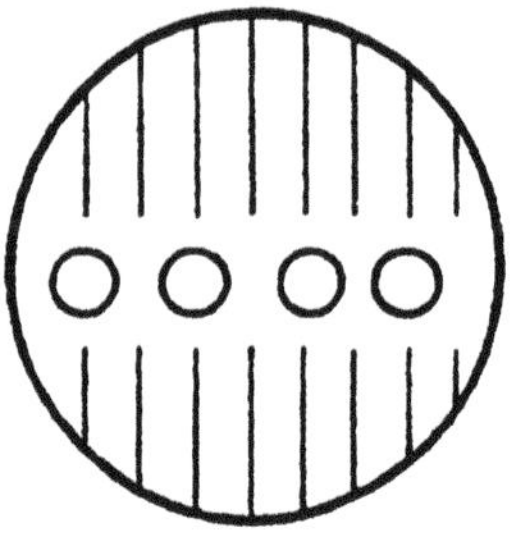

D_4

(17.15) *Exercise.* Replace E^2 in (17.14) by the n-disc E^n, $n \geq 3$. Let D_k^n be obtained from E^n by removing k small disjoint open n-discs. Show $D_k^n \simeq S^{n-1} \vee \ldots \vee S^{n-1}$, k-copies. Let M_k^n be obtained by identifying two copies of D_k^n along their boundaries. Show $H_0(M_k^n) \cong H_n(M_k^n) \cong R$, $H_1(M_k^n) \cong H_{n-1}(M_k^n) \cong R \oplus \ldots \oplus R$ (k-copies) and $H_q(M_k^n) = 0$ for the remaining values of q.

(17.16) *Exercise.* Let T^* be the torus T with a small open disc removed. Let $C \subset T^*$ be the boundary circle. Show that the inclusion $C \to T^*$ induces 0 on H_1. Suggestion: This follows from the calculation in (17.13)(a) and a Mayer-Vietoris sequence computing $H(T)$ in terms of T^* and the disc. Use this observation to calculate $H_q(M_2)$ regarding M_2 as two copies of T^* identified along C. Generalize to calculate $H_q(M_k)$.

(17.17) *Exercise.* Suppose $x_0 \in X$, $y_0 \in Y$. Write $X \bigtriangledown Y = X \times \{y_0\} \cup \{x_0\} \times Y \subset X \times Y$. Obtain a direct sum decomposition for $H_q^*(X \bigtriangledown Y)$. Obtain split short exact sequences $0 \to H_q^*(X \bigtriangledown Y) \to H_q^*(X \times Y) \to H_q^*(X \times Y, X \bigtriangledown Y) \to 0$.

(17.18) *Exercise.* The *join* $X * Y$ of two spaces X, Y is the space obtained from $X \times I \times Y$ by making identifications $(x, 1, y) \sim (x', 1, y)$ and $(x, 0, y) \sim (x, 0, y')$. In $X * Y$ there are the subspaces $C_+X \times Y = \{(x, t, y) \mid \frac{1}{2} \leq t \leq 1\}$ and $X \times C_-Y = \{(x, t, y) \mid 0 \leq t \leq \frac{1}{2}\}$. Then $C_+X \times Y \cap X \times C_-Y = X \times Y$ regarded as the subspace $\{(x, \frac{1}{2}, y)\}$. Show that the triad $(X * Y, C_+X \times Y, X \times C_-Y)$ is exact, and $H_{q+1}^*(X * Y) \cong H_q^*(X \times Y, X \vee Y)$.

(17.19) *Exercise.* Regard the Klein bottle K as two copies of the Möbius strip M identified along their boundaries. Show $H_1(K) \cong R \oplus (R/2R)$ and $H_2(K) \cong R_2 = \{r \mid 2r = 0\}$. Use (12.13) for information on $H_q(M)$. Use homology to prove the nonretraction result in (4.15).

(17.20) *Exercise.* Let A, B be subsets of S^n, $n \geq 2$. Show (a), if A and B are closed, disjoint, and neither separates S^n, then $A \cup B$ does not separate S^n. Show (b) if A, B are connected, open, and $A \cup B = S^n$, then $A \cap B$ is connected.

(17.21) *Exercise.* Let (X, X_1, X_2) be an exact triad, $X = X_1 \cup X_2$ and $B \subset X_1 \cap X_2 = A$. Obtain another relative Mayer-Vietoris sequence

$$\to H_q(A, B) \to H_q(X_1, B) \oplus H_q(X_2, B) \to H_q(X, B) \to \ldots \quad .$$

The cases $A = B$, and $X = Y$ in (17.10), yield direct sum decompositions. Further generalizations appear in Eilenberg-Steenrod [23], Dold [64] or Spanier [52].

(17.22) *Exercise.* Let a denote the antipodal map. Observe that the diagram

$$\begin{array}{ccc} (E_n^+, S^{n-1}) & \to & (S^n, E_n^-) \\ a \downarrow & & \downarrow a \\ (E_n^-, S^{n-1}) & \to & (S^n, E_n^+) \end{array}$$

commutes. Let Γ_1 and Γ_2 be the Mayer-Vietoris connecting homomorphisms associated with the top and bottom sequences respectively. Show $\Gamma_1 = -\Gamma_2$. Suggestion: Use (16.14). Use this relation to obtain another proof of (16.4).

(17.23) *Exercise.* Let $s^{n-1} \subset R^n$ be a homeomorph of s^{n-1} which has an open neighborhood U in R^n homeomorphic to $s^{n-1} \times (-1, 1)$ by a homeomorphism of s^{n-1} with $s^{n-1} \times \{0\}$ ("thickened sphere"). Use Mayer-Vietoris to show $R^n - s^{n-1}$ has two components.

18. The Jordan-Brouwer Separation Theorem

We apply the results of the previous section to compute the homology of the complement of a closed cell inside a sphere (where a *closed cell* is a homeomorph of I^n for some n) and the homology of the complement of a sphere inside a sphere.

(18.1) *Theorem. Let e_r be a closed cell of dimension r in S^n. Then*

$$H_q^{\#}(S^n - e_r) = 0 \quad \textit{for all} \quad q \geq 0$$

Proof: By induction on r. If $r = 0$, e_r is a point, and $S^n - e_r$ is contractible. Suppose $r > 0$ and the theorem true for $r - 1$. Let z be a q-cycle in $S^n - e_r$. Let $\phi: I^r \to e_r$ be a homeomorphism. For $t \in I$, let $e_{r-1}(t) = \phi(t \times I^{r-1})$, a closed $(r-1)$-cell. Since $S^n - e_{r-1}(t) \supset S^n - e_r$, $z = \partial w_t$, where w_t is a $(q+1)$-chain in $S^n - e_{r-1}(t)$. The support[†] $|w_t|$ of w_t is a compact set which doesn't meet the compact set $e_{r-1}(t)$; let ε_t be the distance between these sets, so that $\varepsilon_t > 0$. By uniform continuity, there is a $\delta_t > 0$ such that 2 points in I^r less than δ_t apart have images under ϕ less than ε_t apart. Let I_t be an open interval centered at t of length $< \delta_t$. Let $e_r(t) = \phi(I_t \times I^{r-1})$, an open r-cell; then every point of $e_r(t)$ has distance $< \varepsilon_t$ from $e_{r-1}(t)$, so that $e_r(t)$ doesn't meet $|w_t|$, and $z = \partial w_t$ in $S^n - e_r(t)$.

Now since I is compact and covered by the open intervals I_t, there is a $\rho > 0$ such that every closed interval of length $< \rho$ is contained in some I_t. Choose $m > 0$ so that $1/m < \rho$, and consider the closed intervals

[†] If σ is a singular q-simplex, define its *support* $|\sigma|$ to be $\sigma(\Delta_q)$. For a q-chain $c = \Sigma v_i \sigma_i$, define $|c| = \cup_i |\sigma_i|$.

$$I_0 = [0, 1/m], I_1 = [1/m, 2/m], \ldots, I_{m-1} = [(m-1)/m, 1]$$

Let $e_{r,j}$ be the image of $I_j \times I^{r-1}$. Then there is a chain w_j in $S^n - e_{r,j}$ such that $z = \partial w_j$. By induction on j, we are reduced to proving the following sublemma.

Sublemma. Let J_1, J_2 be closed subintervals of I such that $J_1 \cap J_2 = \{t\}$. Let $e' = \phi(J_1 \times I^{r-1})$, $e'' = \phi(J_2 \times I^{r-1})$. Suppose there are $(q+1)$-chains w', w'' in $S^n - e'$, $S^n - e''$ respectively such that $\partial w' = z = \partial w''$. Then there is also a $(q+1)$-chain w in $S^n - (e' \cup e'')$ such that $z = \partial w$.

Proof: Let $X = S^n - e_{r-1}(t)$, $X_1 = S^n - e'$, $X_2 = S^n - e''$, $A = X_1 \cap X_2 = S^n - (e' \cup e'')$. Since X_1, X_2 are open, (X, X_1, X_2) is an exact triad. Moreover, since $e' \cup e''$ is contractible and S^n is not, A is non-empty; thus the triad is exact for reduced homology also. By inductive assumption, $H_{q+1}(X) = 0$. Since z bounds in X_1 and X_2, Corollary (17.8) tells us that z bounds in A. ■

(18.2) *Corollary. S^n cannot be disconnected by removing a closed cell.*
For if $S^n - e_r$ were disconnected, $H_0^*(S^n - e_r) \neq 0$, contradiction. ■

(18.3) *Theorem. Let s_r be a subspace of S^n which is a homeomorph of S^r. Then $r \leq n$. If $r = n$, then $s_n = S^n$. If $r < n$, then*

$$H_q^*(S^n - s_r) = \begin{cases} R & q = n - r - 1 \\ 0 & \text{otherwise} \end{cases}$$

Proof: Let e_r^+, e_r^- be the images of E_r^+, E_r^-, s_{r-1} the image of $E_r^+ \cap E_r^-$. In case $r = n$ assume $s_n \neq S^n$. Let $X = S^n - s_{r-1}$, $X_1 = S^n - e_r^-$, $X_2 = S^n - e_r^-$, $A = S^n - s_r$. For $r \neq n$, s_r and S^n have different homology, so that $A \neq \phi$. (We also have this for $r = n$ by assumption.) By (18.1), $H_q^*(X_1) = 0 = H_q^*(X_2)$ for all q. By the Mayer-Vietoris sequence,

$$H_{q+1}^*(X) \cong H_q^*(A) \quad \text{all } q$$

By decreasing induction on r, we get

$$H_q^*(S^n - s_r) \cong H_{q-r}^*(S^n - s_0)$$

Now it is easy to see that $S^n - s_0 = S^n$-{two points} is homotopically equivalent to S^{n-1}. Thus

$$H_q^*(S^n - s_r) = \begin{cases} R & q + r = n - 1 \\ 0 & \text{otherwise} \end{cases}$$

For $r = n$ we get $H_{-1}^*(S^n - s_n) = R$, which is absurd. Thus $s_n \neq S^n$ cannot occur. It follows at once that for $r > n$ there is no s_r inside S^n (nor any e_r either, for that matter!). ■

(18.4) *Remark.* For $r = 1$, s_1 is called a *knot.* One is usually interested in $s_1 \subset \mathbf{R}^3$. Regard $\mathbf{R}^3$ as S^3 minus a point P. Let us compute $H_q^*(\mathbf{R}^3 - s_1)$: We have the commutative diagram

$$\begin{array}{ccccccc} \to H_q^*(S^3 - s_1) & \to & H_q^*(S^3) & \to & H_q^*(S^3, S^3 - s_1) & \to \\ \uparrow & & \uparrow & & \uparrow & \\ \to H_q^*(\mathbf{R}^3 - s_1) & \to & H_q^*(\mathbf{R}^3) & \to & H_q^*(\mathbf{R}^3, \mathbf{R}^3 - s_1) & \to \end{array}$$

By excision the thir∨vertical arrow is an isomorphism, and since the reduced homology of a 3-space is 0, we get

$$H_q^*(\mathbf{R}^3 - s_1) \cong H_{q+1}(S^3, S^3 - s_1)$$

By (18.3) and (15.5), we get

$$H_q^*(\mathbf{R}^3 - s_1) = \begin{cases} R & q = 1, 2 \\ 0 & \text{otherwise} \end{cases}$$

In particular, the homology of the complement is the same for all knots. However, the fundamental group of the complement is definitely not the same, and is an important invariant of the way the knot is imbedded (see Crowell and Fox [16]).

(18.5) *Corollary. If $r \leq n$, then removing an s_r from S^n disconnects S^n if and only if $r = n - 1$.*

Once again, we just look at $H_0^*(S^n - s_r)$. ■

In case $n - 1 = r$ we can say much more.

(18.6) *Jordan-Brouwer Separation Theorem. For any s_{n-1} inside S^n, $S^n - s_{n-1}$ consists of two connected components, both having s_{n-1} as frontier.*

Proof: Since $H_0^*(S^n - s_{n-1})$ is isomorphic to R, $S^n - s_{n-1}$ has exactly 2 components K_1, K_2. These are open subsets of S^n, so the frontier of each component is contained in s_{n-1}. Conversely, let $x \in s_{n-1}$, and let U be any open neighborhood of x in S^n. We must show U meets both K_1 and K_2.

Choose any $y_1 \in K_1, y_2 \in K_2$. We can choose a set A such that

$$x \in A \subset U \cap s_{n-1}$$

and $s_{n-1} - A$ is a closed $(n-1)$-cell e_{n-1}. By (18.1), $H_0^*(S^n - e_{n-1}) = 0$, so $S^n - e_{n-1}$ is connected, in fact arcwise-connected. Let σ be a closed arc in $S^n - e_{n-1}$ from y_1 to y_2. Since y_1 and y_2 are in distinct components of $S^n - s_{n-1}$, σ crosses s_{n-1}, and in fact, $\sigma \cap s_{n-1} \subset A$. Since $\sigma \cap s_{n-1}$ is a closed subset of σ, there is a first point x_1 and a last point x_2 of $\sigma \cap s_{n-1} = \sigma \cap A$. Now all points of the part σ_1 of σ from y_1 to x_1 lie in K_1 (except x_1) and U meets σ_1 in a non-empty open subset of σ_1, so U meets K_1; similarly U meets K_2.

(18.7) *Corollary. Let $n \geq 2$, s_{n-1} a homeomorph of S^{n-1}* in $\mathbf{R}^n$. *Then $\mathbf{R}^n - s_{n-1}$ has two connected components, both having s_{n-1} as frontier.*

(Regard $\mathbf{R}^n$ as S^n minus a point.) The bounded component of $\mathbf{R}^n - s_{n-1}$ is called the *inside* of s_{n-1}, the unbounded component (including the point at infinity) the *outside*. Note that the case $n = 1$ is quite different!

For $n = 2$, Corollary (18.7) is the famous Jordan Curve Theorem. Trying to prove this "intuitively obvious" fact is a good way to learn the difference between mathematics and drawing. For $n = 2$, there is also a stronger result: Any homeomorphism of S^1 onto a subspace of the plane extends to a homeomorphism of the plane onto itself. This theorem is due to Schoenflies, and is the key to the triangulability of compact surfaces (Ahlfors and Sario [2], p. 105). Schoenflies' Theorem implies that the inside and outside of an s_1 are open 2-cells. For $n = 3$ this statement is false, as is shown by the Alexander horned sphere (Hocking and Young [32], p. 176). Recently, Mazur [39] and Brown [9] have shown that a "nicely embedded" $s_{n-1} \subset \mathbf{R}^n$ has inside and outside which are open n-cells.

Some other famous separation theorems on S^n are those of Borsuk (Hocking and Young [32], Section 6–17) and Phragmen-Brouwer ([32], Section 8–8). The source of most separation theorems on S^n is the Alexander Duality Theorem (Section 27).

(18.8) *Corollary. Let $n \geq 2$ and $f: E^n \to \mathbf{R}^n$ a one-to-one map. Then f is a homeomorphism of E^n onto $f(E^n)$; if $s_{n-1} = f(S^{n-1})$, then f maps the interior of E^n onto the inside of s_{n-1}.*

Proof: Let $e_n = f(E^n)$. Consider $\mathbf{R}^n$ as S^n minus a point. Then $S^n - e_n$ is connected (18.2), in fact pathwise connected; since $n \geq 2$, $\mathbf{R}^n - e_n$ is also path connected (any path through the point at infinity can be moved slightly so as to avoid that point). $\mathbf{R}^n - e_n$ is unbounded, so is contained in the outside B of s_{n-1}. Hence $e_n \supset \mathbf{R}^n - B = s_{n-1} \cup A$, where A is the inside. Now

$$e_n = f(\dot{E}^n) \cup s_{n-1}$$

(disjoint union), so $A \subset f(\dot{E}^n)$. Since $\dot{E}^n$ is connected, $A = f(\dot{E}^n)$. (That f is a homeomorphism onto e_n is a result of elementary point-set topology). ■

(18.9) *Corollary. Assume $n \geq 2$. Let $U \subset \mathbf{R}^n$ be open connected, $f: U \to \mathbf{R}^n$ a one-to-one map. Then $f(U)$ is connected open and f is a homeomorphism onto $f(U)$. (Invariance of Domain.)*

Proof: Take $x \in U$, and let $V \subset U$ be any open neighborhood of x. Take a closed ball e_n centered at x, $e_n \subset V$. Then by (18.8), $f(\dot{e}_n)$ is an open neighborhood of $f(x)$. ■

(18.10) *Exercise.* Let X, Y be n-dimensional manifolds (6.8), U open in X, $f: U \to Y$ a one-to-one map. Then $f(U)$ is open in Y (hence f maps U homeomorphically onto $f(U)$).

(18.11) *Exercise.* Let X be an n-dimensional manifold, Y an m-dimensional manifold, U open in X. If $m < n$, there is no one-to-one map of U into Y. (Do the case $X = \mathbf{R}^n$, $Y = \mathbf{R}^m$ first.) In particular, X is not homeomorphic to Y.

(18.12) *Exercise.* Let X, Y, U be as in (18.11), but this time assume $m > n$. If $f: U \to Y$ is a one-to-one map, then $f(U)$ is not open in Y, in fact the interior of $f(U)$ is empty. (Hint: Use the Baire-Moore Theorem (Hocking and Young, Theorem 2–79) to reduce to the case of a compact set whose image has nonempty interior, then apply 18.11.) Give an example where f does not map U homeomorphically onto $f(U)$.

(18.13) *Exercise.* Corollary (18.9) is also true for $n = 1$.

(18.14) *Exercise.* Calculate $H_q^*(\mathbf{R}^n - e_r)$, $H_q^*(\mathbf{R}^n - s_r)$ for all q, n, r.

(18.15) *Exercise.* Let x be a point in the unbounded component of $\mathbf{R}^n - s_{n-1}$, $n \geq 1$. Show the inclusion $s_{n-1} \to \mathbf{R}^n - \{x\}$ induces the zero map in homology. It can be proved that if x lies in the bounded component, then the inclusion induces an isomorphism in homology. However, without further assumptions on the embedding of S^{n-1} in $\mathbf{R}^n$, the proof needs the results of section 27.

(18.16) *Exercise.* Let $t_{r,k} \subset S^n$ be a homeomorph of $S^r \times E^k$ and s_r the image of $S^r \times \{0\}$. Show the inclusion $S^n - t_{r,k} \to S^n - S_n$ induces an isomorphism in homology.

(18.17) *Exercise.* We denote the homotopy classes of maps $W \to X$ by $[W, X]$. A map $f:X \to Y$ induces $f_*:[W, X] \to [W, Y]$, $f_*(g) = fg$. Use the idea of support to prove if $f:X \to Y$ induces $f_*:[W, X] \to [W, Y]$ which is a set equivalence for all compact W, then f induces an isomorphism in homology.

Surprisingly, there are examples where f is not a homotopy equivalence, G. W. Whitehead [88], p. 228.

(18.18) *Exercise.* Let s_1 be a knot in R^3. Construct a map $f: S^2 \vee S^1 \to R^3 - s_1$ inducing an isomorphism in homology. If s_1 is the standard circle, construct a homotopy equivalence f. A picture of a possible f is drawn below.

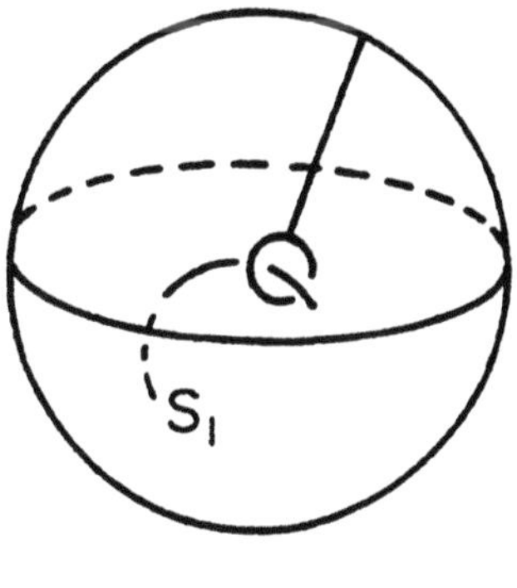

(18.19) *Remark.* A parallel development of the results of this section can be made using the Borsuk Separation Criterion [32] and a clever proof, due to P. H. Doyle, of an analogue of 18.3 (see [85] p. 13 for the generic case).

19. Construction of Spaces: Spherical Complexes

We now consider an important technique for constructing topological spaces, and for spaces constructed in this way, we will derive formulas for computing the homology modules. This method applies to all the familiar spaces, such as projective spaces, compact surfaces, etc.

Suppose we are given a subspace A of a space X and a map f of A into a space Y. In the disjoint union $X \amalg Y$ of X and Y (with the obvious topology), identify each point $x \in A$ with its image $f(x) \in Y$. The quotient space $Z = X \cup_f Y$ of $X \amalg Y$ by the equivalence relation these identifications determine is called *the adjunction space* of the system $X \supset A \xrightarrow{f} Y$. Clearly the quotient map $g: X \amalg Y \to Z$ sends Y homeomorphically onto a subspace of Z; we will identify Y with this subspace. If $\bar{f}: X \to Z$ is the restriction of g to X, then with our identification, $\bar{f}|A = f$.

In what follows, we will concern ourselves only with pairs (X, A) satisfying the following conditions:

(1) X is Hausdorff (2) A is closed in X

(3) Points in $X - A$ can be separated from A, i.e., for any $x \in X - A$, there are disjoint opens U, V such that $x \in U$ and $A \subset V$.

(4) A has a *collaring* B in X, i.e., there is an open neighborhood B of A in X such that A is a strong deformation retract of B, and $A \neq B$.

In that case we say (X, A) is a *collared pair*.

(19.1) *Example*: (E^n, S^{n-1}) is a collared pair (13.14). In this case, the space $Z = E^n \cup_f Y$ is said to be obtained from Y *by adjoining an n-cell via f*.

(19.2) *Example*: More generally, if X is any manifold with boundary (Section 28), and A is its boundary, then (X, A) is a collared pair; in fact, A has an open neighborhood B such that (B, A) is homeomorphic to $(A \times [0, 1), A \times 0)$. (For the proof, see Brown [10] or Vick [75].)

(19.3) *Proposition. Given a collared pair* (X, A) *and a map* $f: A \to Y$, *where* Y *is Hausdorff. Let* $Z = X \cup_f Y$. *Then* (Z, Y) *is a collared pair; in fact, if* B *is a collaring of* A, *then* $Y \cup \bar{f}(B)$ *is a collaring of* Y. *Moreover,* $\bar{f}$ *maps* $X - A$ *homeomorphically onto* $Z - Y$.

A map of pairs $f:(X, A) \to (Y, B)$ such that $X - A$ is mapped homeomorphically to $Y - B$ is called a *relative homeomorphism.*

Proof. Since $g^{-1}(Z - Y) = X - A$, which is open in $X \amalg Y$, $Z - Y$ is open, hence Y is closed, and $\bar{f}$ maps $X - A$ homeomorphically onto $Z - Y$. Let B be a collaring of A in X. Since $B \amalg Y$ is open in $X \amalg Y$ and saturated under the equivalence relation, $Y \cup \bar{f}(B)$ is open in Z. Let $D: B \times I \to B$ be a map such that $D(a, t) = a$ for all $a \in A$, $D(b, 0) = b$ for all $b \in B$, $D(b, 1) \in A$ for all $b \in B$. Define a map $\bar{D}$ on $(Y \cup \bar{f}(B)) \times I$ by

$$\bar{D}(z, t) = \begin{cases} z & \text{if } z \in Y, \\ \bar{f}(D(b, t)) & \text{if } z = \bar{f}(b) \text{ with } b \in B - A. \end{cases}$$

$\bar{D}$ is continuous because it is obtained by passage to the quotient from a continuous map on $(B \amalg Y) \times I$, and $\bar{D}$ exhibits Y as a strong deformation retract of $Y \cup f(B)$. Suppose $z \in Z - Y$, $z = \bar{f}(x)$. Let U, V be disjoint opens in X such that $x \in U$ and $A \subset V$. Then the disjoint Z-opens $Y \cup \bar{f}(V)$ and $\bar{f}(U)$ separate Y from z. Thus condition (3) holds.

To prove Z Hausdorff, given distinct points z_1, z_2, there are three cases.

Case 1. Both points are in $Z - Y$. Since $Z - Y$ is open and homeomorphic to $X - A$, which is open in X Haudorff, we can separate the points.

Case 2. One point is in Y, the other outside Y. This case follows from condition (3).

Case 3. Both points are in Y. Let Y_1, Y_2 be disjoint Y-opens which separate z_1 and z_2. Let $r: B \to A$ be a retraction, and let $B_i = r^{-1}(f^{-1}(Y_i))$, $i = 1, 2$. Then B_i is open in X, and $Y_1 \cup \bar{f}(B_1)$ and $Y_2 \cup \bar{f}(B_2)$ are disjoint Z-opens which separate z_1 and z_2. ∎

(19.4) *Exercise.* Conversely, suppose Y is a closed subspace of a Hausdorff space Z and $\bar{f}: E^n \to Z$ is a map which takes S^{n-1} into Y and sends the open n-cell $\dot{E}^n$ homeomorphically onto $Z - Y$. Then Z is obtained from Y by adjoining an n-cell via $\bar{f} \mid S^{n-1}$. (Use the fact that $\bar{f}$ is a closed mapping.)

Definition. Start with a finite discrete set of points, and successively attach cells, possibly of varying dimensions, but finite in number. We get a compact Hausdorff space, and any space which can be obtained in this way will be called a *spherical complex*. We will obtain some explicit inductive formulas (19.16–19.18) for the homology of a spherical complex.

Some examples:

(19.5) Let f map S^{n-1} onto a point Y. Then $Z \approx S^n$.

(19.6) Let $Y = S^{n-1}$, $f: S^{n-1} \to Y$ the identity. Then $Z = E^n$.

(19.7) Let $Y = S^1$, $f: S^0 \to Y$ a constant map on point P. Then Z consists of two circles with the point P in common. Carrying out this operation $(r-1)$ times (with the same point P) gives the r-leaved rose G_r (17.12).

(19.8) Let $Y = G_2$, denoting the two loops by α and β. Regard E^2 as I^2, S^1 as the perimeter of the square. The diagram describes a map f which attaches a 2-cell to G_2:

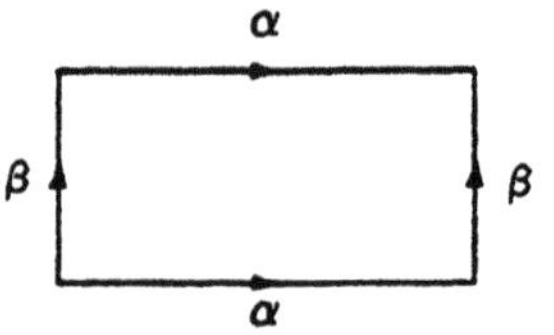

Then Z is homeomorphic to a torus (use 19.4 and the restriction to a square of the canonical map of the plane onto the torus).

(19.9) Every graph (17.12) is obtained from a finite set by attaching 1-cells.

We have already defined real projective n-space $\mathbf{P}^n$ as the quotient space of S^n obtained by identifying antipodal points. We now define complex projective n-space $\mathbf{CP}^n$ analogously: in complex $(n+1)$-space $\mathbf{C}^{n+1}$, consider the subspace defined by $|z| = 1$ (where if $z = (z_0, \ldots, z_n)$, we define $|z|^2 = |z_0|^2 + \ldots + |z_n|^2$); clearly this space is just S^{2n+1}. We identify $z \sim z'$ on S^{2n+1} if $z' = cz$, where c is a complex number of absolute value 1; the resulting quotient space is called $\mathbf{CP}^n$; note that the fibres of the map $f: S^{2n+1} \to \mathbf{CP}^n$ are circles. (For $n = 1$, $\mathbf{CP}^1 \approx S^2$, and f is the Hopf map (Hu [33], p. 66).) We claim:

(19.10) *Proposition.* $\mathbf{CP}^n$ (*resp.* $\mathbf{P}^n$) *is obtained from* $\mathbf{CP}^{n-1}$ (resp. $\mathbf{P}^{n-1}$)

by attaching a 2*n*-cell (*resp. an n-cell*) *via the canonical map* $f: S^{2n-1} \to \mathbf{CP}^{n-1}$ (*resp.* $f: S^{n-1} \to \mathbf{P}^{n-1}$).

Proof. We give the proof for $\mathbf{CP}^n$ and leave $\mathbf{P}^n$ as an exercise. In the diagram

$$\begin{array}{ccc} S^{2n-1} & \subset & E^{2n} \\ \downarrow f & & \downarrow \bar{f} \\ \mathbf{CP}^{n-1} & \xrightarrow{i} & \mathbf{CP}^n \end{array}$$

i is the map $[z_1, \ldots, z_n] \to [0, z_1, \ldots, z_n]$. We define $\bar{f}$ extending f by

$$\bar{f}(z) = [\sqrt{1 - |z|^2}, z_1, \ldots, z_n]$$

where $z = (z_1, \ldots, z_n) \in E^{2n} \subset \mathbf{C}^n$, $|z| \leq 1$. Suppose $|z| < 1$ and $|w| \leq 1$ for $w \in E^{2n}$. If $\bar{f}(z) = \bar{f}(w)$, then

$$(\sqrt{1 - |z|^2}, z_1, \ldots, z_n) = e^{i\theta}(\sqrt{1 - |w|^2}, w_1, \ldots, w_n).$$

Since the zero-th coordinate on the left is real > 0, and $1 - |w|^2$ is real ≥ 0, we must have $e^{i\theta} = 1$, $z = w$. Thus $\bar{f}$ restricted to the open cell $\dot{E}^{2n}$ maps injectively, and its image is clearly $\mathbf{CP}^n - \mathbf{CP}^{n-1}$. Moreover, $\bar{f}$ is a closed map of E^{2n} into $\mathbf{CP}^n$, whence its restriction to $\dot{E}^{2n}$ is a closed map onto $\mathbf{CP}^n - \mathbf{CP}^{n-1}$, i.e., is a homeomorphism. By (19.4) we are done. ■

(19.11) *Remark.* We know that $\mathbf{P}^n$, having S^n as covering space, is a compact connected n-dimensional manifold. We claim that $\mathbf{CP}^n$ is a compact connected $2n$-dimensional manifold: Since it is a continuous image of S^{2n+1}, it is compact connected. Now $\bar{f}(\dot{E}^{2n})$ is an open $2n$-cell. But the other points of $\mathbf{CP}^n$ look locally the same, since $\mathbf{CP}^n$ is *homogeneous*, i.e., given two points x, y, there is a homeomorphism of $\mathbf{CP}^n$ onto itself taking x into y: Write $x = f(x')$, $y = f(y')$, $x', y' \in S^{2n+1}$. There is a nonsingular linear transformation λ of $\mathbf{C}^{n+1}$ taking x' into y', and since λ commutes with multiplication by complex numbers, it passes to the quotient to give the desired homeomorphism (projective transformation) of $\mathbf{CP}^n$ (we regard $\mathbf{CP}^n$ as a quotient space of $\mathbf{C}^{n+1}$ minus the origin).

(19.12) *Exercise.* Let p, q be positive relatively prime integers. The map

$$h:(z_0, z_1) \to (e^{2\pi i/p} z_0, e^{2\pi i q/p} z_1)$$

is a homeomorphism of S^3 whose p-th power is the identity. This gives an operation of $\mathbf{Z}/p$ on S^3 which is properly discontinuous (5.10). The quotient space by this action is denoted $L(p, q)$. It is also a compact connected 3-dimensional manifold which can be shown to be a spherical complex (Artin [3], p. 162) or 21.27. From this its homology can be determined (we know that its fundamental group is $\mathbf{Z}/p$). These "lens spaces" are an important source of examples, e.g., $L(7, 1)$ and $L(7, 2)$ are homotopically equivalent but nonhomeomorphic (see Hilton and Wylie [30], p. 225). A conjecture of Poincaré, as yet unproved, states that a compact 3-dimensional manifold homotopically equivalent to the 3-sphere must be homeomorphic to S^3. (The same statement for S^n instead of S^3 has been proved by Smale [51] and subsequently by Stallings [54a] and Newman [Annals of Math. 1966, pp. 555–571] when $n \geq 5$.)

(19.13) *Exercise.* Using the division ring $\mathbf{H}$ of quaternions instead of $\mathbf{C}$, construct quaternionic projective space $\mathbf{HP}^n$. Show that it is a compact $4n$-dimensional manifold, a quotient space of S^{4n+3} with fibres 3-spheres, and that $\mathbf{HP}^n$ is obtained from $\mathbf{HP}^{n-1}$ by adjoining a $4n$-cell. $\mathbf{HP}^1 \approx S^4$ and we get another Hopf map $S^7 \to S^4$ with fibres S^3. (One can also use the Cayley numbers to obtain the Hopf fibration $S^{15} \to S^8$ with fibres 7-spheres—see Steenrod [55], pp. 105–110; however, there are no analogous $8n$-dimensional projective spaces for $n \geq 3$ because the Cayley numbers are not associative.

Suppose now Z is the adjunction space of a system $X \supset A \xrightarrow{f} Y$, and $\bar{f}: X \to Z$ is the canonical extension of f. Then $\bar{f}$ induces a homomorphism of the homology sequence of the pair (X, A) into that of the pair (Z, Y).

(19.14) *Theorem. Assume (X, A) is a collared pair. Then*

$$H_q(f): H_q(X, A) \to H_q(Z, Y)$$

is an isomorphism for all q.

Proof. Let B be a collaring of A in X. Consider the commutative diagram

$$\begin{array}{ccc} H_q(X, A) & \xrightarrow{\ i\ } & H_q(X, B) \\ {\scriptstyle f_1}\downarrow & & \downarrow{\scriptstyle f_2} \\ H_q(Z, Y) & \xrightarrow[\ j\ ]{} & H_q(Z, Y \cup \bar{f}(B)) \end{array}$$

where the horizontal homomorphisms are induced by inclusions and the vertical ones by $\bar{f}$. We will show that i, j, and f_2 are isomorphisms, hence f_1 is.

The commutative diagram

$$\begin{array}{ccc} H_q(X-A, B-A) & \longrightarrow & H_q(X, B) \\ \downarrow f_3 & & \downarrow f_2 \\ H_q(Z-Y, \bar{f}(B-A)) & \longrightarrow & H_q(Z, Y \cup \bar{f}(B)) \end{array}$$

in which the horizontal arrows are excisions and f_3 induced by a homeomorphism, shows that f_2 is an isomorphism. To see that i (resp. j) is an isomorphism, we use the fact that A (resp. Y) is a deformation retract of B (resp. of $Y \cup \bar{f}(B)$ (19.3)). This implies, e.g., that we have a commutative diagram

$$\begin{array}{ccccccccc} H_q(A) & \to & H_q(X) & \to & H_q(X, A) & \to & H_{q-1}(A) & \to & H_{q-1}(X) \\ \downarrow & & \downarrow & & \downarrow & & \downarrow & & \downarrow \\ H_q(B) & \to & H_q(X) & \to & H_q(X, B) & \to & H_{q-1}(B) & \to & H_{q-1}(X) \end{array}$$

in which the four outside vertical arrows are isomorphisms and the horizontal lines exact. The result follows by the five lemma (14.7). ■

(19.15) Application of the Barratt-Whitehead lemma (17.4) to the ladder

$$\begin{array}{ccccccccccc} \to & H_q(A) & \to & H_q(X) & \to & H_q(X, A) & \to & H_{q-1}(A) & \to & H_{q-1}(X) & \to \dots \\ & \downarrow H_q(f) & & \downarrow H_q(\bar{f}) & & \downarrow \cong & & \downarrow & & \downarrow & \\ \to & H_q(Y) & \to & H_q(Z) & \to & H_q(Z, Y) & \to & H_{q-1}(Y) & \to & H_{q-1}(Z) & \dots \end{array}$$

gives a Mayer-Vietoris exact sequence

$$\dots \to H_q(A) \to H_q(Y) \oplus H_q(X) \to H_q(Z) \to H_{q-1}(A) \to \dots \quad .$$

This sequence will be used for many calculations.

Consider now the special case of adjoining an n-cell, $(X, A) = (E^n, S^{n-1})$. We know that

$$\partial : H_q(E^n, S^{n-1}) \to H^{*}_{q-1}(S^{n-1}).$$

is an isomorphism for all q (14.3). Using the commutative diagram

$$\begin{array}{ccc} H_q(E^n, S^{n-1}) & \longrightarrow & H^*_{q-1}(S^{n-1}) \\ \downarrow & & \downarrow H_{q-1}(f) \\ H_q(Z, Y) & \longrightarrow & H^*_{q-1}(Y) \end{array}$$

we can replace the relative homology modules in the sequence for the pair (Z, Y) and obtain the exact sequence

$$\to H_q(Y) \to H_q(Z) \to H^*_{q-1}(S^{n-1}) \xrightarrow[H_{q-1}(f)]{} H^*_{q-1}(Y) \to H^*_{q-1}(Z) \to$$

Since $H^*_{q-1}(S^{n-1})$ is zero except for $q = n$, this sequence yields the following formulas.

Corollary. We have

(19.16) $H^*_q(Z) \cong H^*_q(Y)$ *for* $q \neq n$ *and* $q \neq n - 1$

(19.17) $H^*_{n-1}(Z) \cong H^*_{n-1}(Y)/\text{Image } H_{n-1}(f)$

(19.18) *an exact sequence*

$$0 \to H^*_n(Y) \to H^*_n(Z) \xrightarrow{\psi} \text{Kernel } H_{n-1}(f) \to 0$$

(19.19) *Remark.* If Kernel $H_{n-1}(f)$ is a free module, this exact sequence splits and we can write

$$H^*_n(Z) \cong H^*_n(Y) \oplus \text{Ker } H_{n-1}(f)$$

("Splits" means there is a homomorphism Ker $H_{n-1}(f) \to H^*_n(Z)$ which is a right inverse to ψ.) This happens for $R = \mathbf{Z}$ or $R = a$ field, for example, since Ker $H_{n-1}(f)$ is then a free R-module.

(19.20) *Corollary. Assume R is a Noetherian ring. If Z is a spherical complex, then $H_q(Z)$ is a finitely generated R-module for every q. If n is the highest dimension of a cell used in constructing Z, then $H_q(Z) = 0$ for $q > n$.*

This follows from the previous formulas by induction on the number of cells attached. ∎

If we take

$$Z = \bigcup_{i=0}^{\infty} C_i$$

where C_i is the circle of radius $1/i$ with center the origin, $i > 0$, and C_0= the origin, we get a compact Hausdorff space which is not a spherical complex, since $H_0(Z)$ is not finitely generated.

(19.21) *Theorem. The homology of complex projective space is given by*

$$H_q(\mathbf{CP}^n) \cong \begin{cases} 0 & q > 2n \text{ or } q \text{ odd}, \\ R & q \text{ even such that } 0 \leq q \leq 2n. \end{cases}$$

Proof. For $n = 0$ it is the homology of a point. Use induction on n. We get $\mathbf{CP}^n$ from $\mathbf{CP}^{n-1}$ by attaching a $2n$-cell. By (19.16) we need only consider $q = 2n$ and $q = 2n - 1$. By (19.17), $H_{2n-1}(\mathbf{CP}^n) = 0$, since $H_{2n-1}(\mathbf{CP}^{n-1}) = 0$. By (19.18), $H_{2n}(\mathbf{CP}^n) = Ker\ H_{2n-1}(f)$, where $f{:}S^{2n-1} \to \mathbf{CP}^{n-1}$ is the canonical map. Since $H_{2n-1}(f)$ is the zero homomorphism, we do get $H_{2n}(\mathbf{CP}^n) \cong R$. ■

(19.22) *Exercise.* Compute the homology of quaternionic projective space (19.13).

(19.23) *Theorem. Let $f{:}S^n \to \mathbf{P}^n$ be the canonical map. If n is even, $H_n(f) = 0$. If n is odd, there are isomorphisms $H_n(\mathbf{P}^n) \cong R \cong H_n(S^n)$ such that $H_n(f)$ is multiplication by 2, i.e.,*

$$\begin{array}{ccc} R & \xrightarrow{\times 2} & R \\ \cong\downarrow & & \downarrow\cong \\ H_n(S^n) & \xrightarrow{H_n(f)} & H_n(\mathbf{P}^n) \end{array} \quad \textit{commutes.}$$

Proof. $\mathbf{P}^n$ is obtained from $\mathbf{P}^{n-1}$ by attaching E^n via the canonical map $f{:}S^{n-1} \to \mathbf{P}^{n-1}$. Hence $\bar{f}{:}(E^n, S^{n-1}) \to (\mathbf{P}^n, \mathbf{P}^{n-1})$ induces isomorphisms in homology (19.14). Consider the ladder obtained from $f{:}(S^n, S^{n-1}) \to (\mathbf{P}^n, \mathbf{P}^{n-1})$

(19.24)
$$\begin{array}{ccccccccc} 0 \to & H_n(S^n) & \xrightarrow{j} & H_n(S^n, S^{n-1}) & \xrightarrow{\partial} & H_{n-1}(S^{n-1}) & \to 0 & & \\ & \downarrow f_1 & & \downarrow f_2 & & \downarrow f_3 & & & \\ 0 = H_n(\mathbf{P}^{n-1}) \to & H_n(\mathbf{P}^n) & \xrightarrow{j'} & H_n(\mathbf{P}^n, \mathbf{P}^{n-1}) & \xrightarrow{\partial'} & H_{n-1}(\mathbf{P}^{n-1}) & \to & H_{n-1}(\mathbf{P}^n) & \to 0. \end{array}$$

The zeros are by (15.5) for the top row, (19.20) and induction on n for the bottom left, and (15.5) for the bottom right. The vertical maps are induced by f. The top line is split exact for algebraic reasons, but to compute f_2 we need a splitting compatible with induced maps. The idea is that collapsing $S^{n-1} \subset S^n$ to a point produces $S^n \vee S^n$, each sphere mapped homeomorphically to $\mathbf{P}^n/\mathbf{P}^{n-1}$. Exact sequences formalize the notion. By the long exact sequence of triples (14.6), excision (15.3) and the direct sum lemma (14.13), we have a direct sum decomposition

(19.25)
$$\begin{array}{ccccc} H_n(S^n, E^-_n) & & & & H_n(S^n, E^-_n) \\ \cong\uparrow & \nwarrow & H_n(S^n, S^{n-1}) & \nearrow & \uparrow\cong \\ H_n(E^-_n, S^{n-1}) & \nearrow & & \nwarrow & H_n(E^+_n, S^{n-1}) \end{array}$$

where $E^\mp_n \subset S^n$ are the upper and lower hemispheres, Let $e^\pm:(E^n, S^{n-1}) \to (E^\pm_n, S^{n-1})$ be homeomorphisms inverse to projecting on the first n-coordinates, $e^\pm(x) = (x, \pm\sqrt{1 - |x|})$, in cartesian coordinates. Then $fe^\pm:(E^n, S^{n-1}) \to (\mathbf{P}^n, \mathbf{P}^{n-1})$ induce isomorphisms in homology by (19.14) and (13.6), and f_2 is an epimorphism. The maps fe^+ and fe^- are not the same. Let $\phi:(E^n, S^{n-1}) \to (E^n, S^{n-1})$ be given by $\phi(x) = -x$. Note $\phi | S^{n-1}$ is the antipodal map. Then

(19.26)
$$fe^- = fe^+\phi$$

since $e^-\phi(x) = (-x, \sqrt{1 - |x|}) = -e^-(x)$. We now use exact sequences to specify generators of various homology modules.

Let $\iota_n \in H_n(S^n)$ and $\iota_{n-1} \in H_{n-1}(S^{n-1})$ be generators mapped to each other in the sequence of isomorphisms (15.4)

$$\begin{array}{ccccccc} H_n(S^n) & \to & H_n(S^n, E^-_n) & \leftarrow & H_n(E^n, S^{n-1}) & \xrightarrow{\partial} & H_{n-1}(S^{n-1}) \\ & & & \nwarrow & \downarrow H(e^+) & & \\ & & & & H_n(E^+_n, S^{n-1}) & & \end{array}$$

Define $K \in H_n(E^n, S^{n-1})$ by $\partial(K) = \iota_{n-1}$, $K^\pm \in H_n(E^\mp_n, S^{n-1})$ by $K^\pm = H_n(e^\pm)(K)$, and $\eta \in H_n(\mathbf{P}^n, \mathbf{P}^{n-1})$ by $\eta = f_2K^-$. By construction η generates $H_n(\mathbf{P}^n, \mathbf{P}^{n-1})$. We regard $K^\pm$ as generators of the summands of $H_n(S^n$, $H_n(S^n, S^{n-1})$ by (19.25). We next calculate the maps in (19.24).

(a) $j(\iota_n) = (K^+, -K^-)$.

By naturality (14.5), $\partial(K^{\pm}) = \iota_{n-1}$ in the top row of (19.24). Hence a generator for ker ∂ is $(K^-, -K^-)$. By exactness $j(\iota_n) = m(K^+, -K^-)$ for some $m \in R$. Evaluating in the upper right corner of (19.25) (and tacitly identifying the groups mapped isomorphically) yields $mK^+ = K^+$ or $m = 1$.

(b) $f_2(K^-) = (-1)^n\eta$.

$$f_2(K^-) = H(fe^-)(K) = H(fe^+\phi)(K) = (-1)^n H(fe^-)(K)$$
$$= (-1)^n f_2(K^+), \quad \text{using (19.26) and (16.4).}$$

Combining (a), (b) we have $f_2 j(\iota_n) = (1-(-1)^n)\eta$. When n is even, $f_2 j(\iota_n) = 0$. Since j' is monic, $H_n(f) = f_1$ is the zero map. When n is odd, we obtain $\partial' = 0$ since f_2 is epic and $f_3 = H_{n-1}(f) = 0$. Hence j' is an isomorphism and $H_n(\mathbf{P}^n) \cong R$ generated by $(j')^{-1}\eta$. Then $H_n(f)(\iota_n) = f_1\iota_n = (j')^{-1}f_2 j\,\iota_n = 2(j')^{-1}\eta$. ■

(19.27) *Theorem. The homology of real projective space is given by*

$$H_q(\mathbf{P}^n) \cong \begin{cases} 0 & q > n, \\ R_2 & q \text{ even such that } 1 < q \le n, \\ R/2 & q \text{ odd such that } 1 \le q \le n-1, \\ R & q = 0 \text{ and } q = n \text{ if } n \text{ is odd} \end{cases}$$

(where R_2 is the submodule of R annihilated by multiplication by 2).

Thus the homology with coefficients in the field of rational numbers looks quite different from the homology with coefficients in a field of characteristic 2.

Proof. By induction on n, $n = 0, 1$ being trivial ($\mathbf{P}^1 \cong S^1$). From (19.16) we have $H_q(\mathbf{P}^n) \cong H_q(\mathbf{P}^{n-1})$ for $q \le n-2$, and $H_q(\mathbf{P}^n) = 0$ for $q > n$ by (19.20). By (19.18) we have an exact sequence

$$0 \to H_n(\mathbf{P}^n) \to H_{n-1}(S^{n-1}) \xrightarrow{H_{n-1}(f)} H_{n-1}(\mathbf{P}^{n-1}) \to H_{n-1}(\mathbf{P}^n) \to 0.$$

If n is even, the middle map is multiplication by 2 so $H_n(\mathbf{P}^n) \cong R_2$ and $H_{n-1}(\mathbf{P}^n) \cong R/2R$. If n is odd the middle map is 0 so $H_n(\mathbf{P}^n) \cong R$ and $H_{n-1}(\mathbf{P}^n) \cong R_2$. ■

Remark. Note that the result for $R = \mathbf{Z}/2\mathbf{Z}$ is immediate after (19.24) and the observation concerning collapsing $S^{n-1} \subset S^n$, since $H_n(f)$ is then the 0 map for all n.

(19.28) *Examples*. For $R = \mathbf{Z}$ we get

$$H_q(\mathbf{P}^2) \cong \begin{cases} \mathbf{Z} & q = 0, \\ \mathbf{Z}/2 & q = 1, \\ 0 & q \geq 2, \end{cases}$$

$$H_q(\mathbf{P}^3) \cong \begin{cases} \mathbf{Z} & q = 0, \\ \mathbf{Z}/2 & q = 1, \\ 0 & q = 2, \\ \mathbf{Z} & q = 3, \\ 0 & q \geq 4 \end{cases}$$

and so on. For $R = \mathbf{Z}/2$ we get

$$H_q(\mathbf{P}^n) \cong \begin{cases} \mathbf{Z}/2 & q \leq n, \\ 0 & q > n. \end{cases}$$

(19.29) *Theorem. The homology of the torus T is*

$$H_q(T) \cong \begin{cases} R & q = 0 \text{ and } q = 2, \\ R \times R & q = 1, \\ 0 & q > 2. \end{cases}$$

Proof. T is obtained from $Y = G_2$ by attaching a 2-cell via the map indicated as

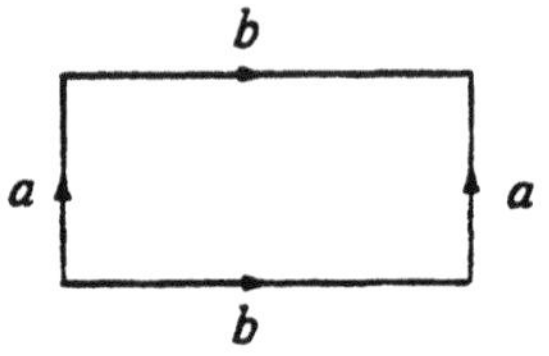

where the sides marked a, b are mapped onto the loops a, b of G_2 (for we know that the unit square I^2 maps onto T by

$$\phi:(x, y) \to (e^{2\pi ix}, e^{2\pi iy})$$

with the identifications indicated). Now $H_1(G_2)$ is the free R-module generated by the homology classes α, β of the loops $\phi \cdot a, \phi \cdot b$ (17.12), and the adjoining map $f:S^1 \to G_2$ sends the generator of $H_1(S^1)$ onto $\alpha + \beta - \alpha - \beta = 0$, so that $H_1(f)$ is the zero-homomorphism. The theorem then follows from (19.16–19.18) compare (17.13). ∎

(19.30) *Example.* Consider the double torus

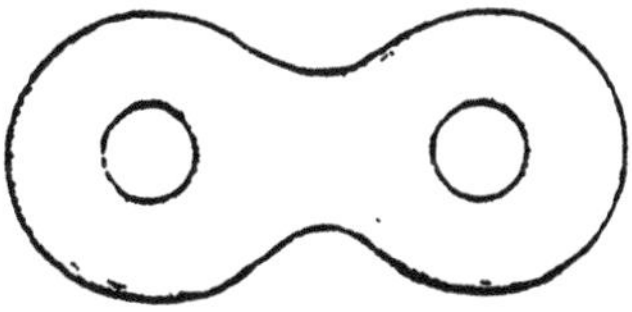

We can think of this as follows. Cut out a small open 2-cell from each of two copies of the torus, then paste these together along the circles which are the frontiers of those open 2-cells. Now the torus with the open 2-cell removed can be described by

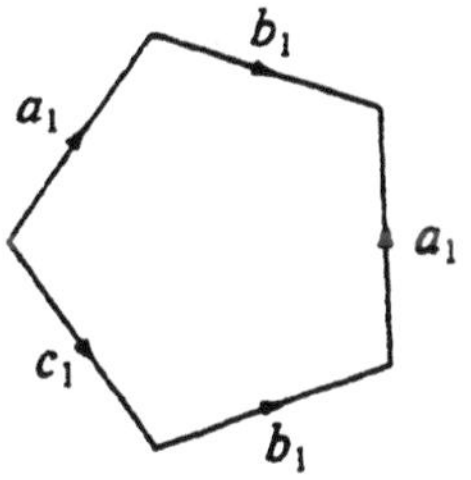

identifying sides of the pentagon as indicated, the loop c_1 being the frontier of the cut-out cell. In pasting another copy, the c's will cancel and we are left with

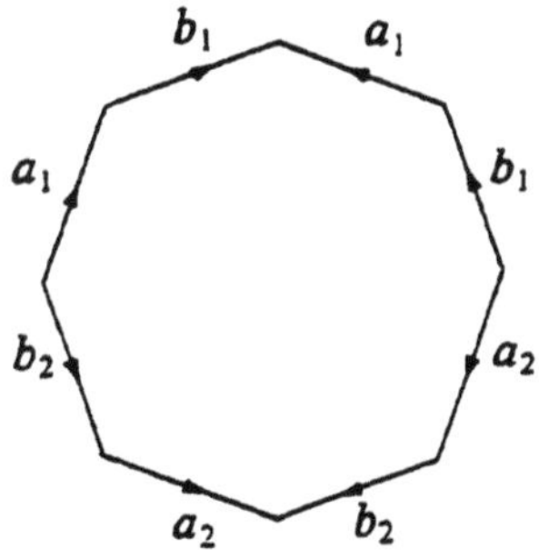

Thus we can describe the double torus precisely as obtained from G_4 by attaching a 2-cell via the map of the perimeter of an octagon described by the expression $a_1b_1a_1^{-1}b_1^{-1}a_2b_2a_2^{-1}b_2^{-1}$. More generally, attaching a 2-cell to G_{2g} via the map of the perimeter of a $4g$-gon described by the expression $a_1b_1a_1^{-1}b_1^{-1}\ldots a_gb_ga_g^{-1}b_g^{-1}$ gives *the g-fold torus* T_g. A calculation as in (19.26) gives

$$H_q(T_g) \cong \begin{cases} R & q = 0 \text{ and } q = 2, \\ R^{2g} & q = 1, \\ 0 & q > 2. \end{cases}$$

The number g is called the *genus* of T_g.

(19.31) *Example.* If we start with the projective plane $\mathbf{P}^2$ and "double it" as we did the torus, we obtain a surface which can be described by attaching a 2-cell to G_2 via the map of the perimeter of a square described by the expression $a_1^2 a_2^2$

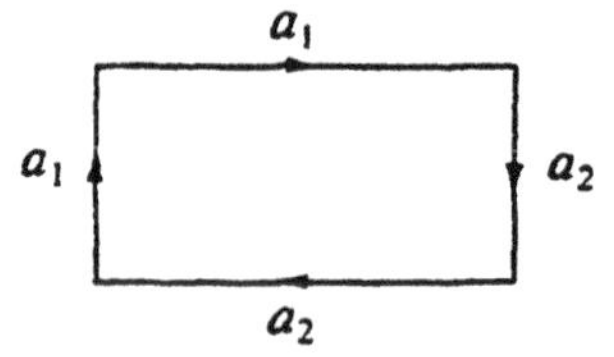

More generally, mapping the perimeter of a $2h$-gon via the expression $a_1^2 \ldots a_h^2$ yields a surface U_h (let $U_1 = \mathbf{P}^2$). Then the homology of U_h is given by

$$H_q(U_h) = \begin{cases} R & q = 0, \\ R^{h-1} \times R/2 & q = 1, \\ R_2 & q = 2, \\ 0 & q > 2 \end{cases}$$

(to see this, note that the homomorphism

$$H_1(f){:}H_1(S^1) - H_1(G_h) \cong R^h$$

sends a generator of $H_1(S^1)$ into $2(\bar{a}_1 + \ldots + \bar{a}_n)$, where $\bar{a}_1, \ldots, \bar{a}_n$ are free generators of $H_1(G_h)$. The corkernel of this homomorphism is isomorphic to $R^{h-1} \times R/2$, since we can use $\bar{a}_1 + \ldots + \bar{a}_h, \bar{a}_2, \ldots, \bar{a}_h$ as free generators, while the kernel is R_2.)

(19.32) *Exercise.* The *Klein bottle* is defined by pasting the two ends of a cylinder after a 180° twist, i.e., by the identification

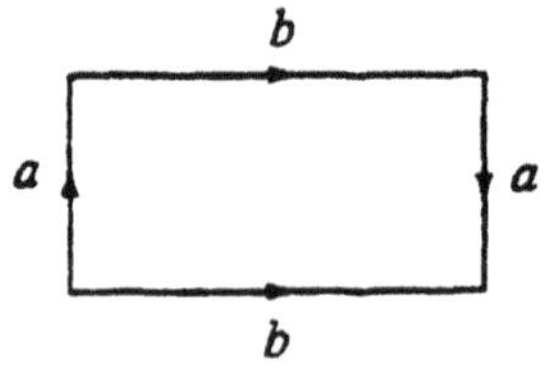

Show that it is homeomorphic to the space U_2 (construct both spaces by pasting two triangles along a side in two different ways).

(19.33) *Exercise.* Show that if the torus is attached to the projective plane (in the same way that we attached two tori to construct the double torus), the resulting surface is homeomorphic to U_3.

(19.34) *Note.* It is a nontrivial theorem that every compact connected 2-dimensional manifold is homeomorphic to either S^2, one of the T_g's or one of the U_h's (see Ahlfors and Sario [2]; the proof uses triangulability and the Jordan-Schoenflies Theorem). Obviously no two of these are homeomorphic (look at their homology groups), so a complete classification of these surfaces is obtained. No such classification is known in dimension 3, and Markov proved it is impossible in dimensions ≥ 4 because the resulting classification of π_1 would solve the word problem for groups, contradicting the theorem of Novikov-Boone (see Massey [67], p. 144).

(19.35) *Example.* If a point is removed from a spherical complex, the resulting space has finitely many path components (do induction on the number of cells attached). Hence the compact path-connected space drawn below is not a spherical complex:

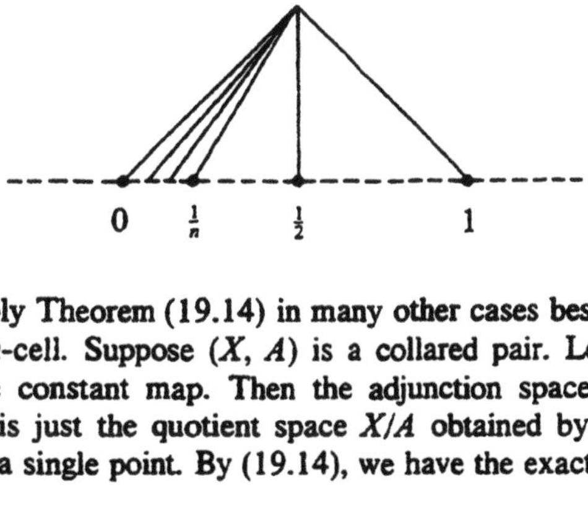

We can apply Theorem (19.14) in many other cases besides the case of adjoining an n-cell. Suppose (X, A) is a collared pair. Let P be a point $f: A \to P$ the constant map. Then the adjunction space of the system $X \supset A \xrightarrow{f} P$ is just the quotient space X/A obtained by identifying the subspace A to a single point. By (19.14), we have the exact sequence

$$\to H_q(P) \to H_q(X/A) \to H_q(X, A) \to H^*_{q-1}(P) \to \ .$$

Since the augmented homology of a point is zero, we obtain the following result.

(19.36) *Proposition. If (X, A) is a collared pair, then the space X/A obtained from X by collapsing A to a point is a Hausdorff space in which* *the distinguished point has a contractible open neighborhood. Moreover, there is a canonical isomorphism*

$$H_q^*(X/A) \cong H_q(X, A) \quad \text{all } q$$

so that we obtain an exact sequence

$$\to H_q(A) \to H_q(X) \to H_q(X/A) \to H_{q-1}^*(A) \to H_{q-1}^*(X) \to .$$

(19.37) *Example* cf. (15.24). Let X be a Hausdorff space. In the space $X \times I$, identify the closed subspace $X \times 0$ to one point and $X \times 1$ to another. The quotient space SX under these identifications is called *the (unreduced) suspension* of X. For example, $S(S^n) \approx S^{n+1}$. Applying the previous proposition twice, we obtain

$$H_q(SX) \cong \begin{cases} H_{q-1}(X) & q > 0, \\ R & q = 0. \end{cases}$$

(19.38) *Example* cf. (17.17). Let (X, x), (Y, y) be pointed Hausdorff spaces in which the points x, y have contractible open neighborhoods (e.g., manifolds. Then the pair $(X \amalg y, \{x, y\})$ is collared. The quotient space obtained by identifying x and y is called the wedge or bouquet $X \vee Y$ of X and Y at these points. Using the preceding proposition, we obtain

$$H_q(X \vee Y) \cong \begin{cases} H_q(X) \oplus H_q(Y) & q > 0, \\ R^{c+d-1} & q = 0, \end{cases}$$

where X has c path components and Y has d.

(19.39) *Example.* Suppose a Hausdorff space Z is the union of two closed subspaces X and Y, and suppose that (X, A) is a collared pair, where $A = X \cap Y$. Then $Z = X \cup_f Y$, where $f{:}A \to Y$ is the inclusion map, and (19.14, 19.36) give the exact sequence

$$\to H_q(Y) \to H_q(X \cup Y) \to H_q(X/X \cap Y) \to H_{q-1}^*(Y) \to .$$

Contrast this with Van Kampen's theorem for $\pi_1(X \cup Y)$ (4.12). There is no general result for $\pi_q(X \cup Y)$ for $q > 0$, at present.

(19.40) *Exercise.* Show that a product of spheres is a spherical complex. Suggestion: Let $f_n{:}(E^n, S^{n-1}) \to (S^n, pt)$ be the relative homeomorphism in

(19.5). Then $f_n \times f_m$ maps $\partial(E^n \times E^m) = S^{n-1} \times E^m \cup E^n \times S^{m-1}$ to $S^n \vee S^m$ and $E^n \times E^m$ is homeomorphic to E^{n+m}; call a homeomorphism ϕ. Then $S^n \times S^m$ is homeomorphic to $Z = E^{n+m} \cup_{(f_n \times f_m)\phi} S^n \vee S^m$ obtained by adjoining an $(n+m)$-cell to $S^n \vee S^m$ via $(f_n \times f_m)\phi$. Generalize to k-fold products by induction on k.

(19.41) *Exercise.* If X is the topologist's sine curve (p. 56) and A is the interval on the y-axis together with the lower arc, prove that (X, A) is not a collared pair by applying (10.13) and (19.36).

The next two exercises use (19.23) to derive some facts about maps of spheres.

(19.42) *Exercise.* If $f: S^n \to S^n$ satisfies $f(-x) = f(x)$ for all $x \in S^n$, then $\deg f$ is even, in particular, if n is even then $\deg f = 0$.

A partial converse of (19.42) is true: If $\deg f$ is even then $f(-x) = f(x)$ for some $x \in S^n$. The proof will have to wait for (26.13).

(19.43) *Exercise.* If $f: S^n \to S^n$ has odd degree, then $f(-x) = -f(x)$ for some $x \in S^n$. Suggestion: suppose not and consider

$$h(x) = (f(x) + f(-x))/| f(x) + F(-x) .$$

Use (19.42).

Remark. A theorem of Borsuk asserts that $f(-x) = -f(x)$ *for all* $x \in S^n$ *implies* $\deg f$ *is odd* (26.25). The partial converse to (19.42) can be proved using (26.25) and a construction similar to (19.43).

(19.44) *Exercise.* Show $\mathbf{P}^k$ is not a retract of $\mathbf{P}^n$ if either n is odd and $n-k$ even, or n even and $n-k$ odd. Could this be done using results only for $R = \mathbf{Z}/2\mathbf{Z}$?

Remark. $\mathbf{P}^k$ is never a retract of $\mathbf{P}^n$ but more structure than (19.27) is needed. Let P_k^n be the space obtained by collapsing $\mathbf{P}^k \subset \mathbf{P}^n$ to a point. The homotopy type of $\mathbf{P}_k^n$ is intimately related to the vector field problem for S^n, [69], [65].

20. Betti Numbers and Euler Characteristic

If we take homology with integer coefficients, for certain spaces, such as spherical complexes (19.20), the homology groups will be finitely generated. If A is a finitely generated Abelian group, a basic theorem (see Lang [35], p. 45) states that the elements of finite order in A form the torsion subgroup T and that the quotient group A/T is free Abelian. The minimal number of generators of A/T is called the *rank* of A. The rank of $H_q(X; \mathbf{Z})$ is called the q-th *Betti number* β_q of the space X, and we also define the *Euler characteristic* $\chi(X)$ by the formula

$$\chi(X) = \sum_q (-1)^q \beta_q$$

when this sum is finite. These numbers are of course topological invariants.

(20.1) *Example*: For S^n, $\beta_0 = \beta_n = 1$, all other $\beta_q = 0$, and we get

$$\chi(S^n) = \begin{cases} 0 & n \text{ odd} \\ 2 & n \text{ even} \end{cases}$$

(20.2) *Example*: For the r-leaved rose G_r, $\beta_0 = 1$, $\beta_1 = r$, all other $\beta_q = 0$, and we get

$$\chi(G_r) = 1 - r$$

(20.3) *Example*: For $\mathbf{CP}^n$ (19.21), we have $\beta_q = 0$ for q odd or $q > 2n$, $\beta_q = 1$ for q even such that $0 \leq q \leq 2n$. Hence

$$\chi(\mathbf{CP}^n) = n + 1$$

(20.4) *Example*: For $\mathbf{P}^n$ (19.27), we have $\beta_0 = 1$, $\beta_n = 1$ if n is odd, and $\beta_q = 0$ in all other cases. Thus

$$\chi(\mathbf{P}^n) = \begin{cases} 1 & n \text{ even} \\ 0 & n \text{ odd} \end{cases}$$

(20.5) *Example*: For the g-fold torus T_g(19.30), $\beta_0 = \beta_2 = 1$, $\beta_1 = 2g$, and $\beta_q = 0$ $q > 2$. Thus

$$\chi(T_g) = 2 - 2g$$

(20.6) *Example*: For the surfaces U_h(19.31), $\beta_0 = 1$, $\beta_1 = h - 1$, $\beta_q = 0$ for $q \geq 2$. Thus

$$\chi(U_h) = 2 - h$$

We can also define in a similar manner the relative Betti numbers and relative Euler characteristic $\chi(X, A)$ for a pair (X, A).

(20.7) *Lemma. When these numbers are defined,*

$$\chi(X) = \chi(A) + \chi(X, A)$$

Proof: Using the exact homology sequence, we are reduced to a purely algebraic lemma.

(20.8) *Lemma. Given an exact sequence of finitely generated Abelian groups*

$$0 \to A_1 \xrightarrow{i_1} A_2 \xrightarrow{i_2} \dots \xrightarrow{i_{r-1}} A_r \to 0$$

then

$$\text{rank}\, A_1\text{-rank}\, A_2 + \dots + (-1)^{r+1}\, \text{rank}\, A_r = 0$$

Proof: *By* induction on r: The cases $r = 1, 2$ are trivial. For $r = 3$, let $\bar{A}_i$ be the quotient group of A_i by its torsion subgroup. We get induced homomorphisms $\bar{i}_1, \bar{i}_2$, hence an induced sequence of free Abelian groups

$$0 \to \bar{A}_1 \xrightarrow{\bar{i}_1} \bar{A}_2 \xrightarrow{\bar{i}_2} \bar{A}_3 \to 0$$

which is not in general exact (take $A_1 = 2\mathbf{Z}, A_2 = \mathbf{Z}$). However, we have the following facts.

(20.9) *Sublemma.* $\bar{i}_1$ *is a monomorphism,* $\bar{i}_2$ *is an epimorphism. and* Kernel $\bar{i}_2$/Image $\bar{i}_1$ *is a torsion group.*

Proof: Exercise.

Hence rank $A_1 =$ rank (Kernel $\bar{i}_2$). Now the exact sequence of free Abelian groups

$$0 \to \text{Kernel } \bar{i}_2 \to \bar{A}_2 \to \bar{A}_3 \to 0$$

splits, so that $\bar{A}_2 \cong \bar{A}_3 \oplus$ Kernel $\bar{i}_2$, whence rank $\bar{A}_2$ = rank $\bar{A}_3$ + rank (Kernel $\bar{i}_2$). Since rank A_i = rank $\bar{A}_i$ by definition, the lemma is proved for $r = 3$.

For $r > 3$, consider the two exact sequences

$$0 \to A_1 \to A_2 \to Im\ i_2 \to 0$$

$$0 \to Im\ i_2 \to A_3 \to \ldots \to A_r \to 0$$

Since each sequence contains fewer than r terms, we are done by induction.

(20.10) *Corollary. If Z is obtained from Y by attaching an n-cell, and* $\chi(Y)$ *is defined, then*

$$\chi(Z) = \chi(Y) + (-1)^n$$

Proof: By (19.14), $H_q(Z, Y) \cong H_q(E^n, S^{n-1})$ for all q, so by the lemma, we need only compute $\chi(Z, Y) = \chi(E^n, S^{n-1})$. Applying the lemma again, $\chi(E^n, S^{n-1}) = 1 - [1 + (-1)^{n-1}] = (-1)^n$. ∎

(20.11) *Corollary. Let X be a spherical complex, obtained from* α_0 *points by attaching* α_q *q-cells for* $q = 1, \ldots, n$ *(in any order). Then*

$$\chi(X) = \sum_0^n (-1)^q \alpha_q$$

This follows from (20.10) by induction on $\Sigma_0^n \alpha_q$.

(20.12) *Exercise.* The q-th Betti number of X is equal to the dimension of the vector space $H_q(X; \mathbf{Q})$ over the rational numbers. (Show that if $z_1, \ldots, z_\beta$

are cycles with integer coefficients whose homology classes form a basis for $H_q(X; \mathbf{Z})$ modulo torsion, then the homology classes of these same cycles form a basis of $H_q(X; \mathbf{Q})$.)

(20.13) *Remark.* Consider the following special case of (20.11). Take a regular polyhedron in 3-space,,(homeomorphic to S^2). Divide its surface into triangles (or quadrilaterals, etc.) so that if two of them meet, they meet in a common edge or a common vertex. Let F be the number of faces, E the number of edges, V the number of vertices. Then always

$$V - E + F = 2.$$

This is an old theorem of Euler. See *Proofs and Refutations* by I. Lakatos for the fascinating history of this formula and its implications for understanding mathematical discovery.

(20.14) *Exercise.* If X, Y are spherical complexes, then so is $X \times Y$. Moreover,

$$\chi(X \times Y) = \chi(X)\chi(Y).$$

Thus for example

$$\chi(S^2 \times S^2) = 4,$$

$$\chi(\mathbf{P}^2 \times \mathbf{P}^2) = 1,$$

etc. Numerical results like these suffice to deduce that certain spaces are homotopically inequivalent, e.g., $S^2 \times S^2$ and S^4.

(20.15) *Exercise.* If X, Y are connected n-dimensional manifolds having well-defined Euler characteristics, then $\chi(X + Y)$ is defined (see 19.39), and

$$\chi(X + Y) = \begin{cases} \chi(X) + \chi(Y) & n \text{ odd}, \\ \chi(X) + \chi(Y) - 2 & n \text{ even}. \end{cases}$$

(Let $X * Y$ be the space obtained from the disjoint union $X \perp\!\!\!\perp Y$ by identifying a small closed n-cell in X with one in Y. By Mayer-Vietoris, $\chi(X * Y) = \chi(X) + \chi(Y) - 1$. Now $X * Y$ is obtained from $X + Y$ by attaching an n-cell so (20.10) applies.)

(20.16) *Note*. If R is any principal ideal domain (PID), the results on Abelian groups we have quoted generalize to modules over R (Lang [35], Chapter XV, Section 2). Thus we can define the Euler characteristic $\chi(X; R)$ by

$$\chi(X; R) = \sum_q (-1)^q \operatorname{rank}_R H_q(X; R)$$

and all the results on Euler characteristic generalize.

(20.17) *Example*: For $\mathbf{P}^n$ (19.28) we have

$$\chi(\mathbf{P}^n; \mathbf{Z}/2) = \begin{cases} 0 & n \text{ odd} \\ 1 & n \text{ even} \end{cases}$$

(20.18) *Example*: For the surfaces U_h (19.31) we have

$$\chi(U_h; \mathbf{Z}/2) = 2 - h$$

(20.19) *Remark*. These examples suggest that the Euler characteristic $\chi(X; R)$ may be independent of the PID R. this is in fact true, provided that X is a spherical complex; for the same proof as before shows that

$$\chi(X: R) = \sum_q (-1)^q \alpha_q$$

for all R. Actually the independence of R holds for arbitrary spaces, as follows from the Universal Coefficient Theorem (see 29.12).

(20.20) *Note on Euler characteristic*: The Euler characteristic of X is a very powerful invariant which has many applications outside of topology, for example:

(1) We have seen that only the odd-dimensional spheres admit non-vanishing vector fields (16.5). These are the spheres of Euler characteristic zero. Now one can define the notion of vector field on any *differential manifold* X (Bishop and Crittenden [7]; differentiability is needed in order to be able to talk about tangent vectors). For X compact connected, there exists a non-vanishing vector field if and only if $\chi(X) = 0$ (Steenrod [55], p. 201). Thus for compact surfaces, only the torus and Klein bottle have them (construct them explicitly-exercise).

If X is odd-dimensional compact, it always admits one (see (26.10)). Differential-geometric techniques show that if X is non-compact, it always admits one.

(2) A more general kind of Euler characteristic plays the key role in the Riemann-Roch Theorem for non-singular projective algebraic varieties (see Hirzebruch [31] as well as the Atiyah-Singer Index Theorem [92]).

(20.21) *Exercise.* Calculate the homology of the two-skeleton of the n-simplex. Suggestion; use (20.11). Generalize to k-skeletons of Δ_n.

21. Construction of Spaces: Cell Complexes and more Adjunction Spaces

(21.1) We consider now a special class of spherical complexes called *finite cell complexes*. These are compact Hausdorff spaces Z such that Z has a finite collection of closed subsets c_j^q (were $q = 0, 1, \ldots$, represents the dimension, and j ranges over some index set J_q) with the following properties: Let

$$Z^q = \cup \{c_j^p | \text{ all } j \in J_p, \text{ all } p \leq q\},$$

$$Z^{-1} = \text{empty set},$$

and put

$$f_j^q = c_j^q \cap Z^{q-1}.$$

Then we require

(1) the only time $c_i^p - f_i^p$ meets $c_j^q - f_j^q$ *is when* $p = q$ and $i = j$;

(2) $Z = \cup_q Z^q$;

(3) for each c_j^q, there is a map $\phi_j^q : E^q \to Z$ which sends S^{q-1} onto f_j^q and maps $E^q - S^{q-1}$ homeomorphically onto $c_j^q - f_j^q$ (S^{-1} is the empty set).

The sets $c_j^q - f_j^q$ are therefore open q-cells for $q > 0$ and their disjoint union over all q, j is Z (note that f_j^0 is empty, so c_j^0 is closed 0-cell. The map ϕ_j^q is called *the characteristic map* of c_j^q; clearly Z is obtained from $(Z - c_i^q) \cup f_i^q$ by attaching a q-cell via the map $\phi_j^q | S^{q-1}$, hence Z is a spherical complex. The closed subspace Z^q is called *the q-skeleton of* Z.

(21.2) *Examples*: All the spherical complexes considered in Section 19 are finite cell complexes. Thus for $n > 0$

$$\mathbf{P}^n = c^0 \cup c^1 \cup c^2 \ldots \cup c^n$$

$$\mathbf{CP}^n = c^0 \cup c^2 \cup c^4 \ldots \cup c^{2n}$$

$$\mathbf{HP}^n = c^0 \cup c^4 \cup \ \ldots \cup c^{4n}$$

$$S^n = c^0 \cup c^n$$

$$E^n = c^0 \cup c^{n-1} \cup c^n \quad n \geq 2$$

$$T_g = c^0 \cup c_1^1 \cup \ldots \cup c_{2g}^1 \cup c^2$$

$$U_h = c^0 \cup c_1^1 \cup \ldots \cup c_h^1 \cup c^2$$

(Of course, these are not disjoint unions!)

(21.3) *Example*: Consider S^2 as obtained from its north pole by attaching a 2-cell via the constant map of S^1 onto the north pole. Let x be a point other than the north pole, and attach a 1-cell to S^2 via the constant map of S^0 on x. The resulting spherical complex Z is not a finite cell complex because the frontier f^2 of its c^2 consists of x and the north pole, hence $c^2 - f^2$ is not an open 2-cell.

One advantage of finite cell complexes over spherical complexes is that we can prove the following nice result:

(21.4) *Theorem. Let X be a finite cell complex, $E \xrightarrow{p} X$ a d-fold covering space, $d > 0$ (i.e., all the fibres $p^{-1}(x)$ have d points). Then E has a structure of finite cell complex for which the map p is cellular. Moreover*

$$\chi(E) = d\chi(X)$$

(We say p is *cellular* if p maps the q-skeleton of E into the q-skeleton of X, for all $q \geq 0$.)

Proof: Given any point $e \in E$, let $x = p(e)$. Then $x \in c_j^q - f_j^q$ for uniquely determined q, j; Let $y = (\phi_j^q)^{-1}(x)$, so that $y \in E^q - S^{q-1}$. Since E^q is contractible, the lifting theorem (6.1) tells us there is a unique map $\psi_j^q: E^q \to E$ such that (1) $\psi_j^q(y) = e$, (2) $p\,\psi_j^q = \phi_j^q$. Clearly p maps $\psi_j^q(E^q - S^{q-1})$ one-to-one onto $c_j^q - f_j^q$; since p is open, ψ_j^q maps $E^q - S^{q-1}$ homeomorphically onto its image.

If we fix x and let e vary in the fibre $p^{-1}(x)$, we get d such maps and we shall denote them by ψ^q_{ij}, where the index i runs from 1 to d. It is clear that the entire system of maps $\psi^q_{1j}, \ldots \psi^q_{dj}$ depends only on ϕ^q_j, not on the point x.

Let c^q_{ij} be the image of E^q under ψ^q_{ij}. Clearly these closed subsets give a representation of E as a finite cell complex for which p is a cellular map. Moreover, for each q, the number of sets c^q_{ij} is d times the number of sets c^q_j. Hence (20.11), $\chi(E) = d\chi(X)$. ■

Note. The formula $\chi(E) = d\chi(X)$ can be proved without assuming that X is a finite cell compex, but the proof requires spectral sequences. (See Spanier [52], Chapter 9.)

(21.5) *Note*. In the theory of Riemann surfaces, one is usually concerned with *branched covering spaces* $E \xrightarrow{p} X$: Roughly, over an open $U \subset X$, $p^{-1}(U) \to U$ is a covering space, but over X-U (*the branch locus*) there may be *ramification*. For example, consider the map of the extended complex plane S^2 onto itself defined locally by $z \to z^2$. The branch locus consists of the two points 0, ∞, and outside this set we have a 2-fold (unramified) covering space. Thus formula (21.4) no longer holds for branched covering spaces; however one can subtract a correction term from the right side which is a function of the "ramification indices" to get a correct formula (for the special case $X = S^2$, see Springer [54], p. 275).

(21.6) *Note*. A subspace X of a Euclidean space $\mathbf{R}^n$ is called a (finite) *polyhedron* if it can be represented as a finite union of geometric simplexes such that the intersection of two such simplexes is either empty or an iterated face of each. (For example, two geometric 2-simplexes which meet must meet in a common edge or a common vertex.) Any such representation is called a *triangulation* of X. Thus figure (b) is a triangulation, while (a) is not.

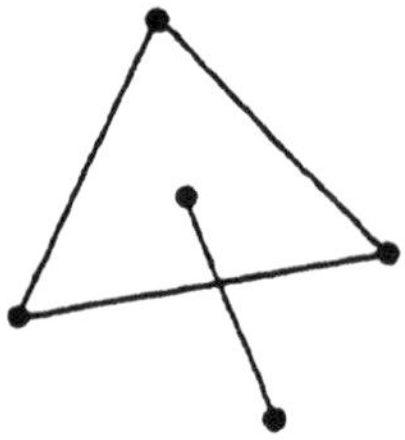

(a)

(b)

A space X is called (finitely) *triangulable* (sometimes called a *finite simplical complex*, although this terminology is gradually being used only for a purely algebraic notion) if it is homeomorphic to a (finite) polyhedron. Clearly such spaces are finite cell complexes.

For triangulable spaces there is a systematic combinatorial procedure for calculating the homology groups. (See Wallace [59], Chap. IX.) One can also compute explicitly the fundamental groups of the compact connected surfaces (Ahlfors and Sario [2], p. 99). However, to triangulate even so simple a surface as the torus requires at least 7 vertices, 21 edges, and 14 triangles, which seems unnaturally complicated. The triangulation approach to homology is due to Poincaré [45], who started the whole subject in 1895.

If X and Y are polyhedra, a map $f:X \to Y$ is called *simplicial* (with respect to given triangulations) if the restriction of f to any geometric simplex s in X is an affine map of s onto a geometric simplex in Y. The great importance of polyhedra is that an arbitrary map can be approximated by a simplicial map (on a suitably fine triangulation) in the same homotopy class (cf. any classical book).

(21.7) *Note.* To treat non-compact spaces, there are several types of infinite complexes used. For homotopy questions the most useful type are the *CW-complexes* of J. H. C. Whitehead [61]. The definition is the same as (21.1) except that infinitely many c_j^q are allowed, subject to an additional requirement on the topology of Z: A subset Y of Z is closed if and only if $Y \cap c_j^q$ is closed in c_j^q for all q, j (where c_j^q has the quotient topology from E^q). One can prove that for any space X whatever, there is a CW-complex Z and a map $f:Z \to X$ such that (i) f induces a one-to-one correspondence between the path components of Z and those of X, and (ii) for every $z \in Z$, $q > 0$, the induced homomorphism

$$(f_*)_q:\pi_q(Z, z) \to \pi_q(X, f(z))$$

is an isomorphism. (Such maps f are called *weak homotopy equivalences.*) This reduces the problem of determining the homotopy groups of X to the same problem for a CW-complex; for the latter, some powerful techniques are available (see Spanier [52], Chapters 7 and 8).

(21.8) *Note.* Let $X = \cup_n X^n$ be a cell complex with X^n the n-skeleton. Define $C_n = H_n(X^n, X^{n-1})$ for $n \geq 1$, $C_0 = H_0(X^0)$ and $\partial_n: C_n \to C_{n-1}$ as the composition

$$H_n(X^n, X^{n-1}) \xrightarrow{\partial} H_{n-1}(X^{n-1}) \to H_{n-1}(X^{n-1}, X^{n-2}).$$

Then $\partial_{n-1}\partial_n = 0$ and

$$H_n(X) \cong \frac{\text{Ker } \partial_n: C_n \to C_{n-1}}{Im\ \partial_{n+1}: C_{n+1} \to C_n}.$$

Proof. Using (19.36) we have $H_q(X^n, X^{n-1}) \cong H_q^*(X^n/X^{n-1}) \cong H_q^*(S^n \vee \ldots \vee S^n)$ for $n \geq 1$. Thus for all n, $H_q(X^n, X^{n-1})$ is nonzero only for $q = n$. By (19.20), $H_n(X^{n-1}) = 0$ and $H_n(X^{n+1}) \cong H_n(X)$. We embed this information in a diagram made by piecing together the long exact sequences for the pairs (X^n, X^{n-1}) for all n.

$$
\begin{array}{ccccccccc}
 & & 0 \to H_n(X^{n-1}) & & & & 0 \to H_{n-1}(X^{n-2}) & & \\
 & & \downarrow & & & \swarrow & & & \\
\to H_{n+1}(X^{n+1}, X^n) & \xrightarrow{\partial'} & H_n(X^n) & \xrightarrow{j} & H_n(X^n, X^{n-1}) & \xrightarrow{\partial} & H_{n-1}(X^{n-1}) & \to & H_{n-1}(X^{n-1}, X^{n-2}) \to \\
 & & \downarrow k & & & & \downarrow & & \\
 & & H_n(X^{n+1}) \to 0 & & & & H_{n-1}(X^n) \to 0 & &
\end{array}
$$

The path from 0 in the upper left corner to 0 in the lower right corner is part of one long exact sequence. Define $\phi{:}H_n(X^{n+1}) \to \ker \partial_n / \operatorname{im} \partial_{n+1}$ by $\phi = jk^{-1}$. Since $k^{-1}(X)$ is a coset of $\partial' H_{n+1}(X^{n+1}, X^n)$, ϕ is well defined. From $H_n(X^{n-1}) = 0$ we have j monic, hence ϕ is monic. Since $H_{n-1}(X^{n-2}) = 0$, we have $\ker \partial_n = \ker \partial = \operatorname{im} j$. Since k is epic, it follows that ϕ is epic, hence an isomorphism. ∎

(21.9) *Note. On the uniqueness theorem for homology.* A fundamental discovery of Eilenberg and Steenrod [23] is the formulation of axioms for homology functors, of which singular homology is an example. Abstractly, a *homology theory* is a sequence $\{H_q\}$ of functors from topological pairs and maps of pairs to R-modules and homomorphisms along with natural transformations $\{\partial_q \mid \partial_q{:}H_q(X, A) \to H_{q-1}(A)\}$ such that

a) $$\to H_{q+1}(X, A) \xrightarrow{\partial_{q+1}} H_q(A) \to H_q(X) \to H_q(X, A) \xrightarrow{\partial_q}$$

is exact.

b) If $g, f{:}(X, A) \to (Y, B)$ are homotopic maps of pairs then $H_q(f) = H_q(g)$.

c) If U is a subset of A such that $\bar{U} \subset \operatorname{int} A$, then inclusion

$$(X - U, A - U) \to (X, A)$$

induces isomorphisms $H_q(X - U, A - U) \cong H_q(X, A)$ for all q.

Of course, these are just the fundamental theorems, excluding the values $H_q(pt)$. Eilenberg and Steenrod prove

(21.10) *Uniqueness Theorem. If H and $\bar{H}$ are homology theories defined on the category of finite CW pairs and $\eta{:}H(pt) \to \bar{H}(pt)$ is a homomorphism, then there exists a unique natural transformation $\bar{\eta}{:}H \to \bar{H}$*

extending η. In particular if η is an isomorphism then $H_q(X,A) \cong \bar{H}_q(X,A)$ for all finite CW pairs (X,A).

The proof is not difficult, essentially induction over skeleta of the sort encountered earlier. We refer to [23], p. 100 for details (in the language of simplicial complexes or [66], p. 86 in the language of finite *CW* pairs).

Thus the different ways of constructing homology must give the same values on finite *CW* complexes, though they may differ on more general spaces.

There are theories satisfying a), b), c) which are different from singular homology because their values at points differ. One of these, ***bordism***, due to Thom [57] takes its intuitive basis in a more direct interpretation of the bounding relation than is taken by the singular (or simplicial) construction. Roughly one considers all maps $f:(M, \partial M) \to (X, A)$ where M is an n-manifold (compact). Two such maps f and f' are equivalent if there is an $(n+1)$-manifold W with $\partial W = M \amalg M'$ (disjoint union) and $F:(W, \partial W) \to (X, A)$ extending $f \amalg f'$. The set of equivalence classes $\mathcal{N}_n(X, A)$ can be given the structure of an abelian group and turns out to satisfy a), b), c). However $\mathcal{N}_n(pt)$ consists of equivalence classes of manifolds. For example $\mathcal{N}_2(pt)$ has two equivalence classes, S^2, $\mathbf{P}^2$. The reader is referred to Conner and Floyd [15] and G. W. Whitehead [88].

(21.11) *Chain complexes for homology theories.* The role of chain complexes in the theoretical understanding of homology theories has been analyzed by Burdick, Conner, and Floyd and we quote their result.[1] Consider a covariant functor L from finite *CW* pairs to chain complexes such that for each pair (X, A) we have the chain complex $\{L_n(X, A), \partial\}$ and for each map $f:(X,A) \to (Y,B)$ we have a chain map $L(f):L(X,A) \to L(Y,B)$. We write $L_n(X, \phi) = L_n(X)$. *Suppose L satisfies*

(1) $0 \to L_n(A) \to L_n(X) \to L_n(X, A) \to 0$

is exact for all pairs (X, A),

(2) $$h_n(X, A) = \frac{\operatorname{Ker} \partial:L_n(X, A) \to L_{n-1}(X, A)}{\operatorname{Im} \partial:L_{n+1}(X, A) \to L_n(X, A)}$$

is a homology theory.
Then, there is a natural equivalence

$$h_n(X, A) \cong \sum_{p+q=n} H_p(X, A; h_q)$$

[1] In Proc. Amer. Math. Soc. *19* (1968), 1115–1118.

where H_p is singular theory and $h_q = h_q(pt)$.

The proof will be accessible after the material on acyclic models in chapter 29.

Construction of spaces: More adjunction spaces. We consider adjunction spaces for some cases where a space larger than a cell is attached. This material is not essential to subsequent developments, but does provide interesting examples.

We continue with the assumptions in force at the beginning of chapter 19. The construction of continuous maps is based on the principle stated in the proof of (10.13) (piecing together maps defined on closed subsets) and the following

(21.12) *Extension Principle.* Let X be attached to Y by $f{:}A \to Y$. Given $g{:}X \to Z$ and $h{:}Y \to Z$ such that $g(a) = hf(a)$ for all $a \in A$, there exists a unique continuous $k{:}X \cup_f Y \to Z$ such that

$$\begin{array}{ccc} X \amalg Y & \xrightarrow{g \amalg h} & Z \\ \downarrow & \nearrow_{k} & \\ X \cup_f Y & & \end{array}$$

commutes. The proof is straightforward, or see Dugundji [20], p. 129.

Mapping Cones. After analyzing spaces, we have the problem of analyzing maps. An important tool is the mapping cone. Recall

$$CX = X \times I / X \times \{0\}$$

is the *cone* on X. The image of $X \times \{0\}$ in CX is the *vertex*. Points in CX are written (x, t) with $x \in X$, $t \in I$. We regard X as the subspace $X \times \{1\}$ in CX. The pair (CX, X) is collared, for example $U = X \times (\frac{1}{2}, 1)$ is a collaring of X. Motivation for arguments using CX is often provided by figure 4,

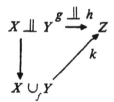

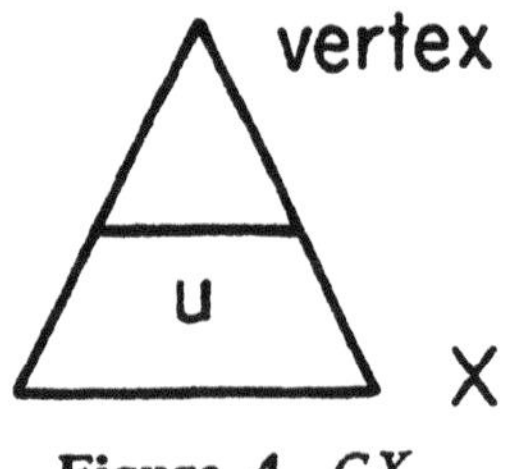

Figure 4. *CX*

Given $f:X \to Y$, we obtain $f:CX \to CY$, $\hat{f}(x,t) = (f(x), t)$.

(21.13) *Lemma. The vertex is a strong deformation retract of CX.*

Proof. The homotopy $H:CX \times I \to CX$ is given by $H(x,t,s) = (x, ts)$. ■

(21.14) *Proposition. $f:X \to Y$ is null-homotopic if and only if f extends to $F:CX \to Y$.*

Proof. If f extends, then it factors through a contractible space, hence is null-homotopic. Conversely, let $F:X \times I \to Y$ be a null-homotopy with $F(\ ,1) = f$. Then F extends to a well-defined map $F:CX \to Y$, since $F(x,0)$ is constant. ■

(21.15) *Definition*. The *mapping cone* Cf of $f:X \to Y$ is the adjunction space $CX \cup_f Y$. Y is embedded as a closed subset of Cf, we write the embedding $e:Y \to Cf$, and $CX - X$ is an open subset, Dugundji [20], p. 128. From (21.14) we have

(21.16) *$ef:X \to Cf$ is null-homotopic.*

Given a commutative diagram

$$\begin{array}{ccc} X & \xrightarrow{f} & Y \\ {\scriptstyle\alpha}\downarrow & & \downarrow{\scriptstyle\beta} \\ X' & \xrightarrow{f'} & Y' \end{array}$$

the extension principle (21.12) provides a unique $\gamma:Cf \to Cf'$ such that

$$\begin{array}{ccc} Y & \xrightarrow{e} & Cf \\ {\scriptstyle\beta}\downarrow & & \downarrow{\scriptstyle\gamma} \\ Y' & \xrightarrow[e']{} & Cf' \end{array} \qquad\qquad \begin{array}{ccc} CX & \longrightarrow & Cf \\ {\scriptstyle\hat{\alpha}}\downarrow & & \downarrow{\scriptstyle\gamma} \\ CX' & \longrightarrow & Cf' \end{array}$$

commute.

Motivation for arguments using Cf is often provided by figure 5.

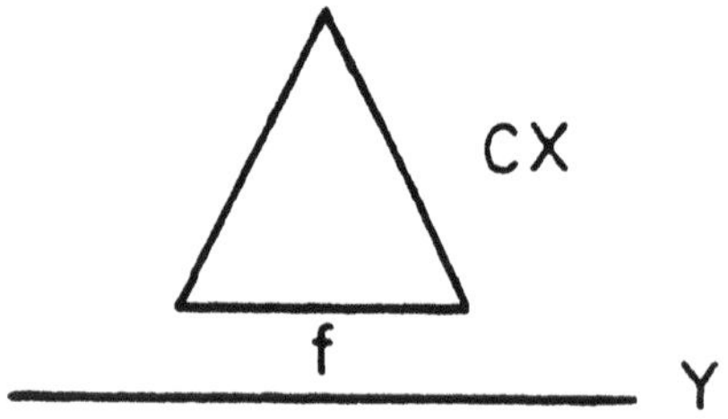

Figure 5. *Cf*

(21.17) *Proposition.* $f:X \to Y$ *is null-homotopic if and only if* Y *is a retract of* Cf.

Proof. If f is null-homotopic, we have an extension $F:CX \to Y$. Then $F \amalg id: CX \amalg Y \to Y$ provides a retraction by (21.12). Conversely, given a retraction $r:Cf \to Y$ we have $f = ref$ null-homotopic by (21.16). ■

As an application of (21.17) we obtain $\mathbf{P}^{n-1}$ *is not a retract of* $\mathbf{P}^n$, (compare exercise (19.44)).

Proof. $\mathbf{P}^n \cong Cf$ where $f:S^{n-1} \to \mathbf{P}^{n-1}$ is the canonical map. By exercise (16.13), f is not null-homotopic. ■

Again, $S^1 \vee S^1$ *is not a retract of* $S^1 \times S^1$.

Proof. The torus is a mapping cone

$$S^1 \xrightarrow{f} S^1 \vee S^1 \to S^1 \times S^1.$$

By the Van Kampen theorem, (4.12), $\pi_1(S^1 \vee S^1)$ is free on two generators a, b and the homotopy class of f is the commutator $[a, b] \neq 1$. ■

Remark. Later applications of (21.17) will show certain maps are not null-homotopic by arguing that the cup product structure of the cohomology of Cf is incompatible with a retraction to Y.

The problem of extending maps to mapping cones has a useful formulation.

(21.18) *Proposition. Let* $f:X \to Y, g:Y \to Z$. *Then* g *extends to* $h:Cf \to Z$ *such that* $he = g$ *if and only if* gf *is null-homotopic.*

Proof. The necessity follows from (21.16). Conversely, if gf is null-homotopic, then Z is a retract of Cgf and the retraction $Cgf \to Z$ composed with the canonical map $Cf \to Cgf$ gives h.

$$\begin{array}{ccccc} X & \xrightarrow{f} & Y & \xrightarrow{e} & Cf \\ \| & & \downarrow g & & \downarrow \\ X & \xrightarrow[gf]{} & Z & = & Cgf. \end{array}$$

■

Remark. h depends on the null-homotopy, and in general is not unique, even up to homotopy. The further analysis will not concern us, but is a standard part of general homotopy theory.

We next look at homological properties of Cf. Since CX is contractible, (19.15) yields a long exact sequence

$$\text{(21.19)} \quad \to H_q^*(X) \xrightarrow{H_q(f)} H_q^*(Y) \xrightarrow{H_q(e)} H_q^*(Cf) \to H_{q-1}^*(X) \to \ldots$$

which is functorial for (strictly) commuting diagrams

$$\begin{array}{ccc} X & \xrightarrow{f} & Y \\ \alpha \downarrow & & \downarrow \beta \\ X' & \xrightarrow[f']{} & Y' \end{array}$$

In practice, however, we often just have commutativity up to homotopy, $f'\alpha \simeq \beta f$.

(21.20) *Proposition. If $f'\alpha \simeq \beta f$ then there exists $\gamma: Cf \to Cf'$, an extension of $e'\beta$, such that*

$$\begin{array}{ccccccccc} \to H_q(X) & \xrightarrow{H_q(f)} & H_q(Y) & \longrightarrow & H_q(Cf) & \longrightarrow & H_{q-1}(X) & \to & \ldots \\ H_q(\alpha) \downarrow & & H_q(\beta) \downarrow & & H_q(\gamma) \downarrow & & H_{q-1}(\alpha) \downarrow & & \\ \to H_q(X') & \xrightarrow[H_q(f')]{} & H_q(Y') & \longrightarrow & H_q(Cf') & \longrightarrow & H_{q-1}(X') & \to & \ldots \end{array}$$

has all squares commutative.

Proof. γ exists, since $e'\beta f \simeq e'f'\alpha \simeq *$. We describe γ more explicitly. Let $H{:}X \times I \to Y'$ be a homotopy such that $H(\ ,0) = f'\alpha$ and $H(\ ,1) = \beta f$. Then we can define γ by

$$(x,t) \longmapsto \begin{cases} (\alpha(x), 2t) & 0 \le t \le \frac{1}{2}, \\ e'H(x, 2t-1) & \frac{1}{2} \le t \le 1, \end{cases}$$

$$e(y) \longmapsto e'\beta(y),$$

and picture γ in figure 6.

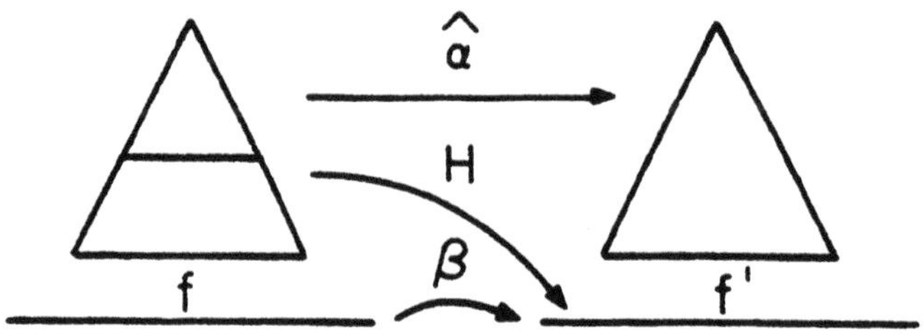

Figure 6.

Our result will follow when we show

$$\begin{array}{ccc} (CX, X) & \xrightarrow{\bar{f}} & (Cf, Y) \\ \hat{\alpha}\downarrow & & \downarrow\gamma \\ (CX', X') & \xrightarrow[\bar{f}']{} & (Cf', Y') \end{array}$$

is a homotopy commutative diagram of pairs, since (21.19) arises from the Barratt-Whitehead lemma applied to the excisions $\bar{f}, \bar{f}'$ and $(\hat{\alpha}, \gamma)$ will induce a map of ladders. The required homotopy $K{:}(CX \times I, X \times I) \to (Cf', Y')$ is given by

$$K(x,t,s) = \begin{cases} \left((\alpha(x)), \dfrac{2t}{s+1}\right) & 0 \le t \le \dfrac{s+1}{2}, \\[2ex] e'H(x, 2t-s-1) & \dfrac{s+1}{2} \le t \le 1. \end{cases}$$

Then $K(x, t, 0) = \gamma\bar{f}(x, t)$ and $K(x, t, 1) = \bar{f}'\hat{\alpha}(x, t)$. ■

Remark. It may be the case that $H_q(\gamma) \neq H_q(\gamma')$ for different extensions $\gamma \neq \gamma'$.

An important property of mapping cones is their homotopy invariance.

(21.21) *Proposition. If $f, g: X \to Y$ are homotopic, then Cf and Cg are homotopy equivalent.*

Proof. We first construct maps $\theta: Cf \to Cg$ and $\psi: Cg \to Cf$. Let $H: X \times I \to Y$ be a homotopy such that $H(\ , 0) = g$ and $H(\ , 1) = f$. Let $\bar{H}(x, t) = H(x, 1 - t)$. Then the formulas

$$\theta(x, t) = \begin{cases} (x, 2t) & 0 \leq t \leq \frac{1}{2}, \\ H(x, 2t - 1) & \frac{1}{2} \leq t \leq 1, \end{cases}$$

$$\theta \mid Y = id$$

and similarly for ψ with $\bar{H}$ replacing H, give well-defined maps. We construct a homotopy $\psi\theta \simeq id(\text{rel } Y)$. Note

$$\psi\theta(x, t) = \begin{cases} (x, 4t) & 0 \leq t \leq \frac{1}{4}, \\ \bar{H}(x, 4t - 1) & \frac{1}{4} \leq t \leq \frac{1}{2}, \\ H(x, 2t - 1) & \frac{1}{2} \leq t \leq 1. \end{cases}$$

Then the required homotopy $K: Cf \times I \to Cf$ can be seen from the picture

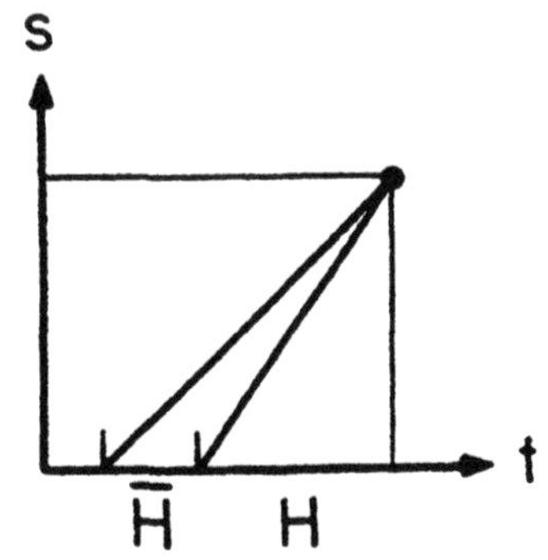

Specifically, K is given by

$$K(x, t, s) = \begin{cases} \left(x, \dfrac{4t}{3s+1}\right) & 0 \leq t \leq \dfrac{3s+1}{4}, \\ H(x, 3s+2-4t) & \dfrac{3s+1}{4} \leq t \leq \dfrac{s+1}{2}, \\ H(x, 2t-1) & \dfrac{s+1}{2} \leq t \leq 1, \end{cases}$$

$$K | Y = id.$$

Similarly for $\theta\psi$. ∎

Remark. The homotopy equivalences provided by (21.21) have some nonobvious consequences. For example, the space made by identifying 3 sides of an (equilateral) triangle according to the edge equation aaa^{-1} is the mapping cone of map $f: S^1 \to S^1$ and $f \simeq id$. But the mapping cone of id is the disc, hence this space (called the *dunce cap*) has the homotopy type of a point. Another example is shown in figure 7. Two tori are interlocked so their surfaces are mutually tangent along the subspaces $S^1 \vee S^1$. The resulting space has the homotopy type of $(S^1 \times S^1) \vee S^2$. (Prove this: consider the attaching map of one torus on the other and make use of the argument following 21.17.)

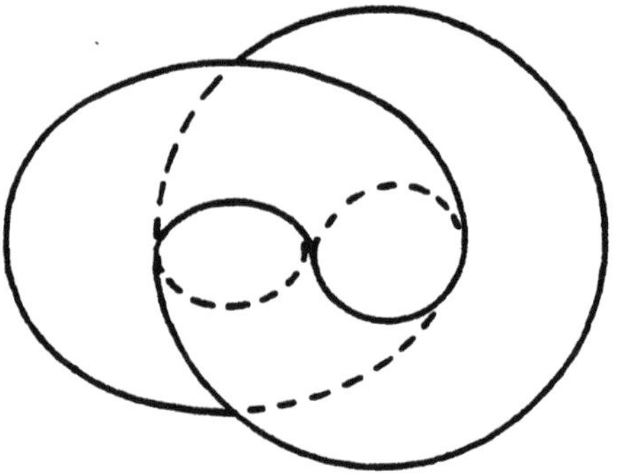

Figure 7.

Some manifolds[1]. Let M, M' be manifolds with homeomorphic boundaries, and $h:\partial M \to \partial M'$ a homeomorphism. The adjunction space $W = W(h, M, M') = M \cup_h M'$ can be proved to be a manifold. One uses the collaring property (19.2) and invariance of domain to construct the local euclidean structure at the points where M is joined to M'. Naturally the first case to consider is $M = M' = E^n$.

(21.22) *Proposition* (Alexander trick). *Let $h:S^{n-1} \to S^{n-1}$ be a homeomorphism. Then $W = E^n \cup_h E^n$ is homeomorphic to S^n.*

Proof. Regarding E^n as the cone on S^{n-1}, we have $\hat{h}:E^n \to E^n$, a homeomorphism extending h. Let $p:E^n \amalg E^n \to S^n$ be the usual identification of S^n in terms of its hemispheres. Then $p(\hat{h} \amalg id):E^n \amalg E^n \to S^n$ extends by (21.12) to $\phi:W \to S^n$. It is apparent that ϕ is a continuous, injective map of a compact space to a Hausdorff space, hence a homeomorphism. ∎

Remark. It is not necessarily the case that W is smoothly equivalent to S^n. The first examples were given by Milnor [83], igniting a significant development in topology.

Lens spaces (19.12). Let $M = E^2 \times S^1$. It turns out that $L(p, q) \cong M \cup_h M$ for certain homeomorphisms of $\partial M = S^1 \times S^1$. We construct the homeomorphisms and do some calculations. Write points in S^1 as $e^{i\theta}$.

(21.23) *Definition.* The *twists* $\lambda, \mu:S^1 \times S^1 \to S^1 \times S^1$ are given by

$$\lambda(e^{i\theta}, e^{i\phi}) = (e^{i(\theta+\phi)}, e^{i\phi}),$$
$$\mu(e^{i\theta}, e^{i\phi}) = (e^{i\theta}, e^{i(\theta+\phi)}).$$

There is no way to distinguish geometrically the factors of $S^1 \times S^1$. But if we regard $S^1 \times S^1 = \partial(E^2 \times S^1)$, then meaningful distinctions can be made. We shall do this only when $E^2 \times S^1 \subset \mathbf{R}^3$ is embedded in the standard (i.e. unknotted) way. In this case the circle $\phi = 0$ is called a *meridian M* and the circle $\theta = 0$ a *longitude L*. The meridian is homotopically trivial in $E^2 \times S^1$ while the longitude represents a generator of H_1. The effect of the twists on M and L is shown in figure 8. (We have actually distorted the images of M, L by a homeomorphism of $\mathbf{R}^3$.)

---

[1] Acknowledgement. This discussion is based on the exposition of Rolfsen [85], to which the reader is referred for further, interesung material.

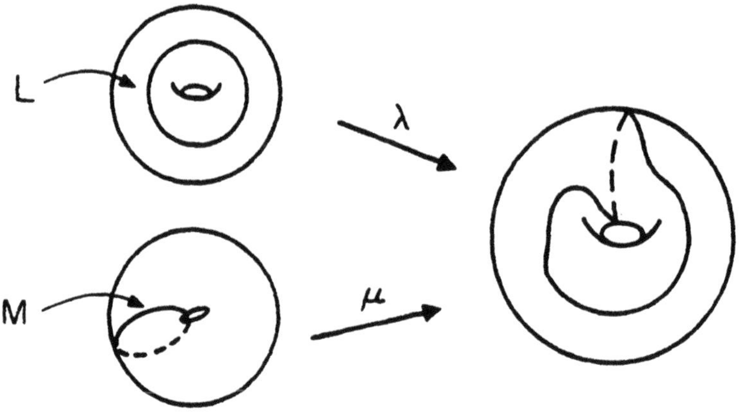

Figure 8.

Of course $\lambda | M = \mu | L = id$.

We now consider homology and take $R = \mathbf{Z}$ for the remainder of this section. The circles M, L determine a basis, m, l for $H_1(S^1 \times S^1)$. More precisely, pick generators of $H_1(M)$, $H_1(L)$ and take their images in $H_1(S^1 \times S^1)$. This amounts to assigning orientations (i.e., directions) to each circle. In terms of this basis, we have

(21.24)
$$H_1(\lambda) = \begin{pmatrix} 1 & 1 \\ 0 & 1 \end{pmatrix}, \qquad H_1(\mu) = \begin{pmatrix} 1 & 0 \\ 1 & 1 \end{pmatrix},$$
$$m = \begin{pmatrix} 1 \\ 0 \end{pmatrix}, \qquad l = \begin{pmatrix} 0 \\ 1 \end{pmatrix}.$$

The inverse homeomorphisms are given by $\lambda^{-1}(e^{i\theta}, e^{i\phi}) = (e^{i(\theta-\phi)}, e^{i\phi})$ and similarly for μ^{-1}. We have

(21.25)
$$H_1(\lambda^{-1}) = \begin{pmatrix} 1 & -1 \\ 0 & 1 \end{pmatrix}, \qquad H_1(\mu^{-1}) = \begin{pmatrix} 1 & 0 \\ -1 & 1 \end{pmatrix}.$$

From linear algebra we have the fact that any 2×2 matrix of integers with det $= 1$ is a product of the four matrices (21.24), (21.25), [89].† The interchange map $i: S^1 \times S^1 \to S^1 \times S^1$, $i(x, y) = (y, x)$ switches m with l and has matrix

$$H_1(i) = \begin{pmatrix} 0 & 1 \\ 1 & 0 \end{pmatrix}.$$

Summarizing,

(21.26) *If A is any 2×2 matrix with integer entries and* $\det A = \pm 1$, *there is a homeomorphism h of $S^1 \times S^1$ with $H_1(h) = A$.*

It turns out that the topological type of $E^2 \times S^1 \cup_h E^2 \times S^1$ depends only on $h(M)$. See Rolfsen [85] for this and further details.

(21.27) *Definition.* Let (p, q) be co-prime integers and h as homeomorphism of $\partial(E^2 \times S^1)$ such that $H_1(h)(m) = qm + pl$ in $H_1(S^1 \times S^1)$. The adjunction space $L(p, q) = E^2 \times S^1 \cup_h E^2 \times S^1$ is a *lens space.*

It is not obvious that this description is the same as (19.12). See M. Cohen [78] for details. The image $h(M)$ is called a (p, q) *torus knot.* There are topological relations among the $L(p, q)$. If $\pm r \equiv q^{\pm 1} \bmod p$, then $L(p, q)$ is homeomorphic to $L(p, r)$. This can be proved by cutting and pasting, see Hilton and Wylie [30] or Rolfsen [85]. Furthermore $L(1, q) \cong S^3$ and $L(0, 1) \cong S^2 \times S^1$. The former derives from the decomposition $S^3 \cong \partial(E^2 \times E^2) = E^2 \times S^1 \cup S^1 \times E^2$ (note the switch of positions for E^2), while the latter follows from $S^2 \times S^1 = E_2^+ \times S^1 \cup E_2^- \times S^1$. We normalize, $0 < q < p$.

(21.28) *Proposition. The integral homology of $L(p, q)$ is given by* $H_3 \cong \mathbf{Z}$, $H_2 = 0$, $H_1 \cong \mathbf{Z}/p\mathbf{Z}$, $H_0 \cong \mathbf{Z}$.

Proof. Using (19.15) we have (and abbreviating $L(p, q) = L$) $H_3(L) \cong H_2(S^1 \times S^1)$ and an exact sequence

† Let $S = \begin{pmatrix} 1 & 1 \\ 0 & 1 \end{pmatrix}$ $R = \begin{pmatrix} 1 & 0 \\ 1 & 1 \end{pmatrix}$ $U = \begin{pmatrix} 0 & 1 \\ -1 & -1 \end{pmatrix}$ Then S and U generate $SL(2, \mathbf{Z})$. But $U = (R^{-1}S)^2$.

$$0 \to H_2(L) \to H_1(S^1 \times S^1) \xrightarrow{\;H_1(j)\,\oplus\,H_1(h)\;} H_1(E^2 \times S^1) \oplus H_1(E^2 \times S^1) \to H_1(L) \to 0$$

where j is an inclusion and h is the attaching map. Denote generators in $H_1(S^1 \times S^1)$ by m, l as in (21.24), and write l_1, l_2 for the generators in the upper and lower terms in the sequence corresponding to longitudes. Then $H_1(j)(m) = 0, H_1(j)(l) = l_1$ while $H_1(h)(m) = pl_2, H_1(h)(l) = xl_2$ for some integer x. Hence the matrix for $(H_1(j) \oplus H_1(h)) \circ \Delta$ with respect to this basis is

$$\begin{pmatrix} 0 & 1 \\ p & x \end{pmatrix}$$

which is a monomorphism. Since $\binom{0}{1}$ and $\binom{1}{x}$ are a basis for $H_1(E^2 \times S^1) \oplus H_1(E^2 \times S^1)$ we obtain the stated result. ■

Remark. In fact $x \equiv \pm q^{-1} \bmod p$, since on $H_1(S^1 \times S^1), H_1(h) = \begin{pmatrix} q & y \\ p & x \end{pmatrix}$ and $qx - py = \pm 1$.

Poincaré homology 3-sphere. We give a construction, due to Dehn, of a space first constructed by Poincaré. We ascertain its homological properties.

Let K be a knot in S^3 with a neighborhood N such that N is the image of $E^2 \times S^1$ via an embedding $\phi: E^2 \times S^1 \to S^3$ such that $\phi(\{0\} \times S^1) = K$ and $X = S^3 - \operatorname{int} N$ is a 3-manifold with ∂X homeomorphic to $S^1 \times S^1$. Note $X \subset S^3 - K$ is a deformation retract as follows: A point $z \in N - K$ has the form $z = \phi(re^{i\theta}, y)$ where $re^{i\theta} \in E^2 - \{0\}$, $y \in S^1$. Then $f: S^3 - K \to X$ given by $f(z) = \phi(e^{i\theta}, y)$, $f \mid S^3 - \operatorname{int} N = id$ has the required properties. Hence by (18.3), X has the homology of S^1.

(21.29) *Exercise.* Let C be a circle of the form $C = \operatorname{im} \phi \mid S^1 \times \{x\}$, where $S^1 = \partial E^2$ and we abuse notation to regard $x \in K$. Show the inclusion $C \to X$ induces an isomorphism in homology. Suggestion: C is an equatorial circle of an embedded 2-sphere obtained by pushing the disc $\operatorname{im} \phi \mid E^2 \times \{x\}$ in opposite directions leaving C fixed. This 2-sphere represents the homology of $R^3 - \{x\}$. Now regard K as the union of 2 arcs intersecting in x and ∞, and compute in the Mayer-Vietoris sequence.

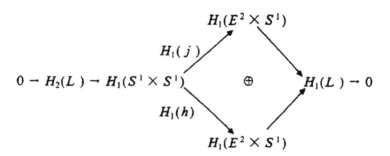

New manifolds $Q = E^2 \times S^1 \cup_h X$ are obtained from homeomorphisms $h: S^1 \times S^1 \to \partial X$. This process is called *surgery* and its elaboration has led to a deep understanding of manifolds.

In figure 9 we have this set up for a trefoil knot K.

(21.30) *Exercise.* Show that J and C in fig. 9 satisfy $[J] + [C] = 0$ in $H_1(X)$. Suggestion: Each under-crossing contributes $-[C]$ and the windings each contribute $+[C]$.

Let $Q = E^2 \times S^1 \cup_h X$ where h is a homeomorphism such that $h(M) = J$, M a meridian of $E^2 \times S^1$. Q is the *Poincaré homology 3-sphere.*

(21.31) *Proposition.* $H_q(Q) \cong H_q(S^3)$.

Proof. Combine exercises (21.29) and (21.30) with the argument in (21.28). ■

Remark. Using the Seifert, Van Kampen theorem (4.12), the fundamental group of Q is $\pi_1(S^3 - K)$ with the additional relation represented by J. Then $\pi_1(Q)$ turns out to be a finite group of order 120 known as the *binary icosahedral* group. Thus Q is not homeomorphic to S^3. The space Q can also be obtained from a properly discontinuous action of the b.i. group on S^3. See F. Klein [81] for the fascinating and extensive interconnections of various parts of mathematics through this group.

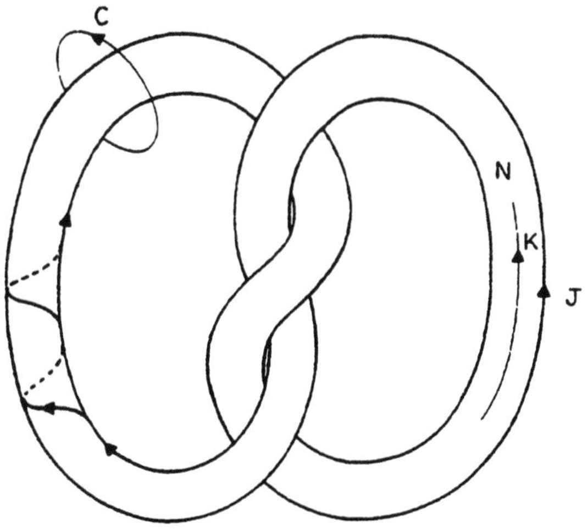

Figure 9.

Part III
ORIENTATION AND DUALITY ON MANIFOLDS

Introduction to Part III

Manifolds are the main geometric topic of part III. A new theoretical tool, cohomology, is introduced. Cup and cap products are introduced by means of the Alexander-Whitney map. The cup product endows cohomology with more algebraic structure than was available for homology in chapter 19. The cap product from cohomology to homology is used for the duality theorems.

The first topic is orientability of manifolds. This discussion illustrates a new use of homology as a means of piecing together "local information" to obtain a grasp on a global property. Orientation is a global property of a manifold M which manifests itself locally in distinctions between "left-handedness" and "right-handedness". This distinction can be described in terms of choosing generators for certain homology modules. The presence of an orientation manifests itself as a coherent system of such choices.

Our treatment of cohomology is divided between this part and part IV. Here we develop material analogous to homology and prove one of the universal coefficient theorems. This theorem is often useful for calculations. Next, products are introduced. We delay a more theoretical treatment of products to part IV. Enough material is developed to prove the duality theorems for manifolds.

Our proof of Poincaré duality follows Milnor [41] (see also [84]). This argument uses Mayer-Vietoris sequences to effect passage from local information to global information.

The Alexander duality theorem of chapter 27 brings to fruition the ideas nascent in chapter 18. The Lefshetz duality theorem of chapter 28 completes the triumvirate of duality theorems.

22. Orientation of Manifolds

Throughout this section, X will be an n-dimensional manifold, $n \geq 1$.

(22.1) *Lemma. For any point $x \in X$,*

$$H_n(X, X - x) \cong R$$

Proof: Let U be an open neighborhood of x homeomorphic to the open unit ball in $\mathbf{R}^n$ (i.e., a "coordinate neighborhood" of x). Excising the closed subset $X - U$ of the open $X - x$, we get

$$H_n(U, U - x) \xrightarrow{\approx} H_n(X, X - x)$$

Since U is contractible, the exact homology sequence of the pair $(U, U - x)$ gives

$$H_n(U, U - x) \xrightarrow{\approx} H^{\#}_{n-1}(U - x)$$

But $U - x$ is homotopically equivalent to S^{n-1}, so $H^{\#}_{n-1}(U - x) \cong R$. ■

Consider the special case $n = 2$, $R = \mathbf{Z}$. Then there are two possible elements of $H_2(X, X - x) \cong H_1(U - x)$ which can generate this infinite cyclic group, namely the ones corresponding to loops winding once around x in opposite directions. Choosing one of these generators corresponds intuitively to "choosing an orientation about the point x."

For $n > 2$, one has to determine the possible generators of $H_{n-1}(U - x) \cong H_{n-1}(S^{n-1})$. Regard S^{n-1} as the frontier of the geometric simplex Δ^n. One can then show the generators are $\pm\partial\delta^n$, where δ^n is the identity singular simplex on Δ^n (see Wallace [59], p. 178), and (22.40).

Definition. A local R-orientation of X at x is a generator of the R-module $H_n(X, X - x)$.

To define the notion of an orientation of X globally, our intuition tells us that we should have orientations of X given at each point such that these local orientations "match up." This need not be possible, as is shown by the open Möbius band: In the space

$$S = \{(s, t) \in \mathbf{R}^2 \mid 0 \leq s \leq 1, 0 < t < 1\}$$

identify $(0, t)$ with $(1, 1 - t)$. (The quotient space is a 2-dimensional manifold.) See (28.17). However, we can always get "matching orientations" throughout a neighborhood of a given point x, in the following sense:

(22.2) *Continuation Lemma. Given an element $\alpha_x \in H_n(X, X - x)$. Then there is an open neighborhood U of x and $\alpha \in H_n(X, X - U)$ such that $\alpha_x = j_x^U(\alpha)$, where*

$$j_x^U : H_n(X, X - U) \to H_n(X, X - x)$$

is the canonical homomorphism induced by inclusion.

Proof: Let a be a relative cycle representing α_x. Then the support $|\partial a|$ of ∂a is a compact subset of X contained in $X - x$, so that $U = X - |\partial a|$ is an open neighborhood of x. Take $\alpha \in H_n(X, X - U)$ to be the homology class of a relative to $X - U$. ■

This lemma tells us we can obtain elements $\alpha_y \in H_n(X, X - y)$ for y near x (i.e., for $y \in U$) from α_x by setting $\alpha_y = j_y^U(\alpha)$. We think of these elements as "matching up" because they all come from one and the same element $\alpha \in H_n(X, X - U)$. Call α a *continuation* of α_x in U. We must show one more point.

(22.3) *Coherence Lemma. If α_x generates $H_n(X, X - x)$, then U and α can be chosen such that α_y generates $H_n(X, X - y)$ for all $y \in U$.*

This follows at once from a stronger result:

(22.4) *Locally Constant Lemma. Every neighborhood W of x contains a neighborhood U of x such that for every $y \in U$, j_y^U is an isomorphism (hence α_x has a unique continuation in U).*

Proof: Let V be a coordinate neighborhood of x contained in W (V is homeomorphic to E^n), and let U be a smaller open set corresponding to an open ball of radius < 1. Then we have the commutative diagram (for any $y \in U$)

$$\begin{array}{ccccc} H_n(X, X-U) & \approx & H_n(V, V-U) & \approx & H^\#_{n-1}(V-U) \\ {\scriptstyle j^U_y}\downarrow & & \downarrow & & \downarrow \\ H_n(X, X-y) & \approx & H_n(V, V-y) & \approx & H^\#_{n-1}(V-y) \end{array}$$

in which the left horizontal isomorphisms are excisions and the right ones are connecting homomorphisms (V is contractible). Now the right vertical arrow is an isomorphism because the inclusion $V - U \to V - y$ is a homotopy equivalence (move out radially from y).

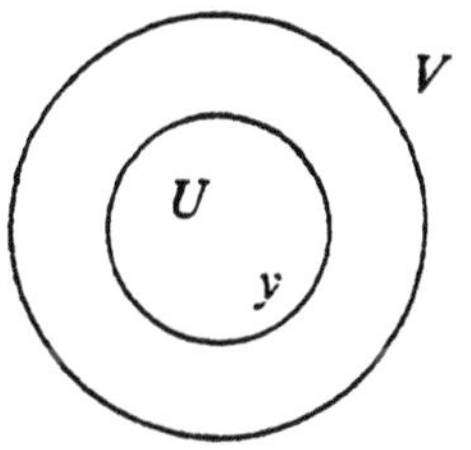

Hence j^U_y is an isomorphism. ■

(22.5) *Remark.* The only properties of U used in the above argument are (a)$(V, V - U) \to (X, X - U)$ is an excision, (b) the inclusion $i: V - U \to V - y$ induces an isomorphism $H^\#_{n-1}(V - U) \to H^\#_{n-1}(V - y)$ for every $y \in U$. These two properties are also satisfied if we take U to be a closed rectangular parallelepiped of dimension $d \leq n$ through x.

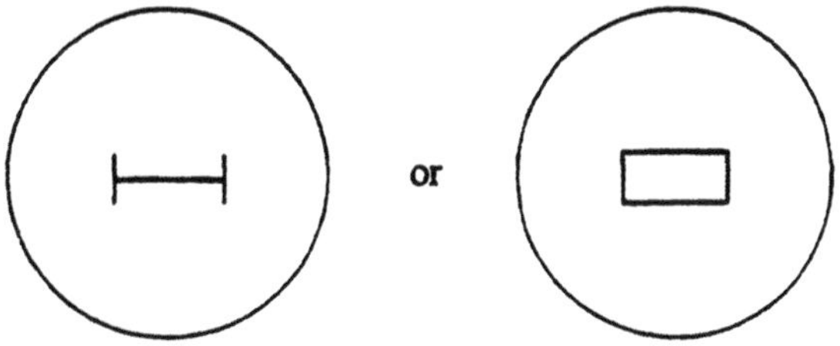

Property (a) is immediate. Now we can choose an $(n - 1)$-sphere $S_{n-1} \subset V - U$. We know the inclusion

$$i' : S_{n-1} \to V - y$$

is a homotopy equivalence, and we can easily show the inclusion

$$i'': S_{n-1} \to V - U$$

to be a homotopy equivalence (the case $d = 1$, $n = 2$ is illustrated);

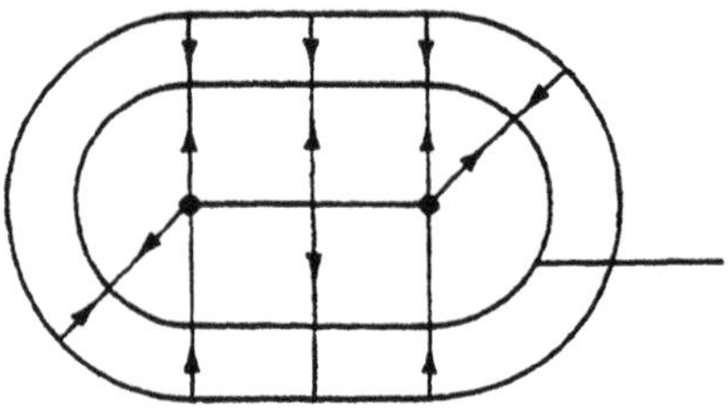

Hence $H_{n-1}(i) = H_{n-1}(i')H_{n-1}(i'')^{-1}$ is an isomorphism.

Definition. Given a subspace $U \subset X$. An element $\alpha \in H_n(X, X - U)$ such that $j_y^U(\alpha)$ generates $H_n(X, X - y)$ for each $y \in U$ will be called a *local R-orientation of X along U.*

(22.6) *Notation.* If $V \subset U$ are subspaces of X,

$$j_V^U: H_n(X, X - U) \to H_n(X, X - V)$$

denotes the homomorphism induced by inclusion. If α is a local R-orientation along U, then $j_V^U(\alpha)$ is one along V, since for any $y \in V$,

$$j_y^V[j_V^U(\alpha)] = j_y^U(\alpha).$$

We now define a *global R-orientation* of X: Suppose we are given
(i) a family of open subspaces U_i which cover X.
(ii) for each i, a local orientation $\alpha_i \in H_n(X, X - U_i)$ of X along U_i.
Call this an *R-orientation system* if the following compatibility condition holds: For any $x \in X$, if $x \in U_i \cap U_{i'}$, then

(iii) $j_x^{U_i}(\alpha_i) = j_x^{U_{i'}}(\alpha_{i'})$

In this case a local R-orientation is unambiguously defined at each point x by

(iv) $\alpha_x = j_x^{U_i}(\alpha_i) \quad x \in U_i$

Given another R-orientation system (V_k, β_k). We say it defines the same R-orientation if

(v) $\alpha_x = \beta_x \quad \text{all } x \in X$

Then *a global R-orientation of X* is by definition an equivalence class of R-orientation systems, the equivalence relation being (v).

We say X is *R-orientable* (resp. *orientable*) if an R-orientation system (resp. a $\mathbf{Z}$-orientation system) exists.

(22.7) *Proposition.* (a) *An open submanifold V of an R-orientable X is R-orientable.* (b) *X is R-orientable if and only if all its connected components are.*

Proof: (a) Let (U_i, α_i) be an R-orientation system for X. For any $x \in V$, let $\beta_x \in H_n(V, V - x)$ correspond to α_x via the excision isomorphism

$$H_n(V, V-x) \approx H_n(X, X-x)$$

By (22.4), there is an open neighborhood V_x of x such that $V_x \subset V \cap U_i$ for some i and such that β_x has a unique continuation to a local R-orientation $\bar{\beta}_x$ of V along V_x; we can further choose V_x so small that $X - V$ is contained in the interior of $X - V_x$. Then for any $y \in V_x$, the diagram

$$\begin{array}{ccccc} H_n(V, V-y) & \approx & H_n(X, X-y) & \searrow & \\ \uparrow & & \uparrow & & H_n(X, X-U_i) \\ H_n(V, V-V_x) & \approx & H_n(X, X-V_x) & \nearrow & \end{array}$$

shows that the local R-orientation of V at y induced by $\bar{\beta}_x$ equals β_y. Thus $(V_x, \bar{\beta}_x)$ is an R-orientation system for V.

Hence (b) follows from (a) and the fact that the connected components of a manifold are open. ■

(22.8) *Proposition. Suppose X is connected. Then two R-orientations of X which agree at one point are equal.*

Proof: Let A be the set of points at which they agree. By (22.4), A and $X - A$ are open, hence $A = X$. ■

(22.9) *Corollary. A connected orientable manifold has exactly two distinct orientations.*

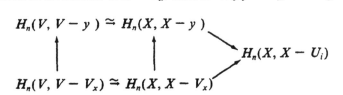

(22.10) *Example*: For $X = S^n$, any $x \in X$, $H_n(S^n) \to H_n(S^n, S^n - x)$ is an isomorphism, since $S^n - x$ is contractible. Taking the open covering consisting of the single set X and α_x a generator of $H_n(S^n)$, we see that S^n is R-orientable.

(22.11) *Example*: $X = \mathbf{R}^n$ is homeomorphic to S^n minus a point, so, by (22.7) X is R-orientable.

(22.12) *Proposition. Every manifold has a unique* $\mathbf{Z}/2$*-orientation.*

For each x, α_x must be the unique non-zero element of $H_n(X, X-x; \mathbf{Z}/2)$. We can choose an open neighborhood U_x of x in which α_x has a unique continuation. Clearly these continuations are compatible. ∎

(22.13) *Remark*. It can actually be shown that if X is orientable, then X is R-orientable for all coefficient rings R. This follows from the universal coefficient theorem which tells how to determine $H_q(X, A; R)$ from $H_q(X, A; \mathbf{Z})$, and R (see Section 29).

(22.14) *Theorem. Let X be a connected non-orientable manifold. Then there is a 2-fold connected covering space $E \xrightarrow{p} X$ such that E is orientable.*

(22.15) *Corollary. Every simply connected manifold is orientable (more generally, every connected manifold whose fundamental group contains no subgroup of index 2 is orientable).*

Proof of theorem: Define E as the set of pairs (x, α_x), where $x \in X$ and α_x is one of the two generators of $H_n(X, X-x; \mathbf{Z})$. Set $p(x, \alpha_x) = x$.

Consider pairs (U, α_U), where U is open in X and α_U is a local orientation of X along U. Let

$$\langle U, \alpha_U \rangle = \{(x, \alpha_x) \mid x \in U, \alpha_x = j_x^U(\alpha_U)\}$$

Suppose

$$(x, \alpha_x) \in \langle U, \alpha_U \rangle \cap \langle U', \alpha_{U'} \rangle$$

By (22.4) there is an open neighborhood $U'' \subset U \cap U'$ of x such that α_x has a unique continuation $\alpha_{U''}$ on U''. We must then have $j_{U''}^U(\alpha_U) = \alpha_{U''} = j_{U''}^{U'}(\alpha_{U'})$, so that

$$\langle U'', \alpha_{U''} \rangle \subset \langle U, \alpha_U \rangle \cap \langle U', \alpha_{U'} \rangle$$

Thus the sets $\langle U, \alpha_U \rangle$ form a base for a topology on E. Since p maps $\langle U, \alpha_U \rangle$ homeomorphically onto U and

$$p^{-1}(U) = \langle U, \alpha_U \rangle \cup \langle U, -\alpha_U \rangle \quad \text{(disjoint)}$$

we do have a 2-fold covering space.

For each $x \in X$, choose open neighborhoods $V \supset U$ as on p. 159 for which an α_V exists; set $\alpha_U = j_U^V(\alpha_V)$. Use the isomorphism

$$H^{\#}_{n-1}(\langle V, \alpha_V \rangle - \langle U, \alpha_U \rangle) \simeq H^{\#}_{n-1}(V - U)$$

to locally orient E along $\langle U, \alpha_U \rangle$. This shows E is orientable.

Suppose E were not connected. Then for each component C, $p \mid C : C \to X$ is a covering space which must be a homeomorphism (since the fibres of E have only 2 points). Hence C is nonorientable, contradicting (22.7(b)). ■

Proof to Corollary. The corollary follows from the fact that $p_*\pi_1(E, e_0)$ has index 2 in $\pi_1(X, x_0)$. ■

(22.16) *Examples*: For $X = \mathbf{P}^2$, we have $E = S^2$. If X is the Klein bottle U_2, we have $E =$ the torus T_1. For let $q : I^2 \to U_2$ be the map given by the identifications (19.32)

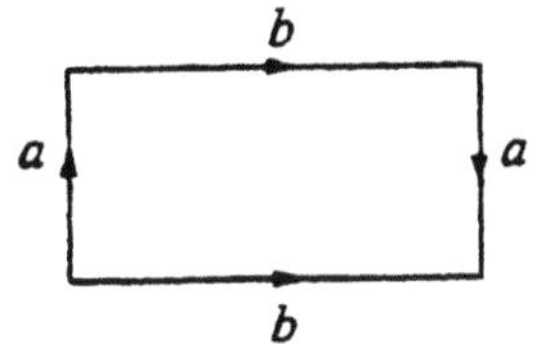

Piecing together two such with the second turned upside down gives the torus

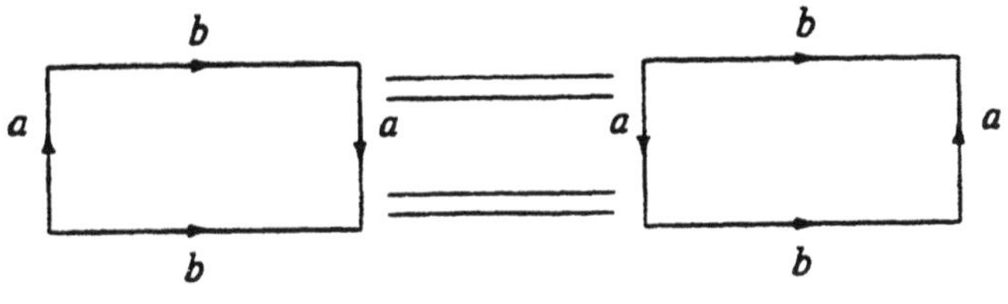

Precisely, define $p : T_1 \to U_2$ by

$$p(e^{2\pi i s}, e^{2\pi i t}) = \begin{cases} q(2s, t) & 0 \le s \le \frac{1}{2} \\ q(2s - 1, 1 - t) & \frac{1}{2} \le s \le 1 \end{cases}$$

Show more generally that for $X = U_h$, we have $E = T_{h-1}$, $h > 1$ (19.32).

(22.17) *Note*. One may ask, conversely, whether each orientable manifold

E is a 2-sheeted covering of a non-orientable X.

The answer is no, for by (21.4), $\chi(E)$ would then be even (when E and X are finite cell complexes), but $E = \mathbf{CP}^{2n}$ is an example where $\chi(E)$ is odd (20.3 and 22.31).

(22.18) It is useful to construct another covering space analogous to E, containing E as an open subspace when $R = \mathbf{Z}$: Let X^0 be the set of all (x, α_x), with $x \in X$, $\alpha_x \in H_n(X, X-x)$. We do not assume α_x is a generator. Set $p(x, \alpha_x) = x$. For any open $U \subset X$, define $\langle U, \alpha_U \rangle$ exactly as before. These sets form a basis for a topology on X^0 such that $p{:}X^0 \to X$ is a covering space. The fibre $p^{-1}(x)$ is in one-to-one correspondence with $H_n(X, X-x)$. X^0 is called *the R-orientation sheaf of X*.

Suppose now $R = \mathbf{Z}$. Define a function v on X^0 with values in the non-negative integers as follows: Given (x, α_x), α_x is an integer multiple of a generator of $H_n(X, X-x)$; the absolute value of that multiple is independent of the generator chosen, and this number we call $v(x, \alpha_x)$.

(22.19) *Lemma. For $q > 0$, $v^{-1}(q)$ is open in X^0 and $v^{-1}(q) \to X$ is a 2-fold covering space.*

This follows at once from (22.4).

Note that $v^{-1}(0)$ is also open, but $v^{-1}(0) \to X$ is a homeomorphism. X^0, being the disjoint union of the $v^{-1}(q)$'s, is never connected. $v^{-1}(1)$ is our previous E.

(22.20) *Exercise*. For general rings R, define an equivalence relation $a \sim b$ if $a = ub$, where u is an element of R having a multiplicative inverse. The set of equivalence classes can be called "R modulo units." Define a map $v{:}X^0 \to$ "R modulo units" and verify the lemma. If 1 denotes the class containing 1, $v^{-1}(1)$ is a covering space of X such that each fibre is in one-to-one correspondence with the multiplicative group of units in R. For $R = \mathbf{Z}/2$, $v^{-1}(1) \to X$ is a homeomorphism, and $X^0 = X \times \mathbf{Z}/2$.

(22.21) *Remark*. For any subspace $A \subset X$, a (continuous!) map $s{:}A \to X^0$ such that $ps =$ inclusion $A \to X$ is called a *section over A*. For $x \in A$, let $s'(x) \in H_n(X, X-x)$ denote the second coordinate of $s(x)$, i.e.,

$$s(x) = (x, s'(x))$$

Let ΓA denote the set of all sections over A. If $s_1, s_2 \in \Gamma A$, then

$$x \to (x, s_1'(x) + s_2'(x)) \quad x \in A$$

defines another section over A, denoted $s_1 + s_2$. If $s \in \Gamma A$, $\lambda \in R$, then

$$x \to (x, \lambda s'(x)) \quad x \in A$$

defines a section over A denoted λs. These operations make ΓA into an R-module. The zero element of this module is the section

$$x \to (x, 0) \quad x \in A$$

Sections over all of X are called *global sections*.

Note that there exists a global section s mapping X into $v^{-1}(1)$ if and only if X is R-orientable, and, in fact, the different R-orientations of X are in one-to-one correspondence with such sections (this is an exercise in the definitions of all these notions). More generally, for any $A \subset X$ we say X is *R-orientable along A* if there is a section over A mapping A into $v^{-1}(1)$.

(22.22) *Proposition. X is R-orientable along A if and only if there is a homeomorphism $\phi: p^{-1}(A) \to A \times R$ (R given the discrete topology) such that the diagram*

$$\begin{array}{ccc} p^{-1}(A) & \xrightarrow{\phi} & A \times R \\ & \searrow \quad \swarrow & \\ & A & \end{array}$$

is commutative. If this is the case, ΓA is isomorphic to the module of all continuous maps $A \to R$; if A has k connected components $k < \infty$, then $\Gamma A \cong R^k$.

Proof: Given a section $s: A \to v^{-1}(1)$. For each $x \in A$, $s'(x)$ is a generator of $H_n(X, X - x)$. If $(x, \alpha_x) \in p^{-1}(A)$, then there is a unique $\lambda_x \in R$ such that $\alpha_x = \lambda_x s'(x)$. Define ϕ by

$$\phi(x, \alpha_x) = (x, \lambda_x) \quad x \in A$$

If U is an open neighborhood of x on which α_x has a unique continuation α_U, then ϕ maps $\langle U, \alpha_U \rangle$ one-to-one onto $U \times \lambda_x$, hence is a homeomorphism. Conversely, given ϕ, we recover s by $s(x) = \phi^{-1}(x, 1)$, $x \in A$. ■

(22.23) *Remark.* There is a canonical homomorphism

$$j_A: H_n(X, X - A) \to \Gamma A$$

defined by $j_A(\alpha)(x) = (x, j_x^A(\alpha))$, $x \in A$ (see (22.6)). (We must verify the continuity of $j_A(\alpha)$: Let a be a relative cycle representing α. If $U =$

$X - |\partial a|$, then U is open and contains A, and if $\alpha_U \in H_n(X, X - U)$ is the homology class of a relative to $X - U$, then α_U induces α under the inclusion of $X - U$ into $X - A$. Now given $x \in A$, consider open neighborhoods V of x such that $V \subset U$ and such that $j_x^A(\alpha)$ has a unique continuation α_V over V (22.4). Since α_U induces α_V, $j_A(\alpha)$ maps $V \cap A$ into $\langle V, \alpha_V \rangle$, but sets of the latter type form a basis of neighborhoods of $(x, j_x^A(\alpha))$ in X^0.)

If $B \subset A$, we have the commutative diagram

$$\begin{array}{ccc} H_n(X, X - A) & \xrightarrow{j_A} & \Gamma A \\ \Big\downarrow j_B^A & & \Big\downarrow r \\ H_n(X, X - B) & \xrightarrow[j_B]{} & \Gamma B \end{array}$$

where the right vertical arrow r is defined by restricting sections over A to the subset B.

(22.24) *Theorem. Suppose $A \subset X$ is closed. Then*

(i) $H_q(X, X - A) = 0$ *for* $q > n$

(ii) j_A *is a monomorphism, and its image is the submodule* $\Gamma_c A$ *of sections with compact support, i.e.,*

$$j_A : H_n(X, X - A) \cong \Gamma_c A$$

In particular, $j_X : H_n(X) \cong \Gamma_c X$, *and* $H_q(X) = 0$ *for* $q > n$.

A section $s \in \Gamma A$ has *compact support if it agrees with the zero-section outside some compact subset of* A (the compact subset will depend on s in general). If A is compact then of course $\Gamma_c A = \Gamma A$.

(22.25) *Corollary. If A is connected and non-compact,* $H_n(X, X - A) = 0$. *In particular,* $H_n(X) = 0$ *if X is connected and non-compact.*

Proof: If $\alpha \in H_n(X, X - A)$, then by connectedness $vj_A(\alpha)$ is constant, and since $j_A(\alpha)$ is zero outside a compact set, $\alpha = 0$. ■

(22.26) *Corollary. If A is compact with k connected components, and X is R-orientable along A, then*

$$H_n(X, X - A) \cong R^k$$

Proof: Apply (22.22). ■

(22.27) *Corollary. If A is a compact subspace of $\mathbf{R}^n$ with k connected components, then k equals the $(n-1)^{st}$ Betti number of the complement of A in R^n (assume $n \geq 2$).*

Proof: $\mathbf{R}^n$, being orientable, is orientable along A, and $H_n(\mathbf{R}^n, \mathbf{R}^n - A) \simeq H_{n-1}(\mathbf{R}^n - A)$. ■

(22.28) *Corollary. Let X be a compact connected manifold. Assume that for any $a \neq 0, a \in R$ and any unit $u \in R$, $ua = a$ implies $u = 1$ (this holds, for example, when R is an integral domain). Then*

$$H_n(X) \cong \begin{cases} R & \textit{if } X \textit{ is } R\textit{-orientable} \\ 0 & \textit{otherwise} \end{cases}$$

Proof: If X is R-orientable, apply (22.22). Suppose there is a global section $s \in \Gamma X$, $s \neq 0$. Then $v(s(X))$ is a constant in R modulo units, i.e., there is $a \in R$, $a \neq 0$, such that $s'(x)$ is a times a generator of $H_n(X, X - x)$ for all $x \in X$, where $s(x) = (x, s'(x))$. The hypothesis on R implies that $s'(x)/a$ is a well-defined generator. The map $x \to (x, s'(x)/a)$ is a section $X \to v^{-1}(1)$, so that X is R-orientable. ■

(22.29) We see that an R-orientation of a compact connected manifold X is determined by a generator of ΓX, or a generator ζ of the top homology module $H_n(X)$; ζ is called *the fundamental class* of the R-orientation. The local R-orientation at each point x is then $j_x^X(\zeta)$.

(22.30) *Corollary. If X is a compact connected manifold then*

$$H_n(X; \mathbf{Z}/2) \simeq \mathbf{Z}/2$$

Use (22.12). ■

(22.31) *Remark*. Comparing with our previous calculations of homology groups, we have the following table:

Orientable		Non-Orientable	
S^n	all $n \geq 1$	U_h	all $h \geq 1$
T_g	all $g \geq 1$	$\mathbf{P}^n$	n even
$\mathbf{P}^n$	n odd		
$\mathbf{CP}^n$	all n		
$\mathbf{HP}^n$	all n		

Proof of Theorem (22.24):

Step 0. A empty. Obvious.

Step 1. If the theorem holds for the closed subsets A_1, A_2 and $A_1 \cap A_2$, then it holds for $A = A_1 \cup A_2$.

We use the relative Mayer-Vietoris sequence (17.10) for the triad $(X, X - A_1, X - A_2)$. It gives $H_q(X, X - A) = 0$ for $q > n$ and for $q = n$ we have the commutative diagram

$$\begin{array}{ccccccc}
0 \to H_n(X, X - A) & \to & H_n(X, X - A_1) \oplus H_n(X, X - A_2) & \to & H_n(X, X - A_1 \cap A_2) \\
\downarrow j_A & & \cong \downarrow j_{A_1} \oplus j_{A_2} & & \cong \downarrow j_{A_1 \cap A_2} \\
0 \to \Gamma_c A & \xrightarrow[(r_1, -r_2)]{} & \Gamma_c A_1 \oplus \Gamma_c A_2 & \xrightarrow[r_1 + r_2]{} & \Gamma_c(A_1 \cap A_2)
\end{array}$$

and chasing around the diagram shows j_A. an isomorphism.

Step 2. A is compact, connected, and contained in a coordinate neighborhood which is evenly covered by p. By excision, we may replace X by $\dot{E}^n$.

Case 1. A is a rectangular parallelepiped (of dimension $\leq n$).

Then (i) follows from $H_q(\dot{E}^n, \dot{E}^n - A) \approx H^*_{q-1}(\dot{E}^n - A) \approx H^*_{q-1}(S^{n-1})$, and (ii) follows from (22.5) and (22.22), since A is connected and E^n is R-orientable.

Case 2. A is a finite union of rectangular parallelepipeds $A_1, \ldots, A_m$ such that each wall of each A_i is parallel to a coordinate hyperplane in $\mathbf{R}^n$. By case 1 we may assume $m > 1$.

We argue by induction on m. Let $A' = A_1 \cup \ldots \cup A_{m-1}$. Then $A' \cap A_m$ is a set of the same type, union of at most $m - 1$ such parallelepipeds (of lower dimension, perhaps, but no matter), so the inductive hypothesis applies to A' and $A' \cap A_m$. By step 1, the theorem holds for A.

Case 3. A compact $\subset U$ evenly covered coordinate patch.

Given $s \in \Gamma A$. We may assume s maps A into one sheet over U (otherwise compactness of A and normality of U give m disjoint opens covering A such that it is true for each $U_i \cap A$). Then s extends to $s^* \in \Gamma U$ (the inverse to p on that sheet). For each point $x \in A$, choose a rectangular parallelepiped of dimension n containing x in its interior, having its walls parallel to the coordinate hyperplanes, and contained in U. Let A' be the union of all these parallelepipeds, a finite union since A is compact. By case 2 and the commutative diagram

$$\begin{array}{ccc} H_n(X, X - A') & \overset{j_{A'}}{\underset{\approx}{\to}} & \Gamma A' \ni s^* \mid A' \\ \downarrow & & \downarrow \\ H_n(X, X - A) & \overset{j_A}{\to} & \Gamma A \ni s \end{array}$$

we see that s is in the image of j_A. Thus j_A is onto.

Given $\alpha \in H_q(X, X - A)$, where $q \geq n$. If $q = n$, assume $j_A(\alpha) = 0$. We want $\alpha = 0$. Let z be a relative cycle representing α. Then $X - |\partial z|$ is an open $V \supset A$. Let α' be the homology class of z in $H_q(X, X - V)$. If $q = n$, since $j_x^V(\alpha') = j_x^A(\alpha) = 0$ for all $x \in A$, (22.4) tells us there is an open V'. $A \subset V' \subset V$, such that $j_x^V(\alpha') = 0$ for all $x \in V'$. Construct A' as above so that $A \subset A' \subset V' \cap U$. Then

$$j_{A'}^V(\alpha') = 0$$

by case 2, so $\alpha = j_A^{A'}(j_{A'}^V(\alpha')) = 0$

Step 3. A compact.

Then A is a finite union of compact sets $A_1, \ldots A_m$ each of which is contained in a coordinate neighborhood evenly covered by p. Use induction on m and steps 1 and 2.

Step 4. $A \subset U$, U open with compact closure $\bar{U}$. Then the theorem is true for U and A.

We use the exact homology sequence (14.6) for the triple

$$(X, U \cup (X - \bar{U}), (U - A) \cup (X - \bar{U})).$$

Note that by excision,

$$H_q(U, U - A) \approx H_q(U \cup (X - \bar{U}), (U - A) \cup (X - \bar{U}))$$

For $q > n$, we get

$$H_{q+1}(X, U \cup (X - \bar{U})) \to H_q(U, U - A) \to H_q(X, (U - A) \cup (X - \bar{U}))$$

The first and third modules are 0 by step 3 applied to the manifold X and the compact subsets $\bar{U} - U$ and $\bar{A} \cup (\bar{U} - U)$, hence the middle term is 0.

For $q = n$ we have the commutative diagram

$$\begin{array}{ccccc} 0 \to H_n(U, U - A) & \to & H_n(X, (U - A) \cup (X - \bar{U})) & \to & H_n(X, U \cup (X - \bar{U})) \\ \downarrow j_A & & \downarrow & & \downarrow \\ 0 \to \Gamma_c A & \xrightarrow{i} & \Gamma(\bar{A} \cup (\bar{U} - U)) & \xrightarrow{r} & \Gamma(\bar{U} - U) \end{array}$$

where the monomorphism i is defined as follows:

Let $s \in \Gamma_c A$ be zero outside a compact $K \subset A$. Then

$$i(s) \mid A = s$$

$$i(s) = 0 \qquad \text{outside } K$$

(we identify U^0 with $p^{-1}(U)$). From this diagram we see that j_A is an isomorphism.

Step 5. General case.

Given $s \in \Gamma_c A$ zero outside compact $K \subset A$. There is an open $U \supset K$ such that $\bar{U}$ is compact (cover K with finitely many coordinate neighborhoods). Consider $A' = A \cap U$, $s' = s \mid A'$. By step 4, applied to $j_{A'}$, and the commutative diagram

$$\begin{array}{ccc} H_n(U, U - A') & \to & H_n(X, X - A) \\ j_{A'} \downarrow \wr & & \downarrow j_A \\ 0 \to s' \in \Gamma_c A' & \xrightarrow{i} & s \in \Gamma_c A \end{array}$$

we see that $s \in \text{Image}(j_A)$.

Given $\alpha \in H_q(X, X - A)$. For $q = n$ assume that $j_A(\alpha) = 0$. We want $\alpha = 0$. Let z be a relative cycle representing α. Applying the above argument to $|z|$, there is an open $U \supset |z|$ such that $\bar{U}$ is compact. Let $A' = A \cap U$. By the same commutative diagram, we are done for $q = n$; for $q > n$ we know the class of z in $H_q(U, U - A')$ is zero by step 4, *a fortiori* $\alpha = 0$. ∎

(22.32) *Remark.* Orientability is a property which is not invariant under homotopy equivalence: For example, the open Möbius band is homotopically equivalent to a circle (28.17).

(22.33) *Exercise.* Let $f: X \to Y$ be an m-fold covering space, where X and Y are n-dimensional compact connected oriented manifolds. Let $R = \mathbf{Z}$. For any $x \in X$, choose a neighborhood U which is mapped homeomorphically onto a neighborhood V of $y = f(x)$ by f. Then f induces an isomorphism

$$H_n(X, X - x) \to H_n(U, U - x) \to H_n(V, V - y) \to H_n(Y, Y - y)$$

f is said to be *orientation-preserving* if for each x, this isomorphism takes the local orientation at x onto the local orientation at y.

Let ζ_X, ζ_Y be the fundamental homology classes for the orientations. Then $H_n(f)(\zeta_X)$ is a multiple of ζ_Y; this multiple is called the *degree* of f. Then f is orientation-preserving if and only if its degree is positive; in that case, degree $(f) = m$. Use the diagram

$$\begin{array}{ccc} H_n(X, X - f^{-1}(y)) \cong \bigoplus\limits_{x \in f^{-1}(y)} H_n(X, X - x) & \to & H_n(Y, Y - y) \\ \uparrow & & \uparrow \\ H_n(X) & \xrightarrow[H_n(f)]{} & H_n(Y) \end{array}$$

(22.34) *Note.* In this section, we have had a glimpse of some techniques of sheaf theory. For a more extended view, see Swan [56].

(22.35) *Note.* Another approach to orientability of a manifold X is through the action of its loop space on the "topological tangent bundle" of X. See E. Fadell [24].

(22.36) *Exercise.* Prove that every connected manifold X is homogeneous, i.e., given $x, x' \in X$, there is a homeomorphism ϕ of X onto itself such that $\phi(x) = x'$. (Hint: First do the case $X = E^n$, and show that ϕ can be chosen to be the identity on a thin shell $1 - \varepsilon \leq |y| < 1$. Then do the case where x and x' both lie in a coordinate neighborhood U (take ϕ to be the identity outside U). In the general case, join x to x' by a path, cover the path with finitely many coordinate neighborhoods, and use the previous case to go from x to x' in finitely many steps.)

(22.37) *Exercise.* Given n-dimensional connected manifolds X, Y, $n \geq 2$. Choose points $x \in X$, $y \in Y$. Let c (resp. d) be a closed n-cell centered at x (resp. y) and contained in a coordinate neighborhood U of x (resp. V of y);

let $\dot{c}\,\dot{d}$ *be the interiors of* c, d, and let s, t be their respective frontiers (so that s and t are homeomorphs of S^{n-1}). In the disjoint union $X - \dot{c} \amalg Y - \dot{d}$, identify s and t by means of a specific homeomorphism h. Show that the resulting space is a connected n-dimensional manifold $X + Y$. To show that the addition just defined depends only on X and Y, up to homeomorphism, is a very difficult problem, related to the Annulus Conjecture [36]. This has recently been proved in dimensions $\neq 4$ by Kirby and Siebenmann (see Bull. Amer. Math. Soc. (1969), 742–749 and Annals of Math. 89 (1970), 575–582).

(22.38) *Exercise.* Show that the addition just defined on the homeomorphism classes of connected n-dimensional manifolds is commutative, associative, and has as identity element the class of S^n. Thus these classes form a monoid, and the classes of compact manifolds form a submonoid. In the case $n = 2$ this submonoid is generated by the classes of T_1 and U_1 subject to the single relation $T_1 + U_1 = U_3$ [e.g., $T_g = T_1 + T_1 + \ldots + T_1$ (g times), $U_h = U_1 + \ldots + U_1$ (h times)].

For some partial generaliztions of this result to higher dimensions, see Smale [51a] and Wall [58a, 73].

(22.39) *Exercise.* Prove $\mathbf{Z}/n\mathbf{Z}$ orientability for some integer $n \neq 2$ (or 1) implies $\mathbf{Z}$-orientability.

(22.40) *Exercise.* Define an equivalence relation on orthonormal frames in $\mathbf{R}^n$ by declaring two frames equivalent if the matrix expressing one in terms of the other has determinant $+1$. Use (15.4) to set up an explicit correspondence between generators of $H_n(\mathbf{R}^n, \mathbf{R}^n - \{0\})$ and equivalence classes of frames.

(22.41) Let $M^{n-1} \subset \mathbf{R}^n$ be embedded such that each $x \in M$ has a closed neighborhood U in $\mathbf{R}^n$ such that $(U, U \cap M)$ and (E^n, E^{n-1}) are homeomorphic pairs. This is called a *locally flat* embedding. Prove that an orientation of M is equivalent to assigning a continuous normal direction $v: M \to S^{n-1}$. Show if v is a local homeomorphism and M is closed, then v induces an isomorphism $H(v): H_{n-1}(M; \mathbf{Z}) \to H_{n-1}(S^{n-1}; \mathbf{Z})$.

(22.42) *Exercise.* An embedding $M^{n-1} \subset N^n$ is *2-sided* if M separates a neighborhood U of M in N into 2 components (it is not required that $N - M$ be separated). Otherwise, for connected M, the embedding is *1-sided.* Construct 3-manifolds by identifying opposite faces of a cube to exhibit 1-sided embeddings of a torus or 2-sided embedding of a Klein bottle. A space formed by identifying opposite faces of a cube is a 3-manifold if and only if its Euler characteristic is 0 (proof in Seifert and Threllfall §60).

(22.43) *Exercise.* If $M^{n-1} \subset N^n$ is a locally flat submanifold of an orientable N, then M is orientable if and only if the embedding is 2-sided.

(22.44) *Exercise.* Show by example that a manifold which bounds need not be orientable. Suggestion: What is the boundary of the Möbius band $\times I$?

(22.45) *Exercise.* Let M be a closed n-manifold. If S^{n-1} is the boundary of a coordinate disc U, prove M is orientable if and only if the inclusion $S^{n-1} \to M - \text{int } U$ induces 0 in homology.

(22.46) *Exercise.* Let M^n be an orientable closed manifold covered by coordinate discs $\{U_i | 1 \leq i \leq k\}$ such that for each i, $\bar{U}_i - U_i$ is homeomorphic to S^{n-1}, and suppose given $\{\alpha_i | \alpha_i \in H_n(M, M - U_i)\}$ forming an R-orientation. Use relative Mayer-Vietoris sequences to construct a generator $\zeta \in H_n(M)$ mapping to α_i for all i.

(22.47) *Exercise.* Prove the connected sum (22.37) of closed orientable manifolds M, N is orientable. Given orientations of M and N and an orientation reversing homeomorphism used in the construction of $M + N$, prove that the connected sum has an orientation compatible with the orientation of M, N.

(22.48) *Exercise.* Let h be a homeomorphism of a torus and $A = H_1(H)$ a 2×2 matrix of integers. Prove h is orientation preserving if and only if $\det A = +1$.

(22.49) *Exercise.* Prove a manifold with a compatible topological group structure is orientable.

(22.50) *Exercise.* Define a closed orientable n-manifold to be *spherical* if there exists $f: S^n \to M$ such that $H(f)\zeta_S = k\zeta_M$ for some $k \neq 0$. Prove that if M is spherical, then $\pi_1(M)$ is finite, provided $n > 1$.

23. Singular Cohomology

The cohomology modules of a topological space were not recognized until quite late in the development of algebraic topology (1930), when Lefschetz formulated a simplified treatment of the duality theorems for manifolds. Cohomology is dual to homology in two senses: (1) There is a bilinear pairing of chains and cochains; (2) H^q is a *contrafunctor*, i.e., a map $X \to Y$ induces a homomorphism $H^q(Y) \to H^q(X)$ in the opposite direction.

Definition. The module $S^q(X)$ of all *singular cochains* on X is $\mathrm{Hom}_R(S_q(X), R) = S_q(X)^*$. Thus a singular cochain of dimension q is an R-linear homomorphism $c: S_q(X) \to R$. If we denote the value of this homomorphism on a chain z by $[z, c]$, we then have the identities

$$[z_1 + z_2, c] = [z_1, c] + [z_2, c],$$

$$[z, c_1 + c_2] = [z, c_1] + [z, c_2],$$

$$[\nu z, c] = \nu[z, c] = [z, \nu c], \qquad \nu \in R$$

so that [,] is a bilinear pairing.

Note: A q-cochain is uniquely determined by its values on the singular q-simplexes, and those values can be assigned arbitrarily. Thus $S^q(X)$ is isomorphic to the direct product of as many copies of R as there are singular q-simplexes in X.

(23.0) *Remark*. Let $S(X)$ be the singular complex for $R = \mathbf{Z}$. Then $R \otimes S(X)$ is canonically isomorphic $(1 \otimes \sigma \to \sigma)$ to the free R-module on singular simplices ($\otimes$ *is over* **Z**), and $\text{Hom}_R(R \otimes S(X), R) \cong \text{Hom}_{\mathbf{Z}}(S(X), R)$ by a canonical isomorphism.

(23.1) *Example.* Take the special case where R is the field **R** of real numbers and X is Euclidean 3-space. A 0-cochain, since it is uniquely determined by its effect on the singular 0-simplexes, is given by an artibrary function $\phi: X \to R$; call it $c^0(\phi)$.

For $q = 1$, suppose we are given a vector field v on X. We assign a 1-cochain $c^1(v)$ to v as follows: Let σ be a singular 1-simplex. If σ is differentiable, define

$$[\sigma, c^1(v)] = \int_\sigma \Omega^1(v)$$

where the right side is the line integral over the path σ of the differential 1-form $\Omega^1(v) = v_1 dx + v_2 dy + v_3 dz$ associated to $v = (v_1, v_2, v_3)$; otherwise define the left side to be zero.

For $q = 2$, associate to v the differential 2-form

$$\Omega^2(v) = v_1\, dy\, dz + v_2\, dz\, dx + v_3\, dx\, dy$$

Define a 2-cochain $c^2(v)$ on differentiable singular 2-simplexes σ by

$$[\sigma, c^2(v)] = \int_\sigma \Omega^2(v)$$

where the right side is the surface integral of the normal component of v.

For $q = 3$, start with a continuous function ϕ and associate the differential 3-form $\Omega^3(\phi) = \phi\, dx\, dy\, dz$. Define a 3-cochain $c^3(\phi)$ on differentiable singular 3-simplexes σ by

$$[\sigma, c^3(\phi)] = \int_\sigma \Omega^3(\phi)$$

where the right side is the mass integral of the density ϕ over the region.

(23.2) *Remark.* Since S_q is a functor from topological spaces to R-modules and $\text{Hom}_R(\ ,R)$ is a contrafunctor on the category of R-modules, the composed functor S^q is a contrafunctor from topological spaces to R-modules. More explicitly, if $f: X \to Y$ is any map, then $S^q(f): S^q(Y) \to S^q(X)$ is defined by the formula

$$[z, S^q(f)c] = [S_q(f)z, c]$$

for any q-chain z, q-cochain c. In case z is a singular q-simplex σ, the formula becomes

$$[\sigma, S^q(f)c] = [f \circ \sigma, c]$$

In other words, $S^q(f)$ is the *transpose* ${}^t S_q(f)$ of $S_q(f)$.

(23.3) *Proposition. There is a unique homomorphism* $\delta: S^q(X) \to S^{q+1}(X)$ *satisfying*

$$[\partial z, c] = [z, \delta c]$$

for all $(q + 1)$*-chains* z *and* q*-cochains* c. *If* $f: X \to Y$ *is any map, then*

$$\delta S^q(f) = S^{q+1}(f)\delta$$

Moreover,

$$\delta\delta = 0$$

Proof: Set $\delta = {}^t\partial$. ■

We call δ the *coboundary operator.*

(23.1)(continued). In our classical example, we can interpret the coboundary as follows: Assume ϕ is a differentiable function on 3-space, and v a differentiable vector field. If σ_q is a differentiable singular simplex of dimension q, we have by definition

(1) $[\sigma_1, \delta c^0(\phi)] = \phi(\sigma_1(E_1)) - \phi(\sigma_1(E_0))$

(2) $[\sigma_2, \delta c^1(v)] = \int_{\partial\sigma_2} \Omega^1(v)$

(3) $[\sigma_3, \delta c^2(v)] = \int_{\partial\sigma_3} \Omega^2(v)$

(4) $[\sigma_4, \delta c^3(\phi)] = \int_{\partial\sigma_4} \Omega^3(\phi)$

On the other hand, we can associate to the function ϕ its gradient vector field $\nabla\phi$ (so that $\Omega^1(\nabla\phi) = d\phi$); then

$$\int_{\sigma_1} \Omega^1(\nabla \phi) = \phi(\sigma_1(E_1)) - \phi(\sigma_1(E_0))$$

Hence

$$\delta c^0(\phi) = c^1(\nabla \phi)$$

So on these cochains, δ corresponds to taking the gradient.

We can associate to the vector field v its curl $\nabla \times v$ (so that $\Omega^2(\nabla \times v) = d\Omega^1(v)$). By Stokes' Theorem,

$$\int_{\sigma_2} \Omega^2(\nabla \times v) = \int_{\partial\sigma_2} \Omega^1(v)$$

Hence

$$\delta c^1(v) = c^2(\nabla \times v)$$

So on these 1-cochains, δ corresponds to taking the curl.

We can also assign to v its divergence $\nabla \cdot v$, which is a differentiable function satisfying $\Omega^3(\nabla \cdot v) = d\Omega^2(v)$. By Gauss' Theorem

$$\int_{\sigma_3} \Omega^3(\nabla \cdot v) = \int_{\partial\sigma_3} \Omega^2(v)$$

Hence

$$\delta c^2(v) = c^3(\nabla \cdot v)$$

and on these cochains, δ corresponds to taking the divergence.

In all cases, if we stick to the language of differential forms, δ corresponds to taking the exterior derivative d. Since $d\Omega^3 = 0$ for any 3-form on 3-space, we see from (4) that $\delta c^3(\phi) = 0$ (General Stokes' Theorem).

(For more details, see Spivak [53].)

We define the modules of cocycles and coboundaries by

$$Z^q(X) = \text{Kernel } \delta: S^q(X) \to S^{q+1}(X)$$

$$B^q(X) = \text{Image } \delta: S^{q-1}(X) \to S^q(X)$$

and the *cohomology module* $H^q(X; R)$ by

$$H^q(X) = Z^q(X)/B^q(X)$$

If $f:X \to Y$ is any map, then by (23.3), $S^q(f)$ respects cocycles and coboundaries, hence induces by passage to the quotient a homomorphism

$$H^q(f):H^q(Y) \to H^q(X)$$

This makes H^q into a contrafunctor $(H^q(gf) = H^q(f)H^q(g))$, so that the cohomology modules are topological invariants.

(23.1) (continued). For $X =$ Euclidean 3-space, we have described certain cohomology classes by differential forms. The cocycles correspond to *closed* forms (ones whose exterior derivative is zero), while the coboundaries are the *exact* forms (forms of type $d\omega$). But for q positive it is classical that a q-form (on $\mathbf{R}^3$!) is closed if and only if it is exact (e.g., curl $v = 0$ iff $v =$ grad ϕ for some ϕ, div $v = 0$ iff $v =$ curl u for some u). This corresponds to the fact that $H^q(\mathbf{R}^3) = 0$ for q positive.

(23.4) *Note.* One can define exterior differentiation of differential forms on any differentiable manifold X. Taking the closed q-forms modulo the exact ones apparently gives another kind of qth cohomology vector space for X. It is a fundamental theorem of de Rham that when X is paracompact, this vector space is canonically isomorphic to $H^q(X; \mathbf{R})$ (see [17, 68]).

(23.5) *Example*: Let X be an open set in the complex plane $\mathbf{C}$, $\phi:X \to \mathbf{C}$ a function which is holomorphic (analytic) in X. For every singular 1-simplex σ_1 in X, define the value of a 1-cochain $c^1(\phi)$ on σ_1 by

$$[\sigma_1, c^1(\phi)] = \int_{\sigma_1} \phi \, dz$$

This 1-cochain is actually a 1-cocycle, for by Cauchy's Theorem

$$[\sigma_2, \delta c^1(\phi)] = \int_{\partial\sigma_2} \phi \, dz = 0$$

for every singular 2-simplex σ_2 in X. Considerations such as these generalize to any complex analytic manifold, and the analogue of de Rham's Theorem, due to Dolbeault, can be proved (Dolbeault [18]).

Suppose we start with a pair (X, A). We define

$$\tilde{S}^q(X, A) = \operatorname{Hom}_R(S_q(X)/S_q(A), R)$$

and define the coboundary $\delta:\tilde{S}^q(X, A) \to \tilde{S}^{q+1}(X, A)$ to be the transpose of the boundary operator $\partial: S_{q-1}(X)/S_{q-1}(A) \to S_q(X)/S_q(A)$. Then we define

$$H^q(X, A) = \text{Kernel } \delta \text{ on } \tilde{S}^q(X, A)/\text{Image } \delta \text{ on } \tilde{S}^{q-1}(X, A)$$

We can interpret this more explicitly as follows: We have an exact sequence of chain complexes

$$0 \to S_q(A) \xrightarrow{i} S_q(X) \xrightarrow{p} S_q(X)/S_q(A) \to 0$$

Applying the functor $\text{Hom}_R(\ , R)$ gives another sequence

$$0 \to \tilde{S}^q(X, A) \xrightarrow{{}^tp} S^q(X) \xrightarrow{{}^ti} S^q(A) \to 0$$

(23.6) *Lemma. This sequence is exact.*

Proof. First show that ${}^ti$ is onto. Let $S_q(X, A)$ be the submodule of $S_q(X)$ generated by all the singular simplexes whose support is not contained in A. Then

$$S_q(X) = S_q(A) \oplus S_q(X, A).$$

Hence any linear functional on $S_q(A)$ can be extended to $S_q(X)$ by setting it equal to zero on $S_q(X, A)$. The remaining steps do not depend on special properties of $S(X)$ but follow formally. An $f: S_q(X) \to R$ such that $fi = 0$ extends to a well-defined $\bar{f}: S_q(X, A) \to R$ such that $\bar{f}p = f$. hence $\text{Ker}\,{}^ti \subset im\,{}^tp$. The other steps are left for the reader. ■

In particular, ${}^tp$ maps $\tilde{S}^q(X, A)$ isomorphically onto the annihiliator $S^q(X, A)$ of $S_q(A)$, a submodule of $S_q(X)$; if we identify $\tilde{S}^q(X, A)$ with $S^q(X, A)$ in this way, the coboundary becomes the restriction of the previous coboundary operator on $S^q(X)$. The relative cocycles $Z^q(X, A)$ are then all cochains which annihilate $S_q(A)$ and $B_q(X)$, i.e., $Z^q(X, A)$ is the annihilator of $B_q(X, A)$. Clearly the relative coboundaries $B^q(X, A)$ are contained in the annihilator of $Z_q(X, A)$, but the reverse inclusion is not generally valid. In any case, we do get a canonical homomorphism

(23.7) $$\alpha: H^q(X, A) \to H_q(X, A)^*.$$

Put another way, we have a bilinear pairing (called the *Kronecker product*)

(23.8) $$H_q(X, A) \times H^q(X, A) \to R$$

given by the formula

$$[\bar{z}, \bar{c}] = [z, c].$$

(23.9) *Proposition. If R is a principal ideal domain* (PID) *then α is an epimorphism.*

$R = \mathbf{Z}$ and $R =$ a field are the main examples of PID's we have in mind.

Proof. We have $H_q(X, A) = \bar{Z}_q(X, A)/\bar{B}_q(X, A)$, where $\bar{Z}_q = p(Z_q)$, $\bar{B}_q = p(B_q)$ are submodules of the free module $S_q(X)/S_q(A)$. Since R is a PID, $\bar{B}_q$ is a free module for all q (Lang [35], p. 387). Hence the exact sequence

(23.10) $$0 \to \bar{Z}_q(X, A) \xrightarrow{i} S_q(X)/S_q(A) \xrightarrow{\partial} \bar{B}_{q-1}(X, A) \to 0$$

splits (there is a homomorphism h such that $\partial h =$ identity, so that $S_q(X)/S_q(A) = h\bar{B}_{q-1} \oplus \bar{Z}_q$). By the same argument as in the lemma above, the dual sequence obtained by applying the functor $\mathrm{Hom}_R(\ ,R)$ is also exact. Thus we have a diagram

$$\begin{array}{ccccc} 0 \to & Z^q(X, A) & \to & S^q(X, A) \\ & \downarrow & & \downarrow t_i \\ 0 \to & H_q(X, A)^* & \to & \bar{Z}_q(X, A)^* \\ & \downarrow & & \downarrow \\ & 0 & & 0 \end{array}$$

with exact rows and columns, and from this the surjectivity of α follows. ■

(23.11) We see that the kernal of α is the quotient module $A^q(X,A)/B^q(X,A)$, where $A^q(X, A)$ is the annihilator of $Z_q(X, A)$. Let us denote this kernel by $E^q(X, A)$. Any map $f:(X, A) \to (Y, B)$ induces a commutative diagram

$$\begin{array}{ccccccc} 0 \to & E^q(X, A) & \to & H^q(X, A) & \to & H_q(X, A)^* \\ & \uparrow & & \uparrow H^q(f) & & \uparrow {}^t H_q(f) \\ 0 \to & E^q(Y, B) & \to & H^q(Y, B) & \to & H_q(Y, B)^* \end{array}$$

so that if $E^q(f)$ denotes the restriction of $H^q(f)$ to $E^q(Y, B)$, we see that E^q becomes a contrafunctor.

(23.12) We now define *the connecting homomorphism*

$$\delta: H^q(A) \to H^{q+1}(X, A).$$

Consider the exact commutative diagram

$$\begin{array}{ccccccccc} 0 & \to & S^q(X, A) & \to & S^q(X) & \xrightarrow{t_i} & S^q(A) & \to & 0 \\ & & \downarrow \delta & & \downarrow \delta & & \downarrow \delta & & \\ 0 & \to & S^{q+1}(X, A) & \to & S^{q+1}(X) & \to & S^{q+1}(A) & \to & 0. \end{array}$$

Let c be a cochain on X such that $'i(c)$ is a cocycle on A, representing a cohomology class $\bar{c}$. Since $'i(\delta c) = 0$, δc is a relative $(q+1)$-cocycle representing a cohomology class $\overline{\delta c} \in H^{q+1}(X, A)$. This class is easily seen to be independent of the choice of c, so we define

$$\delta\bar{c} = \overline{\delta c}.$$

(23.13) *Theorem. The singular cohomology modules have the following properties:*

(1) *Contrafunctoriality.*

(2) *Commutative diagrams*

$$\begin{array}{ccc} H^q(A) & \xrightarrow{\delta} & H^{q+1}(X, A) \\ H^q(f)\uparrow & & \uparrow H^{q+1}(f) \\ H^q(B) & \xrightarrow{\delta} & H^{q+1}(Y, B). \end{array}$$

(3) *Exact cohomology sequence*

$$0 \to H^0(X, A) \to \cdots \to H^q(X) \to H^q(A) \xrightarrow{\delta} H^{q+1}(X, A) \to \cdots.$$

(4) *Homotopy invariant*

$$f \simeq g \Rightarrow H^q(f) = H^q(g).$$

(5) *Excision*

$\bar{U} \subset \dot{A} \Rightarrow H^q(X, A) \to H^q(X - U, A - U)$ an isomorphism.

(6) *For a single point P*

$$H^q(P) \cong \begin{cases} R & q = 0, \\ 0 & q > 0. \end{cases}$$

Proof. (1)–(3) are exercises. (4) follows since homotopic maps induce chain homotopic maps of the singular complex (11.4), (13.13), hence chain homotopic maps of the co-chain complex (change in indices reversed) which induce the same map in cohomology. (5) follows by the same argument using (15.23) to obtain the chain homotopy equivalence of $S(X - U, A - U) \subset S(X, A)$. Alternatively, consider the diagram.

$$\begin{array}{ccccccccc} 0 & \to & E^q(X, A) & \to & H^q(X, A) & \to & H_q(X, A)^* & \to & 0 \\ & E^q(f) = E^q(g) & \uparrow & & \uparrow & & \uparrow & {}'H_q(f) = {}'H_q(g) & \\ 0 & \to & E^q(Y, B) & \to & H^q(Y, B) & \to & H_q(Y, B)^* & \to & 0. \end{array}$$

(5) follows from the excision theorem in homology theory (15.1), (23.15), the above commutative diagam with (X, A) in place of (Y, B) and $(X - U, A - U)$ in place of (X, A) and the five lemma (19.15). (6) follows from the homology of a point, the fact that $E^q(P) = 0$ for all q (23.12), and the canonical isomorphism $R^* \cong R$. ■

(23.13.1) *Note.* Properties (1)–(6) are the axioms of Eilenberg-Steenrod for a cohomology theory. They proved that for triangulable pairs of spaces, there is a unique cohomology theory up to isomorphism. This does not hold for arbitrary spaces (topologist's sine curve). One can use this axiomatic characterization to prove the de Rham Theorem (23.4) (see Schwartz [48]).

(23.14) *Remark.* From the basic properties, many others can be deduced. We list some below and leave the verification as an exercise.

(7) Define augmented cohomology modules for X nonempty by

$$H^{0*}(X) = \text{Cokernel } H^0(P) \to H^0(X)$$

where $X \to P$ is the constant map on a point P. For A nonempty, set

$$H^{0*}(X, A) = H^0(X, A).$$

Then the augmented cohomology sequence is exact.

(8) Exact cohomology sequence for a triple (X', X, A)

$$\cdots \to H^q(X', X) \to H^q(X', A) \to H^q(X, A) \to \cdots.$$

(9) Mayer-Vietoris exact sequence for an exact triad

$$\cdots \to H^q(X_1 \cup X_2) \to H^q(X_1) \oplus H^q(X_2) \to H^q(X_1 \cap X_2) \to \cdots.$$

(10) Relative Mayer-Vietoris Sequence.
(11) X contractible implies $H^{q*}(X) = 0$ for all q.

(23.15) *Exercise.* If (X_k) is the family of path components of X, there is a canonical isomorphism for all q

$$H^q(X) \to \prod_k H^q(X_k)$$

(direct product, not direct sum). If X is nonempty path-connected,

$$H^0(X) = R,$$

$$H^0(X, A) = 0 \quad (A \text{ nonempty}).$$

(Note: These statements do not follow from basic properties (1)–(6), since they are false for Cech-Alexander cohomology.)

(23.16) By the same arguments used in homology theory,

$$\delta: H^{q*}(S^{n-1}) \to H^{q+1}(E^n, S^{n-1}) \quad \text{isomorphism},$$

$$H^{q*}(S^n) \cong \begin{cases} R & q = n, \\ 0 & \text{otherwise.} \end{cases}$$

(23.17) *Remark.* The cohomology analogues of formulas (19.16–19.18) can be proved by the same arguments. Thus if Z is obtained from Y by attaching an n-cell via $f: S^{n-1} \to Y$, we have

(1) $H^q(Z) \to H^q(Y)$ an isomorphism for $q \neq n$, $q \neq n-1$,

(2) $H^{n-1}(Z) \cong \text{Kernel } H^{n-1}(f)$,

(3) an exact sequence

$$0 \to \text{Cokernel } H^{n-1}(f) \to H^n(Z) \to H^n(Y) \to 0.$$

We return to the map in (23.7)

$$\alpha: H^q(X, A) \to H_q(X, A)^*.$$

Example. The projective plane is obtained by attaching a 2-cell to S^1 by a map $f: S^1 \to S^1$ of degree 2. Using (23.17) we have $H^2(\mathbf{P}^2; \mathbf{Z}) \cong \mathbf{Z}/2\mathbf{Z}$. Since $H_2(\mathbf{P}^2; \mathbf{Z}) = 0$ we have $H_2(\mathbf{P}^2; \mathbf{Z})^* = 0$ and Ker $\alpha \neq 0$.

Our approach to this problem is based on the method of *derived functors*. The idea is to study the effect of $\mathrm{Hom}_R(\ ; N)$ on short exact sequences of R-modules

$$0 \to A' \to A \to A'' \to 0$$

where N is a fixed R-module. We follow the treatment in Cartan-Eilenberg [77] but limit the discussion to elementary considerations.

(23.18) *Definition.* A *resolution* of an R-module M is a chain complex $\{C_q, \partial_q\}$ and an epimorphism $\varepsilon: C_0 \to M$ such that im ∂_q = ker ∂_{q-1} and im ∂_1 = ker ε. Recall that our definition of chain complex required each C_q to be a free R-module.

Examples. If R is a field (M a vector space) we can take $C_0 = M$ and $C_q = 0$ for all $q \geq 1$. If R is the integers $\mathbf{Z}$(M is an abelian group) there is a free abelian group F and an epimorphism $\varepsilon: F \to M$ with ker ε a free abelian group. Then a resolution (or presentation) of M is given by $C_0 = F$, C_1 = ker ε, $C_q = 0$, $q \geq 2$. The situation is similar for R a PID, since in this case submodules of free R-modules are free [35].

There is nothing unique about resolutions but they behave pleasantly with respect to maps.

(23.19) *Proposition. Let C, C' be resolutions of M, M'. Given $f: M \to M'$, there is a chain map $\{f_q\}$, $f_q: C_q \to C'_q$ such that $\varepsilon' f_0 = f\varepsilon$*

$$\begin{array}{ccc} C_0 & \xrightarrow{f_0} & C'_0 \\ {\scriptstyle\varepsilon}\downarrow & & \downarrow{\scriptstyle\varepsilon'} \\ M & \xrightarrow[f]{} & M' \end{array}$$

and any two such chain maps are chain homotopic.

Proof. We construct f_q by induction on q. Since ε' is epic, im $f\varepsilon \subset$ im ε', hence f_0 exists because C_0 is free. Given a chain map up to $f_{q-1}:C_{q-1} \to C'_{q-1}$, we have $\partial'_{q-1} f_{q-1}\partial_q = f_{q-2}\partial_{q-1}\partial_q = 0$. Hence $\operatorname{im} f_{q-1}\partial_q \subset \ker \partial'_{q-1} = \operatorname{im} \partial'_q$. Then $f_q:C_q \to C'_q$ such that $\partial'_q f_q = f_{q-1}\partial_q$ exists since C_q is free. Given two chain maps $f, g:C \to C'$, we construct a chain homotopy by induction on q. For $q = 0$, $\varepsilon'(f_0 - g_0) = 0$. Since $\operatorname{im} \partial'_1 = \ker \varepsilon'$ and C_0 is free, $D_0:C_0 \to C'_1$ exists such that $\partial'_1 D_0 = f_0 - g_0$. For the inductive step we have $\partial'_q D_{q-1}\partial_q = (f_{q-1} - g_{q-1} - D_{q-2}\partial_{q-1})\partial_q = \partial'_q(f_q - g_q)$. Hence $\partial'_q(f_q - g_q - D_{q-1}\partial_q) = 0$. Since $\ker \partial'_q = \operatorname{im} \partial'_{q+1}$ and C_q is free, we have $D_q:C_q \to C'_{q+1}$ such that $\partial'_{q+1}D_q + D_{q-1}\partial_q = f_q - g_q$. ∎

Remark. The hypothesis on C can be weakened to $\partial_{q-1}\partial_q = 0$, i.e., C is a complex, and we never used freeness of C'_q. In this form (23.19) is one of the fundamental lemmas of homological algebra.

(23.20) *Proposition. Any two resolutions of M are chain homotopy equivalent.*

Proof. Let C, C' be resolutions of M and $f: C \to C'$, $g:C' \to C$ be chain maps covering $id:M \to M$. Then gf, $id:C \to C$ are chain homotopic and similarly for fg. ∎

We can now define derived functors of $\operatorname{Hom}_R(\ ,N)$. We write $\operatorname{Hom}_R(\ ,N) = H(\ )$ for the moment, to simplify notation. Let C be a resolution of M. Form the co-chain complex $C^* = \{H(C_q), \partial^*_{q+1}:H(C_q) \to H(C_{q+1})\}$. Then the derived functors are the homologies of C^*. More precisely

(23.21) *Definition. The q-th derived functor* $\operatorname{Ext}^q_R(\ ,N)$ *of* $\operatorname{Hom}_R(\ ,N)$ *is given by*

$$Ext^q_R(M, N) = \frac{\ker\{\partial^*_{q+1}:H(C_q) \to H(C_{q-1})\}}{\operatorname{im}\{\partial^*_q:H(C_{q-1}) \to H(C_q)\}}.$$

As things stand, we have a function depending on the choice of resolutions and we have not checked functoriality. We list the steps, leaving their verification for an exercise. Note $Ext^q_R(M, N)$ is an R-module.

Step 1. A chain map $f:C \to C'$ of resolutions of M, M' induces a well-defined homomorphism (note the variance)

$$Ext^q_R(M', N) \to Ext^q_R(M, N).$$

Step 2. Chain homotopic maps in step 1 induce the same map.

Step 3. Up to canonical isomorphism, $Ext_R^q(M, N)$ is independent of resolutions.

Step 4. $f: M \to M'$ induces well-defined homomorphisms

$$f^q: Ext_R^q(M', N) \to Ext_R^q(M, N)$$

such that $(gf)^q = f^q g^q$ and $(id)^q = id$. Thus $Ext^q{}_R(\ ,N)$ is a contra-functor. ∎

(23.22) *Remark*. Application of $H(\ )$ to $C_1 \xrightarrow{\partial_1} C_0 \xrightarrow{\varepsilon} M \to 0$ produces $0 \to H(M) \xrightarrow{\varepsilon^*} H(C_0) \xrightarrow{\partial_1^*} H(C_1)$ which is exact, i.e., ε^* is monic and ker $\partial_1^* = \text{im } \varepsilon^*$. The verification is routine.

Hence ker ∂_1^* can be identified with $\text{Hom}_R(M, N)$ and we use the *Ext* notation only for $q \geq 1$. For the case of abelian groups, we obtain $Ext^q = 0$ for $q \geq 2$ since there are resolutions with $C_q = 0$ for $q \geq 2$. The usual notation then is $Ext(M, N)$ deleting q and $\mathbf{Z}$. The *Ext* notation arises because $Ext(M, N)$ also classifies short exact sequences (extensions) of abelian groups

$$0 \to N \to E \to M \to 0.$$

See MacLane [38].

(23.23) *Exercise*. The biadditivity of $\text{Hom}_R(\ ,\ )$ implies $Ext_R^q(\ ,\ )$ is biadditive. For abelian groups G, $Ext(\mathbf{Z}, G) = 0$, $Ext(\mathbf{Z}/n\mathbf{Z}, G) \cong G/nG$. If G is finitely generated then $Ext(G, \mathbf{Q}) = 0$. $Ext(\mathbf{Q}, \mathbf{Z})$ is uncountable.

To study the effect of $\text{Hom}_R(\ , N)$ on short exact sequences, we mimic the construction of connecting homomorphisms in the topological case. An important feature is the presence of a short exact sequence of singular complexes for the pair (X, A)

$$0 \to S(A) \to S(X) \to S(X, A) \to 0.$$

In our algebraic context we use the construction of a "direct sum" of resolutions to achieve the same set-up.

(23.24) Construction. Given a short exact sequence of R-modules $0 \to A' \xrightarrow{i} A \xrightarrow{j} A'' \to 0$ and resolutions C', C'' of A', A'' set $C_q = C_q' \oplus C_q''$. The projection $\varepsilon: C_0 \to A$ has the form $\varepsilon(x, y) = i\varepsilon'(X) + \bar{j}(y)$ for some $\bar{j}: C_0'' \to A$ such that if $\bar{j} = \varepsilon''$,

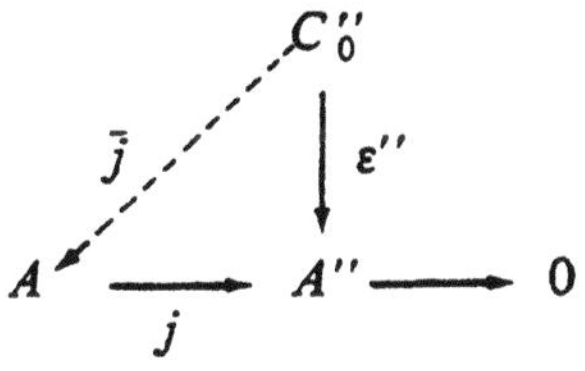

Note a (nonunique) $\bar{j}$ exists since C_0'' is free and j is epic. The boundary operator $\partial_q : C_q \to C_{q-1}$ had the form $\partial_q(x, y) = (\partial_q' x + e_q y,\ \partial_q'' y)$ for some $e_q : C_q'' \to C_{q-1}'$. In particular, since we require $\varepsilon \partial_1 = 0$, we must have $i\varepsilon' e_1 + \bar{j}\partial_1'' = 0$. This equation can be solved for a (nonunique) e_1 because $\operatorname{im} \bar{j}\partial_1 \subset \operatorname{im} i\varepsilon'$, by exactness at A, ε' is epic and C_1'' is free. Translating the requirement $\partial_{q-1}\partial_q = 0$ yields $0 = \partial_{q-1}' e_q + e_{q-1}\partial_q''$ which can be solved inductively for e_q. The verification that ε is epic and $\operatorname{im} \partial_q = \ker \partial_{q-1}$ is an exercise the reader should perform. We call $\{C_q, \partial_q\}$ a *normal extension* of C' and C'' and refer to $\{\bar{j}, e_q\}$ as the *data of the normal extension*.

(23.25) *Lemma. The inclusion $C' \to C$, $x \to (x, 0)$ and projection $(x, y) \to y$ are chain maps and*

$$\begin{array}{ccccccccc} 0 & \to & C_0' & \to & C_0 & \to & C_0'' & \to & 0 \\ & & \downarrow{\scriptstyle \varepsilon'} & & \downarrow{\scriptstyle \varepsilon} & & \downarrow{\scriptstyle \varepsilon''} & & \\ 0 & \to & A' & \xrightarrow{i} & A & \xrightarrow{j} & A'' & \to & 0 \end{array}$$

commutes. Thus $0 \to C' \to C \to C'' \to 0$ is a short exact sequence of resolutions.

Proof. Immediate from the definitions. ■

Note, as in the topological case, C is not necessarily split as a chain complex, but is split as R-modules.

(23.26) *Lemma. Application of* $\operatorname{Hom}_R(\ , N)$ *to a normal extension yields a short exact sequence*

$$0 \to (C'')^* \to C^* \to (C')^* \to 0.$$

Proof. The only nonformal part is the epimorphism on the right, which follows from the R-splitting $C \to C'$. ■

(23.27) *Proposition.* *Let* $0 \to A' \xrightarrow{i} A \xrightarrow{j} A'' \to 0$ *be a short exact sequence. There is a natural homomorphism* $\delta: Ext^q_R(A', N) \to Ext^{q+1}_R(A'', N)$ *and a natural long exact sequence*

$$0 \to \mathrm{Hom}_R(A'', N) \to \mathrm{Hom}_R(A, N) \to \ldots$$

$$\xrightarrow{\delta} Ext^q_R(A'', N) \to Ext^q_R(A, N) \to Ext^q_R(A', N) \xrightarrow{\delta} Ext^{q+1}_R(A'', N) \to .$$

Naturality refers to maps of short exact sequences.

Proof. By (23.20) normal extensions may be used to compute $Ext^q_R(A, N)$. The connecting homomorphism is constructed as in the topological case and the proof of exactness of the long exact sequence follows the topological arguments. The noncanonical construction of normal extensions introduces a complication for naturality not present in the topological case.

Sub lemma. *Let*

$$\begin{array}{ccccccccc} 0 & \to & A' & \xrightarrow{i} & A & \xrightarrow{j} & A'' & \to & 0 \\ & & \downarrow{\scriptstyle \alpha'} & & \downarrow{\scriptstyle \alpha} & & \downarrow{\scriptstyle \alpha''} & & \\ 0 & \to & B' & \to & B & \to & B'' & \to & 0 \end{array}$$

be a commutative diagram with exact rows. Let

$$0 \to C' \to C \to C'' \to 0$$

with data $\{\bar{j}\partial\ e_q\}$ *and*

$$0 \to D' \to D \to D'' \to 0$$

with data $\{\bar{k}, f_q\}$ *be normal extensions over the rows. Given chain maps* $F': C' \to D'$, $F'': C'' \to D''$ *over* α', α'', *there is a chain map* $F: C \to D$ *over* α *such that*

$$\begin{array}{ccccccccc} 0 & \to & C' & \to & C & \to & C'' & \to & 0 \\ & & \downarrow{\scriptstyle F'} & & \downarrow{\scriptstyle F} & & \downarrow{\scriptstyle F''} & & \\ 0 & \to & D' & \to & D & \to & D'' & \to & 0 \end{array}$$

commutes. Furthermore any two such triples (F', F, F''), (G', G, G'') *are chain homotopic via a chain homotopy* $F \simeq G$ *compatible with given chain homotopies* $F' \simeq G'$, $F'' \simeq G''$.

Proof of Sublemma. The required F has the form

$$F(x, y) = (F'x + \lambda_q y, F_q''y)$$

for some $\lambda_q : C_q'' \to D_q'$. The condition that F be a chain map translates to

$$i'\varepsilon'\lambda_0 + \bar{k}F_0'' = \alpha\bar{j}$$

and

$$\partial_q'\lambda_q + f_q F_q'' = F_{q-1}' e_q + \lambda_{q-1}\partial_q''$$

and these are solved inductively for $\{\lambda_q\}$. The part concerning chain homotopies is proved similarly. ■

Naturality of (23.27) follows by diagram chasing. ■

We apply the theory to the problem of ker α. Until further notice, we restrict ourselves to abelian groups ($R = \mathbf{Z}$).

(23.28) *Universal Coefficient Theorem. There is a natural short exact sequence*

$$0 \to Ext(H_{n-1}(X, A; \mathbf{Z}), G) \to H^n(X, A; G) \xrightarrow{\alpha} \mathrm{Hom}(H_n(X, A; \mathbf{Z}), G) \to 0$$

which splits (the splitting is not necessarily natural).

Proof. Write $S_n = S_n(X, A)$, $H_n = H_n(X, A; \mathbf{Z})$, B_n, Z_n for boundaries and cycles. We have natural short exact sequences

$$0 \to B_n \to Z_n \to H_n \to 0,$$

$$0 \to Z_n \to S_n \to B_{n-1} \to 0,$$

and the second splits (not necessarily naturally). Form the diagram

$$
\begin{array}{ccccccccc}
 & & & & 0 & & & & \\
 & & & & & & \mathrm{Hom}(S_{n+1}, G) & & \\
 & & & & \uparrow & \overset{\partial^*_{n+1}}{\nearrow} & \uparrow & & \\
0 & \to & \mathrm{Hom}(H_n, G) & \to & \mathrm{Hom}(Z_n, G) & \to & \mathrm{Hom}(B_n, G) & \to \mathit{Ext}(H_n, G) \to & 0 \\
 & & & & \uparrow & \nearrow & \uparrow & & \\
 & & & & & & 0 & & \\
 & & & & \mathrm{Hom}(S_n, G) & & & & \\
 & & \overset{\partial^*_n}{\nearrow} & & \uparrow & & & & \\
\mathrm{Hom}(S_{n-1}, G) & \to & & & \mathrm{Hom}(B_{n-1}, G) & \to & \mathit{Ext}\,(H_{n-1}, G) & & \\
 & & & & \uparrow & & & & \\
 & & & & 0 & & & &
\end{array}
$$

with exact rows and columns by (23.27). Note the dotted compositions are the boundary operators in the co-chain complex for (X, A) since $\mathrm{Hom}_G(G \otimes S_n, G) \cong \mathrm{Hom}_Z(S_n, G)$. Define

$$F_0 = \mathrm{Ker}\{\mathrm{Hom}(S_n, G) \to \mathrm{Hom}(Z_n, G)\},$$

$$F_1 = \mathrm{Ker}\{\mathrm{Hom}(S_n, G) \to \mathrm{Hom}(B_n, G)\}.$$

Then $F_0 \subset F_1 = \mathrm{Ker}\,\partial^*_{n+1}$. From exactness of the top row we obtain a well-defined epimorphism

$$F_1 \to \mathrm{Hom}(H_n, G)$$

with kernel exactly F_0. From exactness of the left hand column, we obtain $F_0 \cong \mathrm{Hom}(B_{n-1}, G)$. Hence we have an epimorphism

$$F_0 \to \mathrm{Ext}(H_{n-1}, G)$$

with kernel K, and $K \cong \mathrm{im}\,\partial^*_n$. Using $H^n = \ker \partial^*_{n+1}/\mathrm{im}\,\partial^*_n$ we obtain a commutative diagram

$$
\begin{array}{ccccccccc}
 & & K & \xrightarrow{\cong} & \mathrm{im}\,\partial^*_n & & & & \\
 & & \downarrow & & \downarrow & & & & \\
0 & \longrightarrow & F_0 & \longrightarrow & F_1 & \longrightarrow & F_1/F_0 & \longrightarrow & 0 \\
\downarrow & & \downarrow & & \downarrow & & \downarrow\cong & & \downarrow \\
0 & \to & \mathrm{Ext}\,(H_{n-1}, G) & \to & H^n & \to & \mathit{Hom}\,(H_n, G) & \to & 0
\end{array}
$$

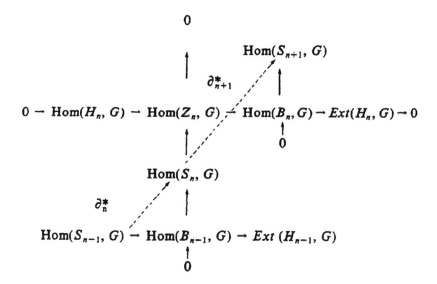

with the desired exact sequence along the bottom. Naturality is a consequence of (23.27) and the naturality of the constructions in the proof. In particular, the epimorphism from F_0 is natural and does not require splitting for its construction. Splitting occurs because a splitting $B_{n-1} \to S_n$ splits the left column giving a map $\mathrm{Hom}(H_n, G) \to F_1$. Composing with the projection $F_1 \to H^n$ splits the universal coefficient sequence. ■

Remark. If R is a PID, (23.28) extends to express $H^n(X, A; N)$ in terms of $H_*(X, A; R)$ and $\mathrm{Hom}_R(\ , N)$, $Ext_R(\ , N)$, where N is an R-module. For more general R an exact sequence is replaced by a more elaborate algebraic device–spectral sequence.

(23.29) *Remark.* We can compute the cohomology of familiar spaces using (23.28) provided we assume R is a PID. Thus for the r-leaved rose G_r,

$$H^q(G_r) \cong H_q(G_r)^* \cong \begin{cases} R & q = 0, \\ R^r & q = 1, \\ 0 & q > 1. \end{cases}$$

For the g-fold torus T_g,

$$H^q(T_g) \cong H_q(T_g)^* \cong \begin{cases} R & q = 0 \text{ and } q = 2, \\ R^{2g} & q = 1, \\ 0 & q > 2. \end{cases}$$

For complex projective space

$$H^q(\mathbf{CP}^n) \cong \begin{cases} R & q \text{ even and } \leq n, \\ 0 & \text{otherwise.} \end{cases}$$

Next consider the nonorientable compact surfaces U_h. Let p be the characteristic of the domain R. There are two cases:

Case 1. $p = 2$. Then $2R = 0, R_2 = R$, all the homology modules are free, and we get

$$H^q(U_h) \cong H_q(U_h)^* \cong \begin{cases} R & q = 0 \text{ and } q = 2, \\ R^h & q = 1, \\ 0 & q > 2. \end{cases}$$

Case 2. $p \neq 2$. Then using (23.28) gives

$$H^q(U_h) \cong \begin{cases} R & q = 0, \\ R^{h-1} & q = 1, \\ R/2 & q = 2, \\ 0 & q > 2. \end{cases}$$

Similarly, for real projective space, if R has characteristic 2,

$$H^q(\mathbf{P}^n) \cong \begin{cases} R & q \leq n, \\ 0 & q > n. \end{cases}$$

(23.30) *Exercise.* Map $\mathbf{P}^2 \to S^2$ by pinching the bottom cell to a point. Compute the induced map on integral and mod 2 cohomology. Use the results to show the splitting in (23.28) is not natural.

(23.31) *Exercise.* Adapt the proof of (23.28) to show $H_q(X; F)$ and $H^q(X; F)$ are dual vector spaces, where F is a field.

(23.32) *Exercise.* If X is a spherical complex with $H_q(X; F) = 0$ for all $q > 0$ where $F = \mathbf{Q}$ for $F = \mathbf{Z}/p\mathbf{Z}$, as p runs over all primes, then $H_q(X; \mathbf{Z}) = 0$. Suggestion: Recall (19.20).

(23.33) *Exercise.* If $f: X \to Y$, a map of spherical complexes, induces an isomorphism $H(f): H_q(X; F) \to H_q(Y; F)$ for all $q > 0$ and F as in (23.32), then f induces an isomorphism of integral homology and cohomology. Suggestion: Consider Cf, the mapping cone. An alternative argument uses naturality in (23.28).

The next series of exercises develops some properties of *Bockstein* homomorphisms. Corresponding to the short exact sequence $0 \to \mathbf{Z} \xrightarrow{n} \mathbf{Z} \to \mathbf{Z}/n\mathbf{Z} \to 0$ we have short exact sequences of chain complexes

$$0 \to S(X) \xrightarrow{n} S(X) \to \mathbf{Z}/n\mathbf{Z} \otimes S(X) \to 0.$$

(23.34) *Exercise.* Obtain natural long exact sequences

$$\to H_q(X; \mathbf{Z}) \xrightarrow{n} H_q(X; \mathbf{Z}) \to H_q(X; \mathbf{Z}/n\mathbf{Z}) \xrightarrow{\beta} H_{q-1}(X; \mathbf{Z}) \to \cdots$$

and

$$\to H^{q-1}(X; \mathbf{Z}) \xrightarrow{\beta} H^q(X; \mathbf{Z}/n\mathbf{Z}) \to H^q(X; \mathbf{Z}) \xrightarrow{n} H^q(X; \mathbf{Z}) \to \cdots$$

by the process of constructing connecting homomorphisms.

(23.35) *Exercise.* Generalize (23.34) to any short exact sequence of abelian groups $0 \to A \to B \to C \to 0$ and examine compatibility of the maps induced by commutative diagrams of short exact sequences.

(23.36) *Exercise.* Prove that the composition of Bocksteins associated with $0 \to \mathbf{Z}/n\mathbf{Z} \xrightarrow{n} \mathbf{Z}/n^2\mathbf{Z} \to \mathbf{Z}/n\mathbf{Z}/n\mathbf{Z} \to 0$ is 0.

(23.37) *Exercise.* Prove $\beta: H^{n-1}(\mathbf{P}^n; \mathbf{Z}/2\mathbf{Z}) \to H^n(\mathbf{P}^n: \mathbf{Z}/2\mathbf{Z})$ is an isomorphism if n is even and the 0 map if n is odd. Prove

$$\beta: H_2(L(p, q); \mathbf{Z}/p\mathbf{Z}) \to H_1(L(p, q): \mathbf{Z}p\mathbf{Z})$$

is an isomorphism where $L(p, q)$ is the lens space (21.28).

(23.38) *Exercise.* Prove the covering projection $S^{2n} \to \mathbf{P}^{2n}$ induces 0 in homology or cohomology with any finitely generated abelian group for coefficients. Recall (16.13) this map is not null-homotopic.

The next series of exercises develops a special case of an important theorem relating cohomology with homotopy classes of maps. We assume spaces X are finite connected CW complexes with X^0 = point and X^1 a bouquet of circles. The general theory of CW complexes shows that any finite connected CW complex is homotopy equivalent to one of this form, [66]. The set of homotopy classes of maps $f: X \to Y$ is denoted $[X, Y]$.

(23.39) *Theorem. There is a natural isomorphism* $\Phi: [X, S^1] \to H^1(X; \pi_1 S^1)$ *given by* $\Phi(f) = H^1(f)(\iota)$ *where* $\iota \in H^1(S^1; \pi_1 S^1)$ *corresponds to the identity in the following sequence of isomorphisms from* (23.28), (12.1).

$$H^1(S^1; \pi_1 S^1) \cong \mathrm{Hom}(H_1(S^1; \mathbf{Z}), \pi_1 S^1)$$
$$\cong \mathrm{Hom}(\pi_1 S^1, \pi_1 S^1).$$

Step 1. Φ factors

$$\begin{array}{ccc} [X, S^1] & \xrightarrow{\Phi} & H^1(X; \pi_1 S^1) \\ \downarrow \Psi & & \downarrow \cong \\ \mathrm{Hom}(\pi_1 X, \pi_1 S^1) & \xleftarrow{\cong} & \mathrm{Hom}(H_1(X; \mathbf{Z}), \pi_1 S^1) \end{array}$$

where $\Psi(f) = f_*$, the induced map of π_1.

Step 2. Give $[X, S^1]$ the structure of an abelian group so that Ψ is a homomorphism;

a) Let $m{:}S^1 \times S^1 \to S^1$ be complex multiplication and $c{:}S^1 \to S^1$ conjugation. Then $m_*{:}\pi_1 S^1 \times \pi_1 S^1 \to \pi_1 S^1$ is addition of homotopy classes, and c_* is multiplication by -1. The latter fact uses $m(1 \times c)\Delta = 0$ where 0 maps $e^{i\theta} \to e^{i0} = (1, 0)$, and Δ is the diagonal.

b) Given $f, g{:}X \to S^1$ define $f + g{:}X \to S^1$ by $f + g = m(f \times g)\Delta$. Show $+$ is well defined on homotopy classes, associative, commutative and $0{:}X \to S^1$ $x \to e^{i0}$ is the unit. Show $cf = -f$ in $[X, S^1]$. Show Ψ is a natural homomorphism with respect to this structure.

Step 3. To prove Ψ is monic, consider $\Psi(f) = 0$ and use the lifting theorem (6.1).

Step 4. To prove Ψ is epic, we use the hypothesis on X for the first time. Using (4.12) show $X^2 \subset X$ induces an isomorphism in π_1 by induction over the skeleta. The cell decomposition $X^2 = (S^1 \vee \ldots \vee S^1) \cup e^2 \cup \ldots \cup e^2$ gives a presentation $1 \to R \to F \to \pi_1 \to 1$, and we regard X^2 as a mapping cone. Using facts from (21.17) obtain $h{:}X^2 \to S^1$ such that $h_* = \phi$ for a prescribed ϕ. To extend h over X, use $\pi_i S^1 = 0$ for $i \geq 2$ and induction on the cells.

(23.40) *Exercise*. Assume R is a *PID* and that H_{q-1} is finitely generated. Let T_{q-1} be the torsion submodule of H_{q-1}. If H_q is also finitely generated and F_q is the quotient module of H_q by its torsion submodule, then

$$H^q \simeq F_q \oplus T_{q-1} \quad \text{(non-canonically)}$$

24. Cup and Cap Products

A key feature of cohomology which distinguishes it from homology is the existence of a natural multiplication called cup product making the direct sum of all the cohomology modules into a graded R-algebra. (Historically, however, multiplication was first defined on the direct sum of the homology groups of a manifold in terms of the intersection of cycles, but the definition is more difficult. See Lefschetz [37], Chap. 4.) This cohomology ring then operates on the direct sum of the homology modules by means of the cap product; this product can then be used to exhibit the duality theorems on manifolds.

Let

$$S^{\cdot}(X) = \bigoplus_{q \geq 0} S^q(X)$$

We first define a *cup product* in $S^{\cdot}(X)$. We require it to be bilinear and also require that $c \cup d \in S^{p+q}(X)$ if $c \in S^p(X)$, $d \in S^q(X)$. To define $c \cup d$, it suffices to specify $[\sigma, c \cup d]$ for any singular $(p+q)$-simplex σ. To do this, consider the affine maps

$$\lambda_p : \Delta_p \to \Delta_{p+q}$$

$$\rho_q : \Delta_q \to \Delta_{p+q}$$

given by $\lambda_p = (E_0 \ldots E_p)$, $\rho_q = (E_p E_{p+1} \ldots E_{p+q})$. Set

$$[\sigma, c \cup d] = [\sigma\lambda_p, c][\sigma\rho_q, d]$$

where the right side is the product of two scalars in R. (Thus we let c operate on the "front" p-face of σ, and let d operate on the "back" q-face, then multiply the resuls.) If $c = \Sigma_p c_p$, $d = \Sigma_q d_q$ are arbitrary elements of $S^{\cdot}(X)$, we have by definition

$$c \cup d = \sum_{p,q} c_p \cup d_q$$

(24.1) *Proposition. The cup product in $S^{\cdot}(X)$ is bilinear, associative, and has as identity element the 0-cochain 1 defined by $[x, 1] = 1$ for every point x in X.*

Proof: Easy. (Note: previously 1 was denoted $\partial\#$ (9.7).) ■

(24.2) Proposition. The coboundary operator is a derivation of the graded ring $S^{\cdot}(X)$, i.e.,

$$\delta(c \cup d) = \delta c \cup d + (-1)^p c \cup \delta d$$

for $c \in S^p(X)$, $d \in S^q(X)$.

Proof: For any $(p + q + 1)$-simplex σ, we have

$$[\sigma, \delta c \cup d] = [\partial(\sigma\lambda_{p+1}), c][\sigma\rho_q, d]$$

$$= \sum_{i=0}^{p+1} (-1)^i [(\sigma\lambda_{p+1})^{(i)}, c][\sigma\rho_q, d]$$

$$= \sum_{i=0}^{p} (-1)^i [\sigma^{(i)}\lambda_p, c][\sigma\rho_q, d]$$

$$+ (-1)^{p+1}[\sigma\lambda_p, c][\sigma\rho_q, d]$$

$$[\sigma, c \cup \delta d] = [\sigma\lambda_p, c][\partial(\sigma\rho_{q+1}), d]$$

$$= \sum_{i=p}^{p+q+1} (-1)^{i-p}[\sigma\lambda_p, c][(\sigma\rho_{q+1})^{(i-p)}, d]$$

$$= [\sigma\lambda_p, c][\sigma\rho_q, d]$$

$$+(-1)^p \sum_{i=p+1}^{p+q+1} (-1)^i [\sigma\lambda_p, c][\sigma^{(i)}\rho_q, d]$$

If we multiply this last by $(-1)^p$, and add to the first, the term $[\sigma\lambda_p, c][\sigma\rho_q, d]$ occurs with opposite signs and cancels, while the remaining terms add up to

$$\sum_{i=0}^{p+q+1} (-1)^i [\sigma^{(i)}\lambda_p, c][\sigma^{(i)}\rho_q, d] = [\sigma, \delta(c \cup d)] \quad \blacksquare$$

(24.3) *Corollary. The direct sum $Z^{\cdot}(X)$ of the cocycle modules is a subring of $S^{\cdot}(X)$ and the direct sum $B^{\cdot}(X)$ of the coboundary modules is a two-sided ideal in $Z^{\cdot}(X)$, hence by passage of the cup product to the quotient, the direct sum $H^{\cdot}(X)$ of the cohomology modules becomes a graded R-algebra.*

Proof: Immediate (note that 1 is a cocyle, since $\partial \# \partial = 0$). ■

Let $f: X \to Y$ be a map. Then f induces homomorphisms $S^q(f): S^q(Y) \to S^q(X)$ for all q, hence a homomorphism $S^{\cdot}(f)$: $S^{\cdot}(Y) \to S^{\cdot}(X)$ defined by

$$S^{\cdot}(f)\left(\sum_p c_p\right) = \sum_p S^p(f)(c_p)$$

Similarly we have a module homomorphism $H^{\cdot}(f): H^{\cdot}(Y) \to H^{\cdot}(X)$.

(24.4) *Proposition. $S^{\cdot}(f)$ and $H^{\cdot}(f)$ are ring homomorphisms.*

Proof: For $c \in S^p(Y)$, $d \in S^q(Y)$, and any $(p+q)$-simplex σ,

$$\begin{aligned}[\sigma, S^{p+q}(f)(c \cup d)] &= [f\sigma, c \cup d] \\ &= [(f\sigma)\lambda_p, c][(f\sigma)\rho_q, d] \\ &= [f(\sigma\lambda_p), c][f(\sigma\rho_q), d] \\ &= [\sigma, S^p(f)(c) \cup S^q(f)(d)] \end{aligned} \quad \blacksquare$$

(24.5) *Corollary. $S^{\cdot}$ and $H^{\cdot}$ are contrafunctors from the category of topological spaces to the category of graded R-algebras.*

(24.6) *Example*: Let X be a single point P. Then as R-module, $H^{\cdot}(P) = H^0(P) = R$, a generator being the cohomology class 1 of the cocycle 1. Since $1 \cup 1 = 1$, we see that $H^{\cdot}(P)$ is ring-isomorphic to R.

(24.7) *Note*. The ring $S^{\cdot}(X)$ does not in general have any good commutativity properties. For example, take X to be the unit interval Δ_1, and define a 0-cochain c and a 1-cochain d by

$$[x, c] = \begin{cases} 1 & x = E_0, \\ 0 & \text{otherwise}, \end{cases}$$

$$[\sigma, d] = \begin{cases} 1 & \sigma = \delta_1, \\ 0 & \text{otherwise}. \end{cases}$$

We have

$$[\delta_1, c \cup d] = 1,$$

$$[\delta_1, d \cup c] = 0.$$

However,

(24.8) *Theorem*. *$H^{\cdot}(X)$ is skew-commutative, i.e.,*

$$a \cup b = (-1)^{pq} b \cup a$$

for $a \in H^p(X)$, $b \in H^q(X)$. In particular, if $a = b$ and p is odd, $a \cup a = 0$, provided R has characteristic $\neq 2$.

Proof.[+] The proof of this theorem is surprisingly complicated, and will involve some new ideas, enumerated in the following six steps.

1. Let $\pi = \pi_p$ be any permutation of the integers $[0, 1, \ldots, p]$. Regard π as an affine map $\Delta_p \to \Delta_p$ by sending the vertex E_i into $E_{\pi(i)}$. Then for any singular p-simplex σ in a space X, σ composed with π will be a new singular p-simplex. Extending by linearity gives an endomorphism $z \to z\pi$ of $S_p(X)$ (*permutation operator*).

2. If $i_0, \ldots, i_q$ are $q + 1$ integers between 0 and p, let $(i_0 \ldots i_q)$ denote the affine map $\Delta_q \to \Delta_p$ sending E_j into E_{i_j} for all j. Then for any singular p-

[+] An alternative proof combines (29.17) and (29.29).

simplex σ in a space X, $\sigma(i_0 \ldots i_q)$ will be a singular q-simplex. (The special case $(0 \ldots j^{-1}\ j\ j\ j^{+1} \ldots p)$ is called the jth *degeneracy operator* of dimension p.) The submodule of $S_q(X)$ generated by all singular simplexes $\sigma(i_0 \ldots i_q)$ will be denoted $C(\sigma)_q$.

3. Clearly the boundary operator sends $C(\sigma)_q$ into $C(\sigma)_{q-1}$. Hence the sequence of these submodules forms an algebraic chain complex, denoted $C(\sigma)$, and we can consider its homology modules $H_qC(\sigma)$ obtained by taking the kernel of ∂ on $C(\sigma)_q$ modulo the image of ∂ on $C(\sigma)_{q+1}$.

(24.9) *Lemma. $C(\sigma)$ is an acyclic complex, i.e.,*

$$H_qC(\sigma) = 0 \qquad \textit{all } q > 0$$

Proof: Define an operator $O{:}C(\sigma)_q \to C(\sigma)_{q+1}$ on the basis elements by

$$O(\sigma(i_0 \ldots i_q)) = \sigma(0 i_0 \ldots i_q)$$

This is just σ composed with the join of E_0 to $(i_0 \ldots i_q)$(15.10). Hence for $q > 0$ and any $z \in C(\sigma)_q$, we have

$$\partial(Oz) = z - O(\partial z)$$

so that if z is a cycle, $z = \partial(Oz)$. ■

4. For any p, let θ_p be the permutation of $[0, 1, \ldots, p]$ which reverses the order: $\theta_p(i) = p - i$. Define an endomorphism θ of the graded module $S.(X)$ by

$$\theta(z) = (-1)^{\frac{1}{2}p(p+1)} z\theta_p$$

for any $z \in S_p(X)$. For brevity let $\varepsilon_p = (-)^{\frac{1}{2}p(p+1)}$ be the sign of θ_p. We have put in this sign because of the following lemma:

(24.10) *Lemma.* $\theta\partial = \partial\theta$.

Proof: If σ is a singular p-simplex, then

$$\partial\theta(\sigma) = \varepsilon_p \partial(\sigma\theta_p) = \varepsilon_p \sum_{i=0}^{p} (-1)^{p-i} \sigma(p \ldots \hat{i} \ldots 0)$$

while

$$\theta(\partial\sigma) = \varepsilon_{p-1} \sum_{i=0}^{p} (-1)^{i} \sigma(p \ldots \hat{i} \ldots 0)$$

Thus we must show

$$\varepsilon_p(-1)^{p-i} = \varepsilon_{p-1}(-1)^i$$

which is easy. ■

It follows from (24.10) that θ induces by passage to the quotient an endomorphism of $H_.(X)$. We wish to show that this endomorphism is the identity. In fact, we get the following:

(24.11) *Lemma. θ on $S.(X)$ is chain homotopic to the identity, i.e., there is an endormorphism J of $S.(X)$ raising degrees by 1 such that*

$$Id - \theta = \partial J + J\partial$$

Proof: One can, if one likes, write out explicitly the complicated expression for J. Instead we use step 5, *the technique of acylic carriers*. Note first that for any p-simplex σ, both $(Id)(\sigma)$ and $\theta(\sigma)$ are in $C(\sigma)_p$ (C "carries" both Id and θ). Moreover, $\theta = Id$ on $S_0(X)$. We can therefore apply the following algebraic lemma to $Id - \theta$. ■

5. Let $\phi: S_.(X) \to S_.(Y)$ be a chain homomorphism (meaning ϕ preserves dimension of chains and commutes with boundary operators; more generally, $S_.(X)$ and $S_.(Y)$ can be replaced by any algebraic chain complexes). Assume $\phi = 0$ on $S_0(X)$. Assume ϕ has an acyclic carrier C (meaning that to each singular simplex σ is associated an acyclic subcomplex $C(\sigma)$ of $S_.(Y)$ such that $\phi(\sigma) \in C(\sigma)$, and such that for any face $\sigma^{(i)}$ of σ, $C(\sigma) \supset C(\sigma^{(i)})$).

(24.12) *Lemma. ϕ is chain homotopic to zero.*

Proof: We construct $J: S_p(X) \to S_{p+1}(Y)$ by induction. Set $J = 0$ for $p = 0$. For p positive, assume J defined in dimensions less than p so as to satisfy $\phi = J\partial + \partial J$ and such that for a simplex τ of dimension less than p, $J(\tau) \in C(\tau)$. It then suffices to define $J(\sigma)$ and verify these conditions when σ is a simplex of dimension p. Now

$$J\partial\sigma \cup_i C(\sigma^{(i)}) \subset C(\sigma)$$

$$\phi\sigma \in C(\sigma)$$

so $\phi\sigma - J\partial\sigma \in C(\sigma)$. But this chain is a cycle:

$$\partial(\phi\sigma - J\partial\sigma) = \phi\partial\sigma - \partial J\partial\sigma = (\phi - \partial J)\partial\sigma$$

but since $\phi = \partial J + \partial J$ in dimension $p - 1$,

$$(\phi - \partial J)\partial\sigma = J\partial\partial\sigma = 0$$

Since $p > 0$ and $C(\sigma)$ is acyclic, there exists $z \in C(\sigma)_{p+1}$ such that

$$\partial z = \phi\sigma - J\partial\sigma$$

So we set $J\sigma = z$ and conclude the induction. ■

6. We return finally to the cohomology modules. The transpose θ' of θ is an endomorphism of $S^{\cdot}(X)$ given by

$$[z, \theta' c] = [\theta z, c]$$

θ' commutes with the coboundary operator, hence induces an endomorphism of $H^{\cdot}(X)$ which must be the identity (just transpose (24.11)). One verifies easily the formulas

$$\theta_{p+q}\lambda_p = \rho_p\theta_p$$

$$\theta_{p+q}\rho_q = \lambda_q\theta_q$$

For $c \in S^p(X)$, $d \in S^q(X)$, we compute $\theta'(c \cup d)$:

$$\begin{aligned}
[\sigma, \theta'(c \cup d)] &= [\theta\sigma, c \cup d] \\
&= \varepsilon_{p+q}[\sigma\theta_{p+q}, c \cup d] \\
&= \varepsilon_{p+q}[\sigma\theta_{p+q}\lambda_p, c][\sigma\theta_{p+q}\rho_q, d] \\
&= \varepsilon_{p+q}[\sigma\rho_p\theta_p, c][\sigma\lambda_q\theta_q, d] \\
&= \varepsilon_{p+q}\varepsilon_p\varepsilon_q[\sigma\rho_p, \theta' c][\sigma\lambda_q, \theta' d] \\
&= \varepsilon_{p+q}\varepsilon_p\varepsilon_q[\sigma, \theta' d \cup \theta' c]
\end{aligned}$$

using at the last step the fact that R is commutative. Now

$$\varepsilon_{p+q}\varepsilon_p\varepsilon_q = (-1)^{pq}$$

so that

$$\theta'(c \cup d) = (-1)^{pq}\theta' d \cup \theta' c$$

Since θ' induces the identity on cohomology, the cohomology classes a, b of c, d (respectively) satisfy

$$a \cup b = (-1)^{pq} b \cup a$$ ■

(24.13) *Note*. The technique of acyclic carrier (and its generalization *acyclic models*) is extremely useful. For example, it is the key to showing that different homology theories are isomorphic (Hilton and Wylie [30], pp. 321–329). See also Section 29, Appendix (29–23A).

(24.14) *Remark*. In the classical example (23.1) of Euclidean 3-space and cochains given by differentiable functions ϕ and differentiable vector fields v, we have the multiplication ϕv of a vector field by a function, or the vector product $v \times w$ of two vector fields; equivalently we have the wedge product of differential forms, since

$$\Omega^1(\phi v) = \phi\Omega^1(v)$$

$$\Omega^2(v \times w) = \Omega^1(v) \wedge \Omega^1(w)$$

This product does not correspond to the cup product on the cochain level, e.g., $c^2(v \times w) \neq c^1(v) \cup c^1(w)$ in general; but the two products do correspond on the level of cohomology (see Goldberg [26], Appendix B).

(24.15) *Example*: $H^{\cdot}(S^n)$ is the graded R-algebra generated by one element a of degree n subject to the single relation $a^2 = 0$.

(24.16) *Example*: To determine the cohomology algebra of the torus T, we need to exhibit explicitly a generator of its second homology module. Examining the argument in (19.29), we see that $H_2(T)$ was determined via the isomorphisms

$$H_2(T) \to H_2(T, G_2) \leftarrow H_2(E^2, S^1) \xrightarrow{\partial} H_1(S^1)$$

Thus it suffices to exhibit a relative 2-cycle on (E^2, S^1) whose boundary represents a generator of $H_1(S^1)$. From the diagram

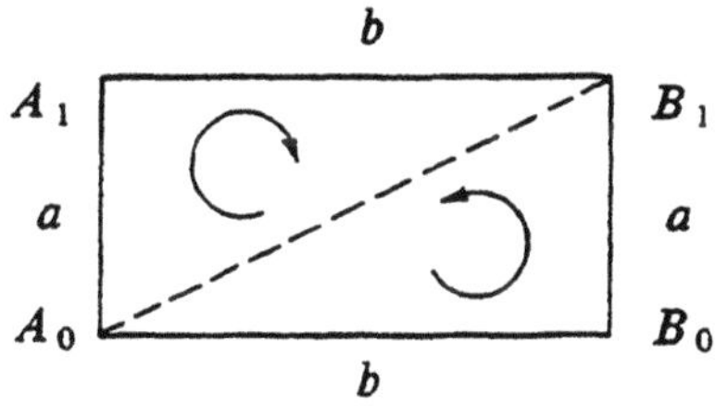

we see that

$$z = (A_0A_1B_1) - (A_0B_0B_1)$$

is such a cycle. If $\phi: I^2 \to T$ is the quotient mapping, then $S_2(\phi)(z)$ is a 2-cycle on T whose homology class ζ generates $H_2(T)$.

We know (23.29) that $H^1(T) \cong R \oplus R$ when R is a PID. Let α be the homology class of the loop $a = \phi(A_0A_1)$, β the class of $b = \phi(A_0B_0)$, so that α, β are generators of $H_1(T)$. Let α^*, β^* be the dual basis of $H^1(T)$, so that

$$[\alpha, \alpha^*] = 1 = [\beta, \beta^*],$$

$$[\alpha, \beta^*] = 0 = [\beta, \alpha^*].$$

Then

$$[\zeta, \alpha^* \cup \beta^*] = [\alpha, \alpha^*][\beta, \beta^*] - [\beta, \alpha^*][\alpha, \beta^*]$$

$$= 1 + 0 = 1$$

$$[\zeta, \alpha^* \cup \alpha^*] = [\zeta, \beta^* \cup \beta^*] = 0 + 0 = 0.$$

Now the isomorphism $H^2(T) \to R$ is given by sending a class $\gamma \in H^2(T)$ onto $[\zeta, \gamma] \in R$. It follows that $\alpha^* \cup \beta^*$ generates $H^2(T)$. Thus $H^{\cdot}(T)$ is the graded R-algebra generated by two elements α^*, β^* of degree 1 subject to the relations

$$(\alpha^*)^2 = 0 = (\beta^*)^2, \qquad \alpha^*\beta^* = -\beta^*\alpha^*.$$

(24.17) *Exercise*. By making explicit a generator of H_2 as above, determine the cohomology algebras of all the compact surfaces T_g, U_h with coefficients in a PID R (19.30–19.31).

(24.18) *Note*. One of the most remarkable applications of cup products in topology is *the Hopf invariant*. We have discussed the Hopf maps $S^3 \to S^2$, $S^7 \to S^4$, $S^{15} \to S^8$ obtained by regarding the image spheres as projective one-space with coefficients in **C**, **H**, and the Cayley numbers (respectively) (19.13). Consider more generally any n and any map

$$f: S^{2n-1} \to S^n.$$

Take $R = \mathbf{Z}$ now. Choose generators $\zeta \in H^{2n-1}(S^{2n-1})$, $\eta \in H^n(S^n)$ of the top dimensional cohomology groups (this amounts to choosing orientations). Let y be a cocycle representing η. Since $\eta \cup \eta = 0$, $y \cup y$ is a coboundary; say $y \cup y = \delta u$. Also, since

$$H^n(f)(\eta) \in H^n(S^{2n-1}) = 0$$

(n at least 2), there is an $(n-1)$-cochain x on S^{2n-1} such that $S^{\cdot}(f)(y) = \delta x$. Now $x \cup S^{\cdot}(f)(y)$ and $S^{\cdot}(f)(u)$ are both $(2n-1)$-cochains on S^{2n-1}, and their difference is actually a cocycle:

$$\delta(x \cup S^{\cdot}(f)(y) - S^{\cdot}(f)(u)) = \delta(x \cup \delta x) - S^{\cdot}(f)(y \cup y)$$

$$= \delta x \cup \delta x - S^{\cdot}(f)(y) \cup S^{\cdot}(f)(y) = 0$$

Hence the cohomology class of this cocycle is a multiple $\gamma\zeta$ of ζ, for a uniquely determined integer γ. It can be shown that γ is independent of all the choices we made (except that changing the orientation on S^{2n-1} changes the sign of γ), and this integer is the Hopf invariant of f. It turns out to depend only on the homotopy class of f, and the assignment $f \to \gamma(f)$ induces a homomorphism

$$\gamma : \pi_{2n-1}(S^n) \to \mathbf{Z}$$

such that

(i) n odd $\Rightarrow \gamma = 0$

(ii) n even $\Rightarrow 2 \in \text{Image}(\gamma)$

(iii) if $n = 2, 4, 8$ and f is the Hopf map, then $\gamma(f) = 1$ (see Hilton and Wylie [30], pp. 379–387, for the proofs). Work of Adem and Frank Adams settles the converse question: Up to homotopy, the Hopf maps are the only ones of Hopf invariant 1. This implies the purely algebraic theorem that **C**, **H**, and the Cayley numbers are the only non-trivial division algebras over **R**! (For a slick treatment of these results using "extraordinary" cohomology, see [6], p. 136–7, after (29.37) see [87].)

We next introduce the adjoint operation, *the cap product*. For each p, q this will be a bilinear pairing

$$\cap : S_{p+q}(X) \times S^p(X) \to S_q(X)$$

For $c \in S^p(X)$, $z \in S_{p+q}(X)$, $z \cap c$ will be the unique q-chain such that

(24.19) $$[z \cap c, d] = [z, c \cup d]$$

for all q-cochains d. Explicitly, for any singular $(p + q)$-simplex σ, set

$$\sigma \cap c = [\sigma\lambda_p, c]\sigma\rho_q$$

and extend to arbitrary $(p + q)$-chains by linearity. Then (24.19) follows from the bilinearity of the Kronecker product and the definition of the cup product. Using a further extension by linearity, we get a pairing

$$\cap : S_{\cdot}(X) \times S^{\cdot}(X) \to S_{\cdot}(X).$$

Remark. There are several variants of the definition of cap product, for example $\sigma \cap c = (-1)^{pq}[\sigma\lambda_p, c]\sigma\rho_q$ or $[\sigma\rho_p, c]\sigma\lambda_q$. In addition some authors write the evaluation pairing with homology on the right. The end result is considerable variation in signs appearing in formulas relating different kinds of products. A systematic approach to all this is offered by J. F. Adams [62].

(24.20) *Proposition. This pairing makes $S_{\cdot}(X)$ a right unitary $S^{\cdot}(X)$-module.*

Proof. Exercise. In particular $z \cap (c \cup d) = (z \cap c) \cap d$. ∎

(24.21) *Proposition. For $z \in S_{p+q}(X)$, $c \in S^p(X)$, we have*

$$\partial(z \cap c) = (-1)^p[(\partial z) \cap c - z \cap \delta c].$$

Proof. It suffices to show that the Kronecker product of each side with an arbitrary cochain gives the same value; this follows from (24.1) and (24.2). ∎

(24.22) *Corollary. By passage to the quotient, the cap product induces a bilinear pairing*

$$\cap : H_{p+q}(X) \times H^p(X) \to H_q(X).$$

(24.23) *Exercise.* Suppose X path-connected, so that $\partial\#: H_0(X) \to R$ is an isomorphism. Then for any $a \in H_p(X)$, $b \in H^p(X)$,

$$\partial\#(a \cap b) = [a, b].$$

The cap product is functorial in the first variable and contrafunctorial in the second.

(24.24) *Propositon. For any map* $f: X \to Y$,

$$H_q(f)[a \cap H^p(f)(b)] = H_{p+q}(f)(a) \cap b$$

$a \in H_{p+q}(X)$, $b \in H^p(Y)$.

Proof. One verifies the same formula on the chain-cochain level by taking the Kronecker product of each side with an arbitrary q-cochain on Y and using (23.3), (24.4) and (24.19). ∎

(24.25) *Products in the relative case.* All the previous results on cup and cap products generalize to the relative case. Thus $H^{\cdot}(X, A)$ becomes a skew-commutative graded R-algebra and $H_{\cdot}(X, A)$ a right unitary $H^{\cdot}(X, A)$-module. These structures are functorial.

We will need a generalization of (24.25), namely cap products

(24.26) $$H_{p+q}(X, A) \times H^p(X, A) \to H_q(X)$$

(24.27) $$H_{p+q}(X, A) \times H^p(X) \to H_q(X, A)$$

induced by the previously defined cap product. Note first that if σ is a singular $(p + q)$-simplex on A, then $\sigma\lambda_p$ is in $S_p(A)$. Hence $c \in S^p(X, A)$, $w \in S_{p+q-1}(A)$ imply $w \cap c = 0$. Thus if $z \in Z_{p+q}(X, A)$, $c \in Z^p(X, A)$ formula (24.21) shows that $z \cap c$ is actually a q-cycle on X (not just a relative q-cycle), and we get the pairing (24.26). Similarly for (24.27).

(24.28) *Exercise.* Fix a class $a \in H_{p+q}(X, A)$. Then the diagram

$$\begin{array}{ccc} H^p(X, A) & \to & H^p(X) \\ \downarrow a\cap & & \downarrow a\cap \\ H_q(X) & \to & H_q(X, A) \end{array}$$

is commutative (where $a\cap$ means cap product with a).

(24.29) *Note.* There are other cohomology operations such as the Steenrod squares that have important applications to homotopy theory (see Mosher and Tangora [69]).

(24.30) *Exercise.* Calculate the cup product structure for orientable surfaces by pinching homologically trivial embedded circles to points (e.g. δ in (12.11)) to obtain a bouquet of tori. Use naturality to induct from known results on tori.

(24.31) *Exercise.* Let $f: S^{2n-1} \to S^n$ be as in (24.18). Form the mapping cone Cf, with $i: S^n \to Cf$ the inclusion. Show $H^q(Cf; \mathbf{Z}) \cong \mathbf{Z}$ for $q = 0, n, 2n$ and 0 otherwise.

(24.32) *Exercise.* Write the generators of the nonzero groups in (24.31) as α, β with $H^n(i)(\alpha) = \eta$ and $\delta(\zeta) = \beta$ where η and ζ are defined in (24.18) and δ is from (21.19). Prove $\alpha \cup \alpha = \gamma(f)\beta$. Hence by (21.21), $\gamma(f)$ is a homotopy invariant. Suggestion: Find suitable representatives in the diagram

$$\begin{array}{ccccccccc}
0 & \leftarrow & S^\cdot(S^{2n-1}) & \leftarrow & S^\cdot(E^{2n}) & \leftarrow & S^\cdot(E^{2n}, S^{2n-1}) & \leftarrow & 0 \\
 & & \uparrow S^\cdot(f) & & \uparrow & & \uparrow & & \\
0 & \leftarrow & S^\cdot(S^n) & \xleftarrow[S^\cdot(i)]{} & S^\cdot(Cf) & \leftarrow & S^\cdot(Cf, S^n) & \leftarrow & 0
\end{array}$$

to calculate δ.

25. Algebraic Limits

For the Alexander duality and Poincaré duality theorems on manifolds, the correct cohomology modules are not, in general, the singular cohomology modules but others obtained from them by passing to the limit. This limiting process is a purely algebraic one which is extremely useful in many other contexts.

Limits will be taken over a *directed (filtering) set*: This is a set I, together with a partial order relation $i \leq i'$ defined for certain pairs of elements of I, such that for any $i, i' \in I$, there is an $i'' \in I$ such that $i \leq i''$ and $i' \leq i''$.

(25.1) *Example*: Let K be a subset of a set X, and let I be the set of those subsets of X which contain K. Define $V \leq V'$ to mean $V \supset V'$. Then for any V, V', the set $V \cap V'$ fulfills the directedness condition. (This is the example of most interest in topology.)

(25.2) *Example*: Let I be the set of non-zero integers. Let $i \leq i'$ mean i divides i'. Then for any i, i', the least common multiple fulfills the condition. (This example and its generalizations are of interest in number theory.)

(25.3) *Example*: Let I be the set of all covering spaces with base points $(E, e_0; p)$ of a pointed space (X, x_0), Define $(E, e_0; p) \leq (E', e_0'; p')$ when there is a map $f:(E', e_0') \to (E, e_0)$ such that $pf = p'$. To satisfy the directedness condition, take the fibre product, given by

$$E'' = \{(e, e') \in E \times E' \mid pe = p'e'\}$$

$$e_0'' = (e_0, e_0')$$

$$p''(e, e') = pe = p'e'$$

where E'' has the topology induced as a subspace of $E \times E'$. (The analogue of this example in algebraic geometry has been used to construct the ètale cohomology [5].)

(25.4) *Definition.* Suppose $(M_i)_{i\in I}$ is a family of R-modules indexed by the directed set I, and that for $i \leq i'$ we are given a homomorphism

$$\phi_{i',i}: M_i \to M_{i'}$$

such that

$$\phi_{i'',i'}\phi_{i',i} = \phi_{i'',i} \qquad \text{if} \quad i \leq i' \leq i''$$

$$\phi_{i,i} = \text{identity}$$

Call this set-up a direct (inductive) system of modules. A *direct (inductive) limit* of this system is a module M together with a family of homomorphisms

$$\phi_i: M_i \to M$$

indexed by I such that

$$\phi_{i'}\phi_{i',i} = \phi_i \qquad \text{if} \quad i \leq i'$$

and such that this collection is universal with respect to the following property. For any module N and any family of homomorphisms

$$\psi_i: M_i \to N$$

satisfying

$$\psi_{i'}\phi_{i',i} = \psi_i \qquad \text{if} \quad i \leq i'$$

there is a unique homomorphism $\psi: M \to N$ such that

$$\dagger \qquad \psi_i = \psi\phi_i$$

for all i.

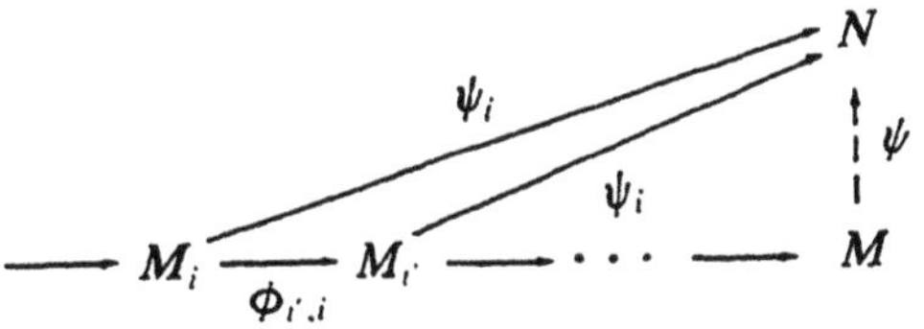

It is clear from this universal property that any two inductive limits M, N are isomorphic by a unique isomorphism satisfying condition †. So we can safely talk about "the" inductive limit and denote it

$$\varinjlim M_i$$

The unique homomorphism ψ will also be denoted $\varinjlim \psi_i$.

(25.5) *Proposition. The inductive limit exists.*

Proof: Let M^+ be the direct sum of all the M_i, with $\phi_i^+ : M_i \to M^+$ the monomorphism sending any $x_i \in M_i$ onto the vector whose i^{th} component is x_i and whose other components are 0. Form the submodule generated by all elements

$$\phi_{i'}^+ \phi_{i',i}(x_i) - \phi_i^+(x_i)$$

for all pairs $i \leq i'$ and all $x_i \in M_i$, and let M be the quotient module of M^+ by this submodule, $\pi : M^+ \to M$ the quotient homomorphism. Then M together with the homomorphisms $\phi_i = \pi\phi_i^+$ is an inductive limit. ■

(25.6) *Example*: Suppose the M_i are all submodules of some module, and $i \leq i'$ means M_i is a submodule of $M_{i'}$, with $\phi_{i',i}$ the inclusion. Then as inductive limit we can take the union

$$\cup_i M_i$$

(which is also a submodule, by the directedness condition), with ϕ_i the inclusion. We have a more general result.

(25.7) *Addendum.* $M = \cup_i \phi_i(M_i)$.

Proof: Let M' be the module on the right. Let $\psi_i : M_i \to M'$ be ϕ_i considered as a homomorphism into M'. Then the system $(M', (\psi_i))$ is clearly another inductive limit. Let $\psi : M \to M'$, $\phi : M' \to M$ be the unique isomorphisms satisfying †. Then ϕ is the inclusion $M' \to M$ and $\psi = \phi^{-1}$. Thus $M' = M$. ■

(25.8) *Exercise.* Suppose the directed set I has a largest element m, so that $i \leq m$ for all $i \in I$. Then the homomorphism

$$\phi_m : M_m \to \varinjlim M_i$$

is an isomorphism.

(25.9) *Exercise.* Turn the arrows around in the definitions of inductive systems and their limits. One gets the notion of *projective (inverse) system and limit*. Construct the projective limit as a submodule of the direct product of the M_i.

We will need later some basic lemmas about inductive limits.

(25.10) *Lemma* (*Additivity*). *Suppose that for each* i *we have a direct sum decomposition* $M_i = N_i \oplus P_i$, *and that for* $i \leq i'$, *the homomorphism* $\phi_{i',i}$ *decomposes accordingly*: $\phi_{i',i} = \psi_{i',i} + \rho_{i',i}$. *Let* $N = \varinjlim N_i$, $P = \varinjlim P_i$, *so that we get induced homomorphisms* $\psi:N \to M$, $\rho:P \to M$ *such that*

$$\psi\psi_i = \phi_i \mid N_i,\ \rho\rho_i = \phi_i \mid P_i$$

Then $\psi \oplus \rho:N \oplus P \to M$ *is an isomorphism.*

Proof: We construct the inverse. Given $x \in M$, choose $x_i \in M_i$ such that $x = \phi_i(x_i)$(25.7). Write $x_i = y_i + z_i$ uniquely, with $y_i \in N_i$, $z_i \in P_i$. Define $\theta(x) = (\psi_i y_i, \rho_i z_i) \in N \oplus P$. One verifies easily that $\theta(x)$ is independent of the choice of x_i and that θ is a homomorphism inverse to $\psi \oplus \rho$. ∎

Secondly, there is an important case in which one need not look at all the M_i's in order to obtain M. A subset $J \subset I$ is called *final* (sometimes "cofinal") if J is a directed set under the induced ordering from I and if for any $i \in I$, there is a $j \in J$ such that $i \leq j$. We can then form the inductive limit over the set J, and we get a canonical homomorphism

$$\lambda:\varinjlim M_j \to \varinjlim M_i$$

by applying the universal property to the family of homomorphisms $\phi_j:M_j \to M$.

(25.11) *Lemma.* λ *is an isomorphism.*

Proof: Let $M' = \varinjlim M_j$, $\phi_j':M_j \to M'$ the canonical homomorphism (so that $\lambda\phi_j' = \phi_j$ for all j). Given $x \in M$ write $x = \phi_i(x_i)$ for some $x_i \in M_i$. Since J is final, there is $j \in J$ such that $i \leq j$. Then also $x = \phi_j(x_j)$, where $x_j = \phi_{j,i}(X_i)$; thus $x = \lambda\phi_j' x_j$, so λ is onto.

Given $x' \in M'$ such that $\lambda x' = 0$. Write $x' = \phi_j'(x_j)$ for some $x_j \in M_j$, so that $\phi_j(x_j) = 0$. We need the following fact.

(25.12) *Sublemma.* If $\phi_i x_i = 0$, there is an i' with $i \leq i'$ such that $\phi_{i',i} x_i = 0$.

Granting this for the moment, there is an $i' \in I$ with $j \leq i'$ such that $\phi_{i',j} x_i =$

0. Since J is final, there is $j' \in J$ with $i' \leq j'$; then $\phi_{j',j}x_j = \phi_{j',i'}\phi_{i',j}x_j = 0$, whence $x' = \phi_{j'}'\phi_{j',j}x_j = 0$.

Proof of sublemma: Go back to the construction of M (proof of (25.5)). Since $\phi_i x_i = 0$, $\phi_i^+ x_i$ is a sum of elements

$$\phi_{k'}^+\phi_{k',k}y_{k',k} - \phi_k^- y_{k',k}$$

where $y_{k',k} \in M_k$. By definition of $\phi_i^{\mp}$ we must have

(i) $$x_i = \sum_{k'=i}\phi_{i,k}y_{i,k} - \sum_{k=i}y_{k',i}$$

while for $h \neq i$,

(h) $$0 = \sum_{k'=h}\phi_{h,k}y_{h,k} - \sum_{k=h}y_{k',h}$$

Choose an index i' greater than all k' which occur. Apply $\phi_{i',i}$ to equation (i), and $\phi_{i',h}$ to each equation (h) which is non-empty, and add: one gets

$$\phi_{i',i}x_i = \sum_{\text{all }(k',k)} \phi_{i',k'}\phi_{k',k}y_{k',k} - \phi_{i',k}y_{k',k}$$

which is zero by definition of inductive system. ■

(25.13) *Exercise.* If $\phi_{i',i}:M_i \to M_{i'}$ is an isomorphism for all $i \leq i'$, then $\phi_i:M_i \to M$ is an isomorphism for all i.

Next we consider the compatibility of inductive limits with exact sequences. Suppose we have three inductive systems over the same directed set I. Suppose further we have homomorphisms

$$M_i^* \underset{\lambda_i}{\to} M_i \underset{\rho_i}{\to} M_i^{**}$$

such that the sequence is exact and such that for $i \leq i'$ the diagram

$$\begin{array}{ccccc} M_i^* & \overset{\lambda_i}{\to} & M_i & \overset{\rho_i}{\to} & M_i^{**} \\ \phi_{i',i}^* \downarrow & & \phi_{i',i}\downarrow & & \downarrow \phi_{i',i}^{**} \\ M_{i'}^* & \overset{\lambda_{i'}}{\to} & M_{i'} & \overset{\rho_{i'}}{\to} & M_{i'}^{**} \end{array}$$

is commutative. Passing to the limit gives homomorphisms

$$M^* \xrightarrow{\lambda} M \xrightarrow{\rho} M^{**}$$

such that $\lambda\phi_i^* = \phi_i\lambda_i$, $\rho\phi_i = \phi_i^{**}\rho_i$, for all i.

(25.14) *Lemma. The limit sequence is exact.*

Proof: Given $x^* \in M^*$, choose $x_i^* \in M_i^*$ such that $x^* = \phi_i^* x_i^*$. Then $\rho\lambda x^* = \phi_i^{**}\rho_i\lambda_i x_i^* = 0$.

Given $x \in M$ such that $\rho x = 0$. Write $x = \phi_i x_i$. Since $\phi_i^{**}\rho_i x_i = 0$, there is an i' with $i \le i'$ such that $0 = \phi_{i'i}^{**}\rho_i x_i = \rho_{i'}\phi_{i'i}x_i$. By exactness at stage i', there is $x_{i'}^* \in M_{i'}^*$ such that $\phi_{i'i}x_i = \lambda_{i'}x_{i'}^*$. Then $\lambda\phi_{i'}^* x_{i'}^* = \phi_{i'}\lambda_{i'}x_{i'}^* = \phi_i x_i = x$. ■

(25.15) *Corollary. If each λ_i is a monomorphism (resp., if each ρ_i is an epimorphism) then so is λ (resp. ρ).*

Finally, we consider iterated limits. Suppose J is another directed set. Suppose that to each $j \in J$ is associated a directed subset $I_j \subset I$ so that if $j \le j'$ we have $I_j \subset I_{j'}$. Assume also that

$$I = \cup_j I_j$$

Then for each j, we can form

$$M_j^* = \varinjlim_{i \in I_j} M_i.$$

If $j \le j'$, there is a homomorphism $\psi_{j'j}: M_j^* \to M_{j'}^*$ defined as follows: Given $x \in M_j^*$, choose $i \in I_j$ and $x_i \in M_i$ such that $x = \phi_i^* x_i$, where $\phi_i^*: M_i \to M_j^*$ is the canonical homomorphism. Choose $i' \in I_{j'}$ such that $i \le i'$. Set $\psi_{j'j}x = \phi_{i'i}^* x_i$ (verify this is independent of the choices). By means of these $\psi_{j'j}$ we get another inductive system and we can take its limit

$$M^* = \varinjlim M_j^*$$

Then there are unique homomorphisms ψ, θ which make the diagrams

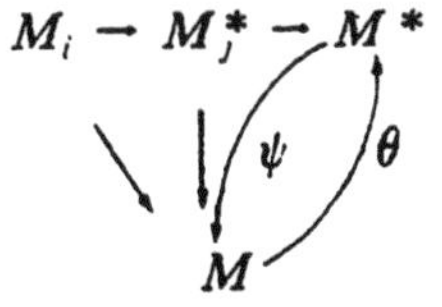

commutative for all $i \in I_j$, all $j \in J$.

(25.16) *Exercise.* ψ and θ are inverse isomorphisms.

(25.17) *Exercise.* Let x be any point in a Hausdorff space X. Then

$$\varinjlim_{x \in U \text{ open}} H_n(X, X - U) \to H_n(X, X - x)$$

is an isomorphism.

Remark. The analogue of (25.14) for projective limits is false, i.e., a projective limit of exact sequences need no longer be exact (see Eilenberg and Steenrod [23], p. 225). It becomes true when we assume that R is a *field*.

(25.18) *Exercise.* The analogues of (25.10, 25.11, and 25.16) hold for projective limits.

26. Poincaré Duality

Throughout this section X will denote an n-dimensional manifold. If X is R-orientable, we choose once and for all an R-orientation of X; otherwise, we take $R = \mathbf{Z}/2$ and take the unique $\mathbf{Z}/2$-orientation (22.12). For each compact $K \subset X$, restricting the R-orientation (regarded as a section of the R-orientation sheaf) to K and using the isomorphism $H_n(X, X-K) \to \Gamma K$ gives a *fundamental class* $\zeta_K \in H_n(X, X-K)$ (22.21 and 22.24). This notation will be used throughout this section.

If X is not compact, the duality theorem we are aiming for requires a different cohomology theory, the *singular cohomology with compact supports*. This is obtained from the ordinary singular cohomology as follows: The compact subspaces of X form a directed system under inclusion ($K \leq K'$ means $K \subset K'$). The modules $H^q(X, X-K)$ then form an inductive system indexed by the compact subspaces (the homomorphism $H^q(X, X-K) \to H^q(X, X-K')$ is induced by inclusion). We define

$$H_c^q(X) = \varinjlim_{K \text{ compact}} H^q(X, X-K)$$

Of course if X is compact, $H_c^q(X) = H^q(X)$ (25.8). A cohomology class in $H_c^q(X)$ is represented by a cochain which "vanishes" off some compact subspace K, i.e., which annihilates all chains with support contained in $X-K$.

(26.1) *Remark.* Given a map $f: X \to Y$, then for any compact $K \subset X$, $f(K)$ is compact. However, f needn't send $X-K$ into $Y-f(K)$, so f does not induce a homomorphism of H_c^q. We do get an induced homomorphism if we assume f is *proper*, i.e., for any compact $L \subset Y$, $f^{-1}(L)$ is compact.

Then f maps $X - f^{-1}(L)$ into $Y - L$, so there is an induced homomorphism

$$H^q(Y, Y - L) \to H^q(X, X - f^{-1}(L)) \to H^q_c(X)$$

As L varies, these homormorphisms are compatible with those induced by inclusion, hence passing to the limit gives an induced homomorphism $H^q_c(f): H^q_c(Y) \to H^q_c(X)$. If X is a subspace of Y and f is the inclusion map, then f is not proper unless X is closed.

(26.2) *Remark.* If U is an open subspace of X, we can nevertheless define a canonical homomorphism

$$H^q_c(U) \to H^q_c(X)$$

(Note that this homomorphism goes in the opposite direction to the one in (26.1).) Namely, for any compact $K \subset U$, we have the inverse of the excision isomorphism $H^q(U, U - K) \to H^q(X, X - K)$. Since these are compatible with the inclusion homomorphisms, passing to the limit gives a unique homomorphism making the diagram

$$\begin{array}{ccc} H^q_c(U) & \longrightarrow & H^q_c(X) \\ \uparrow & & \uparrow \\ H^q(U, U - K) & \to & H^q(X, X - K) \end{array}$$

commutative for all K.

(26.3) *Example*: Let $U = \mathbf{R}^n$ considered as S^n minus a point x. The sets $S^n - K$ for compact $K \subset U$ form a fundamental system of neighborhoods of x which contains a final system of contractible neighborhoods; for the latter,

$$H^q(S^n, S^n - K) \simeq H^{q*}(S^n)$$

is an isomorphism (23.14.7). Hence (25.11)

$$H^q_c(\mathbf{R}^n) \simeq H^{q*}(S^n)$$

is an isomorphism. This isomorphism generalizes to the one-point-compactification of any manifold (see (27.4)).

Now for any compact $K \subset X$, consider the homomorphism from relative cohomology to absolute homology

$$\zeta_K \cap : H^q(X, X-K) \to H_{n-q}(X)$$

given by the cap product $\gamma \to \zeta_K \cap \gamma$ with ζ_K(24.26). If $K \subset K'$, the diagram

$$\begin{array}{ccc} H^q(X, X-K) & & \\ & \searrow & \\ \downarrow & & H_{n-q}(X) \\ & \nearrow & \\ H^q(X, X-K') & & \end{array}$$

is commutative. Passing to the limit gives a homomorphism

$$D: H^q_c(X) \to H_{n-q}(X)$$

(Note that for $q > n$, D is the zero homomorphism.)

(26.4) *Exercise.* Let U be open in X, and give U the induced R-orientation. Then the diagram

$$\begin{array}{ccc} H^q_c(X) & \xrightarrow[D]{} & H_{n-q}(X) \\ \uparrow & & \uparrow \\ H^q_c(U) & \xrightarrow[D]{} & H_{n-q}(U) \end{array}$$

in commutative.

(26.5) *Exercise.* Let $R = \mathbf{Z}$. Let X, Y be compact connected oriented manifolds, and let $f: X \to Y$ be an m-fold covering space. Assume f is orientation preserving. Then the diagram

$$\begin{array}{ccc} H^q(X) & \xrightarrow[D]{} & H_{n-q}(X) \\ {\scriptstyle H^q(f)}\uparrow & & \downarrow{\scriptstyle H_{n-q}(f)} \\ H^q(Y) & \xrightarrow[mD]{} & H_{n-q}(Y) \end{array}$$

is commutative (22.33 and 24.24).

(26.6) *Poincarė Duality Theorem. If X is an R-oriented n-dimensional manifold, the homomorphism*

$$D: H^q_c(X) \to H_{n-q}(X)$$

is an isomorphism (for all q).

Proof. The proof is in several steps. We will prove it simultaneously for the open submanifolds of X.

Before detailing the steps, we outline what is going on (for compact X). As K enlarges to become all of X, the modules $H^q(X, X-K)$ "see" more of the cohomology of X. For example, take X to be the torus and K to be the compact portion above the dashed line. A representative of the dual (in algebraic sense) one dimensional homology is pictured.

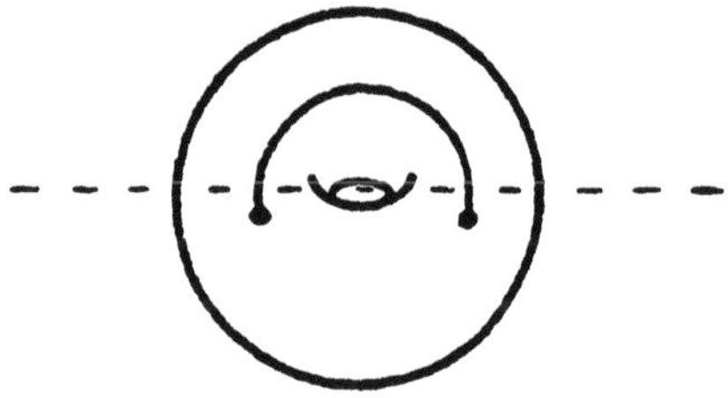

A basic comparison in the proof is a composition of the form

$$H^q(X, X-K) \cong H^q(W, W-K) \xrightarrow{\zeta_K \cap} H_{n-q}(W)$$

where W is open and the inclusion $(W, W-K) \to (X, X-K)$ is an excision. Thus, corresponding to cohomology of X seen by $H^q(X, X-K)$, there is homology of X whose support lies in W. Enlarging K by forming unions produces corresponding enlargement of W. The technical tool is a relative Mayer-Vietoris sequence for cohomology working in tandem with an absolute Mayer-Vietoris sequence for homology. The key step is to check compatibility of cap products with these sequences. Then the proof amounts to observing that one starts with an isomorphism (e.g. $K=$ point) and passes through compatible isomorphisms to reach the theorem.

Step 1. If the theorem holds for open sets U, V and $B = U \cap V$, then the theorem holds for $Y = U \cup V$.

Let K (resp. L) be compact in U (resp. V). We use the Mayer-Vietoris sequences for the triad $(Y, Y-K, Y-L)$ (17.11 and 23.14). Consider the diagram

$$\begin{array}{ccccccccc}
\to H^{q+1}(B, B-K\cap L) & \to & H^q(Y, Y-K\cup L) & \to & H^q(U, U-K)\oplus H^q(V, V-L) & \to & H^q(B, B-K\cap L) & \to \\
\downarrow \zeta_{K\cap L}\cap & & \downarrow \zeta_{K\cup L}\cap & & \downarrow \zeta_K\cap \oplus \zeta_L\cap & & \downarrow \zeta_{K\cap L}\cap & \\
\to H_{n-q-1}(B) & \to & H_{n-q}(Y) & \to & H_{n-q}(U)\oplus H_{n-q}(V) & \to & H_{n-q}(B) & \to
\end{array}$$

The top row is the relative Mayer-Vietoris sequence for $(Y, Y-K, Y-L)$ combined with excisions of the form $(W, W-S) \subset (Y, Y-S)$. The bottom row is the Mayer-Vietoris sequence for (Y, U, V). Commutativity of the right most 2 squares is naturality of the cap product map.

Sublemma. The following diagram commutes up to $(-1)^{q+1}$.

$$\begin{array}{ccc} H^{q+1}(Y, Y-K\cap L) & \xleftarrow{\ \gamma\ } & H^{q}(Y, Y-K\cup L) \\ \downarrow \cong & & \\ H^{q+1}(B, B-K\cap L) & & \downarrow \zeta_{K\cup L}\cap \\ \zeta_{K\cap L}\cap \downarrow & & \\ H_{n-q-1}(B) & \xleftarrow{\ \Gamma\ } & H_{n-q}(Y). \end{array}$$

Proof. We examine the Mayer-Vietoris boundaries γ and Γ on the chain (cochain) level. We use (17.11) for these calculations and (15.22) to have chain homotopy inverses to excisions. The focal point of the calculation is the following commutative diagram

$$\begin{array}{lccccc} & & & (4) & & (6) \\ (1) & S^{\cdot}(Y, Y-K\cup L) & \leftarrow & S^{\cdot}(Y, Y-K) & \leftarrow & S^{\cdot}(Y, Y-K\cap L) \\ & \uparrow(5) \quad S^{\cdot}(U, U-K) & \nwarrow\swarrow & \uparrow(8) \quad S^{\cdot}(B, B-K\cap L) & \nwarrow\swarrow & \uparrow \\ (2) & S^{\cdot}(X, X-K\cup L) & \leftarrow & S^{\cdot}(X, X-K) & \leftarrow & S^{\cdot}(X, X-K\cap L) \\ & & & (3) & & (7) \end{array}$$

where all maps are induced by inclusions, the vertical and oblique maps are excisions, and the horizontal maps are one route around the square used to construct Mayer-Vietoris sequences. To simplify notation, if $R \subset S \subset X$ and $j:(X, X-S) \to (X, X-R)$ is the inclusion, then $j_*(\zeta_S \cap j^*x) = \zeta_R \cap x$ in $S(X, X-R)$ will be abbreviated to $\zeta_S \cap x = \zeta_R \cap x$. Let c be in spot (1), such that $\delta c = 0$. Let a_2, in spot (2), be the image of c via a chain homotopy inverse to excision. Then $\delta a_2 = 0$. Let a_3 in (3) be an element given by (17.11) such that δa_3 in (7) represents the Mayer-Vietoris coboundary of a_2. Let a_4, a_5 in (4) and (5) be the images of a_3. Then the image of a_4 in (1) is $c + \delta Dc$ where D is a chain homotopy, hence $\gamma(c)$ is represented by δa_4 in (6). By commutativity, the images of δa_4 and δa_3 in (8) are equal. We have $\zeta_{K\cap L} \cap \gamma(c) = \zeta_{K\cap L} \cap \delta a_4 = \zeta_{K\cap L} \cap \delta a_3$ and we take the image in $S(B)$.

Furthermore

$$\zeta_{K\cap L}\cap\delta a_3=\zeta_K\cap\delta a_3=(-1)^{q+1}\partial(\zeta_K\cap a_3)=(-1)^{q-1}\partial(\zeta_K\cap a_i),\quad i=4,5.$$

On the other hand

$$\zeta_K\cap a_4=\zeta_{K\cup L}\cap(c+\delta Dc)=\zeta_{K\cup L}\cap c\pm\partial(\zeta_{K\cup L}\cap Dc).$$

Hence $\Gamma(\zeta_{K\cup L}\cap c)$ is represented by $\partial(\zeta_K\cap a_5)$ using (17.11) and $S(Y)\leftarrow S(U)\leftarrow S(B)$. So going around the two ways give representatives in $S(B)$ which are equal (up to sign) in $S(X)$. But $S(B)\to S(X)$ is injective. ∎

Returning to the proof of step 1, we observe that every compact in Y has the form $K\cup L$. Passing to the limit gives a sign-commutative diagram

$$\begin{array}{ccccccccc}
\leftarrow H_c^{q+1}(B) & \leftarrow & H_c^q(Y) & \leftarrow & H_c^q(U)\oplus H_c^q(V) & \leftarrow & H_c^q(B) & \leftarrow \\
D\downarrow & & D\downarrow & & D\oplus D\downarrow & & D\downarrow & \\
\leftarrow H_{n-q-1}(B) & \leftarrow & H_{n-q}(Y) & \leftarrow & H_{n-q}(U)\oplus H_{n-q}(V) & \leftarrow & H_{n-q}(B) & \leftarrow
\end{array}$$

in which the rows are exact and all vertical arrows except those involving Y are given to be isomorphisms (25.10 and 25.14). by the five lemma (14.7), the theorem is true for Y.

Step 2. Let (U_i) be a system of open sets totally ordered by inclusion, and let U be their union. If the theorem is true for all the U_i then the theorem is true for U.

This amounts to verifying the isomorphisms

$$\psi_1:\varinjlim H_{n-q}(U_i)\to H_{n-q}(U),$$

$$\psi_2:\varinjlim H_c^q(U_i)\to H_c^q(U).$$

The point is that for any compact $K\subset U$, we have $K\subset U_i$ for some i (cover K by finitely many U_j's; since the system is totally ordered, all the U_j's are contained in one U_i). It follows that ψ_1 is an isomorphism (consider the compact support of any chain). Similarly, the result on iterated limits (25.16) implies that ψ_2 is an isomorphism.

Step 3. U is contained in a coordinate neighborhood. Regard U as a subspace of $\mathbf{R}^n$.

Case 1. U is convex.

Then U is homeomorphic to $\dot{E}^n$ (exercise). Now in computing the inductive limit

$$\varinjlim_K H^q(\dot{E}^n, \dot{E}^n - K)$$

it suffices to let K run through the final system of closed balls of radius < 1 centered at the origin. But for such K, the modules in question are zero unless $q = n$, and

$$\zeta_K \cap : H^n(\dot{E}^n, \dot{E}^n - K) \to H_0(\dot{E}^n) \cong R$$

is certainly an isomorphism (24.23). Hence the limiting homomorphism D is also an isomorphism.

General Case. We enumerate the dense set of points in U having rational coordinates, and choose a convex open V_j contained in U about the jth point. Let

$$U_i = U_{i-1} \cup V_i \qquad i > 1,$$

$$U_1 = V_1.$$

The inductive hypothesis is that the theorem is true for a union of k convex open sets, $k < i$. Observe that $U_{i-1} \cap V_i$ is the union of at most $i - 1$ convex sets. By case 1, step 1, and induction on i, the theorem is true for U_i. By step 2, the theorem is true for U.

Step 4. Proof for X.

By Zorn's lemma (steps 2 and 3), there is a maximal open subspace U of X for which the theorem is true. For any open V contained in a coordinate neighborhood, the theorem is true for $U \cup V$ (steps 1 and 3). By maximality, $U = X$. ■

(26.7) *Corollary. If X is connected R-orientable, then $H^n_c(X) \cong R$.*

In that case, the generator of $H^n_c(X)$ corresponding to the canonical generator of $H_0(X)$ under the duality isomorphism will be called the *fundamental cohomology class* of the R-orientation. If $f: X \to Y$ is an m-fold covering space satisfying the hypotheses of (26.5), then $H^n(f)$ takes the fundamental class of Y onto m times the fundamental class of X.

(26.8) *Corollary. If X is compact orientable, then the Betti numbers of X satisfy*

$$\beta_q = \beta_{n-q}$$

for all q.

Proof: this follows from (23.40), taking $R = \mathbf{Z}$. ■

For example, if $n = 3$, knowledge of β_1 and the number of connected components of X determines all the Betti numbers.

Of course (23.28) tells us about the torsion in the homology as well: If T_q is the torsion subgroup of $H_q(X; \mathbf{Z})$, then

(26.9) $$\boxed{T_q \cong T_{n-q-1}}$$

Warning: These formulas do not hold without the assumption of orientability—consider $\mathbf{P}^2$, for example.

(26.10) *Corollary. Assume R is a* PID. *If X is odd-dimensional compact R-orientable, then* $\chi(X; R) = 0$.

(26.11) *Corollary. If X is even-dimensional compact orientable and the dimension is not divisible by 4, then* $\chi(X)$ *is even.*

Proof: By (26.8), if $n = 4k + 2$, we must show β_{2k+1} even. According to (20.12) and (23.31), this amounts to showing dim $H^{2k+1}(X; \mathbf{Q})$ even. We may assume X connected and identify $H^n(X; \mathbf{Q})$ with $\mathbf{Q}$ (26.7). Then the cup product is a non-degerate skew-symmetric bilinear form on $H^{2k+1}(X; \mathbf{Q})$ (24.8 and 24.19). Since the determinant of an odd order skew-symmetric matrix is zero, β_{2k+1} must be even. ■

The conclusion of the corollary is false for $4k$-dimensional manifolds (consider $\mathbf{CP}^{2k}$).

Note. We have tacitly been assuming that the Betti numbers of a compact manifold are finite. See Appendix (26.17).

Note. If X is non-orientable, we have proved a Poincaré duality theorem for coefficients in $\mathbf{Z}/2$. A stronger result is often needed, i.e., duality with coefficients in the orientation sheaf (see Swan [56], Chap. XI).

We can apply Poincaré duality to determine the cohomology algebras of projective spaces. Consider for example the complex projective plane. Let $\zeta \in H_4$ be a fundamental class for some R-orientation. Let γ be a generator of H^2 (23.29). Then $\zeta \cap \gamma$ generates H_2. Since the Kronecker product $H_2 \times H^2 \to R$ is non-degenerate, we see that

$$[\zeta \cap \gamma, \gamma] = [\zeta, \gamma \cup \gamma]$$

generates R. Hence $\gamma^2 = \gamma \cup \gamma$ generates the module H^4. Thus γ generates the R-algebra $H^{\cdot}$. A more general statement is possible.

(26.12) *Proposition. If* γ *generates the module* $H^2(\mathbf{CP}^n)$, *then* γ *generates the cohomology algebra* $H^{\cdot}(\mathbf{CP}^n)$.

Proof: By induction on n, using the fact that the inclusion $\mathbf{CP}^{n-1} \to \mathbf{CP}^n$ induces isomorphisms in dimensions $\leq 2n-2$ (19.10). ■

One says that $H^{\cdot}(\mathbf{CP}^n)$ is the *truncated polynomial algebra* generated by an element of degree 2 and height $n+1$ (meaning $\gamma^{n+1}=0$).

(26.13) *Exercise.* Let $R = \mathbf{Z}/2$. Then $H^{\cdot}(\mathbf{P}^n)$ is the truncated polynomial algebra generated by an element of degree 1 and height $n+1$.

(26.14) *Exercise.* R arbitrary. Then $H^{\cdot}(\mathbf{HP}^n)$ is the truncated polynomial algebra generated by an element of degree 4 and height $n+1$.

As an application of (26.13) we have

(26.15) *Borsuk-Ulam Theorem. If* $n > m \geq 1$, *then there is no map* $g{:}S^n \to S^m$ *which commutes with the antipodal maps.*

Proof: Any such map would induce by passage to the quotient a map $f{:}\mathbf{P}^n \to \mathbf{P}^m$ making the diagram

$$\begin{array}{ccc} S^n & \xrightarrow{g} & S^m \\ {\scriptstyle p'}\downarrow & & \downarrow{\scriptstyle p} \\ \mathbf{P}^n & \xrightarrow{f} & \mathbf{P}^m \end{array}$$

commutative (the vertical arrows are the double coverings).

Sublemma. There exists $f'{:}\mathbf{P}^n \to S^m$ *such that* $pf' = f$.

Proof: We use the lifting criterion (6.1). If $m = 1$, it's clear, since the only homomorphism $\pi_1(\mathbf{P}^n) \to \pi_1(\mathbf{P}^1)$ is zero. Suppose $m > 1$. Consider the induced algebra homomorphism

$$H^{\cdot}(f){:}H^{\cdot}(\mathbf{P}^m) \to H^{\cdot}(\mathbf{P}^n)$$

(coefficients $\mathbf{Z}/2$). Let γ_m, γ_n be generators of these algebras. Since

$$0 = H^{\cdot}(f)(\gamma_m^{m+1}) = H^{\cdot}(f)(\gamma_m)^{m+1}$$

and $n > m$, we have $\gamma_n \neq H^\cdot(f)(\gamma_m)$. Hence $H^\cdot(f)(\gamma_m) = 0$ (the only other element of H^1).

Let $i{:}\mathbf{P}^1 \to \mathbf{P}^n, j{:}\mathbf{P}^1 \to \mathbf{P}^m$ be the inclusions obtained by setting all but the first two homogeneous coordinates equal to zero. Now $H^1(j)$ is an isomorphism (induction on m, using the proof of 19.27), so $H^1(j)(\gamma_m) \neq 0$. Hence $H^1(j) \neq H^1(fi)$, so that fi is not homotopic to j. But i and j can be regarded as generators of the fundamental group ($\mathbf{P}^1 \approx S^1$), so

$$f_*{:}\pi_1(\mathbf{P}^n) \to \pi_1(\mathbf{P}^m)$$

is the zero homomorphism. Hence (6.1) applies again. ∎

Getting back to the theorem, we have $pf'p' = pg$. If $x \in S^n$, either $g(x) = f'p'(x)$ or $g(-x) = f'p'(x) = f'p'(-x)$. Thus $f'p'$ and g are two liftings of fp' which agree at a point. By (5.1), $f'p' = g$. But $g(x) \neq g(-x) = -g(x)$, while $p'(x) = p'(-x)$, contradiction. ∎

(26.16) *Project.* For an n-dimensional R-oriented manifold X, the cup product

$$H^p(X, X - K) \times H^q(X, X - L) \to H^{p+q}(X, X - K \cap L)$$

induces by passage to the limit a cup product

$$H^p_c(X) \times H^q_c(X) \to H^{p+q}_c(X)$$

hence the direct sum $H^\cdot_c(X)$ becomes an R-algebra (without an identity element unless X is compact) which is contravariantly functorial with respect to proper maps. Using the duality isomorphisms, we can transport this multiplication to construct the homology algebra $H_\cdot(X)$, so that the resulting "intersection product" of a p-dimensional class and a q-dimensional class has dimension $p + q - n$. This algebra is *not* covariantly functorial with respect to the naturally induced module homomorphisms $H_\cdot(f){:}H_\cdot(X) \to H_\cdot(Y)$. (See Dold [64], p. 335ff. and p. 196.) A special case is treated in chapter 31.

(26.17) *Appendix: Absolute Neighborhood Retracts*

If we identify the boundary of E^n to a point, we obtain a space homeomorphic to S^n, so that

$$\dot{E}^n \approx S^n - \text{point} \approx \mathbf{R}^n$$

(26.17.1) *Lemma. For any point x in an n-dimensional manifold X, there is a map $f{:}X \to S^n$, an open neighborhood U of x, and a point P in S^n such that*

$$f : U \approx S^n - P$$

$$f(X - U) = \{P\}$$

Proof: Choose a coordinate neighborhood V of x and a coordinate neighborhood $U \subset V$ of x such that $\bar{U} \subset V$. Let $g : \bar{U} \approx E^n$ be a homeomorphism sending U onto $\dot{E}^n$, $h : E^n \to S^n$ the quotient map identifying the boundary to the point P. Then define f by

$$f(x') = \begin{cases} P & x' \in X - U \\ hg(x') & x' \in U \end{cases}$$ ■

(26.17.2) *Lemma. If K is a compact subspace of X then there is an open neighborhood V of K and an injective map of V into a* Euclidean space.

Proof: For each $x \in K$, choose U_x and f_x as in (26.17.1). Let finitely many of these open sets, say $U_1, \ldots, U_r$, cover K, and take V to be their union. If $f_1, \ldots, f_r$ are the associated maps, define

$$f : X \to S^n \times S^n \times \cdots \times S^n$$

by $f(x) = (f_1(x), \ldots, f_r(x))$. Then the restriction of f to V is injective. Moreover, the product of spheres is imbedded in Euclidean $r(n + 1)$-space. ■

(26.17.3) *Corollary. A compact manifold can be imbedded in a Euclidean space.*

Recall that a space X is called an *AR* = *absolute retract* (or solid) if it has the following universal property: For any normal space Y and any map $f : B \to X$ of a closed subspace B of Y into X, f extends to a map of Y into X. Tietze's extension theorem states that the closed unit interval is an *AR*; since a cartesian product of *AR*'s is an *AR*, E^n is an *AR* (see Dugundji [20], Chap. VII, No. 5). If in the above definition we require only that f extend to an open neighborhood of B, then X is called an *ANR* = *absolute neighborhood retract.* Clearly an open subspace of an *ANR* (a fortiori of an *AR*) is an *ANR*.

(26.17.4) *Theorem. Every compact manifold is an ANR.*

(26.17.5) *Corollary. If the compact manifold X is imbedded in some Euclidean space, then X is a retract of some open neighborhood.*

Just apply the universal property to $B = X$ and $f =$ identity.

Proof of theorem: Since X can be covered by finitely many coordinate neighborhoods, and each of them is an *ANR*, we are reduced to proving the following lemma:

(26.17.6) *Lemma. If X_1, X_2 are ANR's open in X such that $X = X_1 \cup X_2$, then X is an ANR.*

Proof: Given B closed $\subset Y$ normal, $f: B \to X$. Let $A_i = B - f^{-1}(X_1)$, $i = 1, 2$. Then $A_1 \cap A_2$ is empty, and A_1, A_2 are closed. Since Y is normal, we can separate A_1 and A_2 by open sets Y_1, Y_2.

Let $Y_0 = Y - (Y_1 \cup Y_2)$, closed in Y, hence also normal. Let $B_i = Y_i \cap B$, $i = 1, 2, 0$. Then

$$f(B_i) \subset X_i \qquad i = 1, 2$$

$$f(B_0) \subset X_1 \cap X_2$$

Since $X_1 \cap X_2$ is an *ANR*, $f| B_0$ can be extended to a map g_0 on a neighborhood U_0 in Y_0 (U_0 is relatively open in Y_0). Then $U_0 \cap B = B_0$, so f together with g_0 defines a map $g: U_0 \cup B \to X$ (g is continuous because $U_0 = (U_0 \cup B) \cap Y_0$ is relatively closed in $U_0 \cup B$).

We now apply normality to the disjoint closed sets B_0 and $Y_0 - U_0$: There are disjoint relative open sets $V, W \subset Y_0$ such that $B_0 \subset V$, $Y_0 - U_0 \subset W$. Then $U_0' = Y_0 - W$ is closed and $U_0' \subset U_0$.

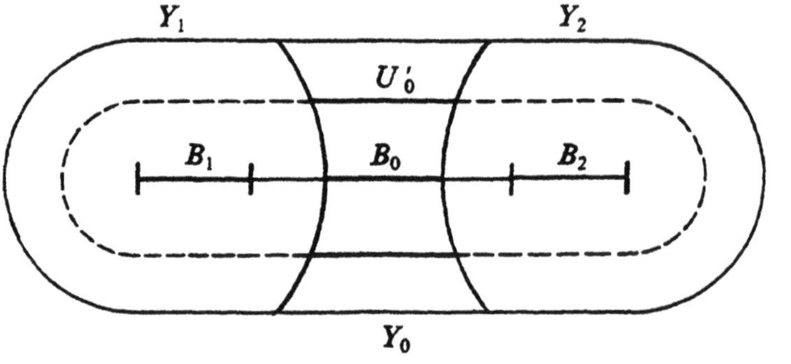

Then $g(U_0' \cup B_i) \subset X_i$, $i = 1, 2$, and $U_0' \cup B_i$ is closed in Y, $i = 1, 2$. Since X_i is an *ANR*, $g| U_0' \cup B_i$ extends to a map $G_i: U_i \to X_i$ on an open neighborhood U_i, $i = 1, 2$. Now

$$U_i' = U_i \cap (U_0' \cup Y_i)$$

is closed in $U'_1 \cup U'_2 = U$, $i = 1, 2$, and

$$U'_0 = U'_1 \cap U'_2$$

(exercise). Hence we can define a map $F: U \to X$ to be G_i on U'_i, $i = 1, 2$, and the definition is licit. Moreover, U contains the open neighborhood

$$(U_1 \cap (V \cup Y_1)) \cup (U_2 \cap (V \cup Y_2))$$

of B (exercise). ■

(26.17.7) *Corollary. If X is a compact manifold and Δ is the diagonal in $X \times X$, then there is an open neighborhood V of Δ such that the identity map of V is homotopic in $X \times X$ to a retraction of V onto Δ.*

Proof: Imbed X in $\mathbf{R}^N$(26.17.3), and let U be an open neighborhood having a retraction $r: U \to X$ (26.17.5). Let $\varepsilon =$ distance from X to $\mathbf{R}^N - U$, and let V be the ε-neighborhood of Δ in $X \times X$. Define

$$F: X \times X \times I \to \mathbf{R}^N$$

by

$$F(x, x', t) = (1 - t)x + tx'.$$

Then F maps $V \times I$ into U (by definition of V). Let

$$G = r(F \mid V \times I): V \times I \to X$$

so that $G(x, x', 0) = r(x) = x$, $G(x, x', 1) = r(x') = x'$. Define

$$H: V \times I \to X \times X$$

by

$$H(x, x', t) = (x, G(x, x', t)).$$

Then H is the required homotopy. ■

(26.17.8) *Theorem. Let the compact subspace K of a Euclidean space be an ANR. Then $H_q(K)$ is a finitely generated module for all q.*

Proof. Let $r: U \to K$ be a retraction of an open neighborhood U onto K. There is a finite cell complex C such that $K \subset C \subset U$ (see the proof of 22.24, case 3), so that K is a retract of C. Hence

$$H_q(r \mid C): H_q(C) \to H_q(K)$$

is an epimorphism for all q. We know that $H_q(C)$ is finitely generated. ∎

(26.17.9) *Corollary. The homology modules of a compact manifold are finitely generated.*

(26.17.10) *Note.* Theorems (26.17.3) and (26.17.4) can be generalized to paracompact manifolds by transfinite induction [29]. Of course (26.17.9) does not generalize. The imbedding theorem we have proved is very weak. Whitney proved that every paracompact differentiable manifold of dimension n can be imbedded (differentiably) in $\mathbf{R}^{2n}$. This is the best possible general result, for we shall see, e.g., that the projective plane cannot be imbedded in 3-space (27.11). (See Lashof [36], for a survey of work on imbeddings. Further developments are surveyed in S. Gitler, Immersions and Embeddings of Manifolds, Proceedings of Symposia in Pure Mathematics Vol. 22).

Manifolds are assumed to be connected and compact for the following exercises.

(26.18) *Exercise.* Let $\dim M = n$. If M is orientable, then $H_{n-1}(M; \mathbf{Z})$ is torsion free. If M is nonorientable, then $H_n(M; \mathbf{Z}/k\mathbf{Z}) = 0$ if k is odd, the torsion subgroup of $H_{n-1}(M; \mathbf{Z}/2\mathbf{Z})$ is cyclic of order 2, and $H_1(M; \mathbf{Z}/2\mathbf{Z}) \neq 0$.

(26.19) *Exercise.* If M is an orientable 3-manifold with $H_1(M; \mathbf{Z}) = 0$, then M has the homology of a 3-sphere.

(26.20) *Exercise.* If M is a nonorientable 3-manifold, then $H_1(M; \mathbf{Z})$ is infinite.

(26.21) *Exercise.* Prove that the cup product pairing $H^p(M) \oplus H^q(M) \to H^n(M)$, $p + q = n$ and coefficients in a field, has the property that for a fixed $x \in H^p(M)$, $x \cup y = 0$ for all $y \in H^q(M)$ only if $x = 0$.

(26.22) *Exercise.* Prove that $\mathbf{CP}^{2n}$ *admits no orientation reversing homotopy equivalence.*

(26.23) *Exercise.* Note that (26.12) and (26.14) combined with exercise (24.32) gives maps $S^3 \to S^2$, $S^7 \to S^4$ with Hopf invariant one. Prove that for these maps

$$\begin{array}{ccc} S^{2n-1} & \longrightarrow & S^n \\ {\scriptstyle -1}\downarrow & & \downarrow{\scriptstyle -1} \\ S^{2n-1} & \longrightarrow & S^n \end{array}$$

cannot commute even up to homotopy, where -1 means a map of degree -1. Suggestion: Use (21.20).

(26.24) *Exercise.* Prove that a necessary condition for the lens space $L(p, q)$ to admit an orientation reversing homotopy equivalence is that -1 be a quadratic residue mod p, i.e., there is an integer λ such that $-1 \equiv \lambda^2 \bmod p$. Suggestion: Consider the mod p Bockstein $\beta : H^1 \to H^2$ and use (26.21). A refinement will appear in (31.10).

(26.25) *Exercise.* Prove a theorem of Borsuk: *if $f: S^n \to S^n$ commutes with the antipodal map then f has odd degree*. Suggestion: Do the case n odd first, then extend to n even using composition of the suspension Σf with reflection of S^{n+1} through the equatorial S^n. Unlike (26.23) this is not a homotopy-theoretic result; indeed $fa \simeq af$ regardless, since homotopy theory proves that homotopy classes of maps of S^n are classified by their degree.

(26.26) *Exercise.* An orientable n-manifold M is *spherical* if there exists $f: S^n \to M$ such that $H(f)\zeta_S = k\zeta_M$ for some $k \neq 0$. Prove that if $p \nmid k$, then f induces an isomorphism of mod p homology. Conclude that $H_i(M; \mathbf{Z})$ is finite $1 \leq i < n$.

(26.27) *Exercise.* Let X be a compact submanifold of a connected manifold M, with $\dim X < n = \dim M$. Apply Theorem (26.17.4) to deduce that $H_n(M, X) \cong (M - X)$ (see Theorem (22.24) for the meaning of Γ_c). If $c_b(A)$ denotes the number of components of a subset of $A \subset M$ which have compact closure in M, deduce from the above and the exact homology sequence that

$$c_b(M - X) = c_b(M) + \dim(\ker[H_{n-1}(X; \mathbf{Z}/2\mathbf{Z}) \to H_{n-1}(M; \mathbf{Z}/2\mathbf{Z})])$$

If M is orientable, then also

$$c_b(M - X) = c_b(M) + \operatorname{rank}(\ker[H_{n-1}(X; \mathbf{Z}) \to H_{n-1}(M; \mathbf{Z})])$$

Note that $c_b(M) = 1$ or 0 according as M is compact or not. These results generalize the Jordan-Brouwer Separation Theorem. Deduce from them that if $\dim X = n - 1$, X is connected, M orientable, and the embedding $i: X \subset M$ induces the zero homomorphism $H_{n-1}(X; \mathbf{Z}/2\mathbf{Z}) \to H_{n-1}(M; \mathbf{Z}/2\mathbf{Z})$, then X is also orientable and i induces the zero homomorphism $H_{n-1}(X; \mathbf{Z}) \to H_{n-1}(M; \mathbf{Z})$ (Dold [64], p. 262).

27. Alexander Duality

In this section X will be an R-oriented n-dimensional manifold, A a closed subset. Let $U = X - A$. We have seen that there is a canonical homomorphism

$$i:H_c^q(U) \to H_c^q(X)$$

when we use cohomology with compact supports. We aim to imbed this homomorphism in a long exact cohomology sequence, and for this purpose we must consider another cohomology theory for the subspace A.

Consider the family of all open neighborhoods V of A, directed by reverse inclusion ($V \leq V'$ means $V' \subset V$). The inclusion homomorphisms make the modules $H^q(V)$ into an inductive system. We define

$$\check{H}^q(A) = \varinjlim_{V} H^q(V)$$

By passing the inclusion homomorphisms $H^q(V) \to H^q(A)$ to the limit we get a canonical homomorphism

$$\kappa:\check{H}^q(A) \to H^q(A)$$

If this is an isomorphism, A is said to be *tautly imbedded* in X.

(27.1) *Proposition. If A is an ANR, then κ is an epimorphism; if also X is an ANR, then κ is an isomorphism.* (See Note below).

Proof: Let $r:V \to A$ be a retraction of a neighborhood V of A onto A,

$i: A \to V$ the inclusion. Then $H^q(ri)$ = identity shows that $H^q(i)$ is an epimorphism, hence κ is.

Suppose X is an ANR. Let U be any neighborhood of A. Let U' be a smaller neighborhood with a retraction $r: U' \to A$. We will find an even smaller neighborhood V such that if $i: A \to U'$, $j: V \to U'$ are the inclusions, then there is a homotopy

$$i(r|V) \simeq j$$

so that $H^q(j) = H^q(r|V)H^q(i)$. Thus we have a factorization

$$\begin{array}{ccccc} H^q(U) & & \longrightarrow & & H^q(A) \\ \downarrow & & & & \uparrow \\ H^q(U') & & & & H^q(V) \\ & \searrow & & \nearrow & \\ & & H^q(A) & & \end{array}$$

so that any class in $H^q(U)$ going to zero in $H^q(A)$ goes to zero in $H^q(V)$; thus κ is a monomorphism

One the closed subset $(U' \times 0) \cup (A \times I) \cup (U' \times 1)$ of $U' \times I$, set

$$F(x, t) = \begin{cases} x & \text{if } t = 0 \quad \text{and} \quad x \in U' \\ r(x) & \text{if } t = 1 \quad \text{and} \quad x \in U' \\ x & \text{if } x \in A \end{cases}$$

Since U' is an ANR, F extends to a map of a neighborhood of this set into U' (see Note below). That neighborhood contains a set of type $V \times I$, where V is a neighborhood of A; this gives the desired homotopy.

Note. We need an extra hypothesis to know that $U' \times I$ is normal, e.g., X paracompact (see *Topology* by H. Schubert, Theorems 2 and 3, pp. 95–6). ■

(27.2) *Remark.* If A is not tautly imbedded, it can still be shown that $\check{H}^q(A)$ depends only on A and not on the imbedding; it is in fact the Alexander-Cech cohomology module of A (Spanier [52], Chap. 6).

Let us now assume the manifold X to be compact, so that A is also. For any open neighborhood V of A, $K = X - V$ is compact and contained in U. The homomorphism

$$H^q(V) \xrightarrow{\delta} H^{q+1}(X, V) \cong H^{q+1}(U, U - K)$$

is compatible with change in V, so passing to the limit gives a connecting homomorphism

$$\delta:\check{H}^q(A) \to H_c^{q+1}(U)$$

(27.3) *Theorem. Assume X is compact. Then the sequence*

$$\cdots \to H_c^q(U) \xrightarrow{i} H^q(X) \xrightarrow{j} \check{H}^q(A) \xrightarrow{\delta} H_c^{q+1}(U) \to \cdots$$

is exact, where A is closed and $U = X - A$.

Proof: Given a class in $H^q(U, U-K)$, its image under $H^q(U, U-K) \to H^q(X, X-K) \to H^q(X) \to H^q(X-K)$ is zero; since $X-K$ is a neighborhood of A we see $ji = 0$. The exactness of

$$H^q(X, X-K) \to H^q(X) \to H^q(X-K)$$

implies that Kernel j = Image i.

For any class in $H^q(X)$, its image under

$$H^q(X) \to H^q(X) \to H^{q+1}(X, X) = H^{q-1}(U, U)$$

is zero, so $\delta j = 0$. Conversely, any class in the kernel of δ is represented by an element of $H^q(V)$ sent to zero in $H^{q+1}(X, V)$, for some V; but that element is in the image of $H^q(X) \to H^q(V)$, hence Kernel δ = Image j.

Similarly Kernel i = Image δ. ∎

(27.4) *Corollary. Let A be a compact ANR in a compact ANR X,* $U = X - A$*. Then the homomorphisms*

$$H^q(U, U-K) \cong H^q(X, X-K) \xrightarrow{\text{Incl.}} H^q(X, A)$$

for K compact $\subset U$ *induce by passage to the limit an isomorphism*

$$H_c^q(U) \cong H^q(X, A)$$

Proof: Apply the 5-lemma to the diagram

$$\begin{array}{ccccccccc}
\to H^{q-1}(A) & \to & H_c^q(U) & \to & H^q(X) & \to & H^q(A) & \to \\
\| & & \downarrow & & \| & & \| & \\
\to H^{q-1}(A) & \to & H^q(X, A) & \to & H^q(X) & \to & H^q(A) & \to
\end{array}$$

using (27.1) to identify $\check{H}^q(A)$ with $H^q(A)$ for all q. ∎

Now let $\zeta_A \in H_n(X, X-A)$ be the fundamental class determined by the R-orientation of X. For any open neighborhood V of A, we have

$$H_n(V, V - A) \cong H_n(X, X - A)$$

by excision; the pre-image of ζ_A under this isomorphism will again be denoted ζ_A. Taking the cap product with ζ_A gives a homomorphism

$$\zeta_A \cap : H^q(V) \to H_{n-q}(V, V - A) \cong H_{n-q}(X, X - A)$$

(24.27). These homomorphisms are compatible with changes in V (24.24), hence passing to the limit gives a homomorphism

$$D_A : \check{H}^q(A) \to H_{n-q}(X, X - A)$$

(Note that for $q > n$ D_A is the zero homomorphism.)

(27.5) *Alexander Duality Theorem. Assume X is R-oriented compact n-dimensional manifold, A closed. Then D_A is an isomorphism for all q.*

Proof: The diagram

$$\begin{array}{ccccccc} \to H^q_c(U) & \to & H^q(X) & \to & \check{H}^q(A) & \to \\ \downarrow D_U & & \downarrow D_X & & \downarrow D_A & \\ \to H_{n-q}(U) & \to & H_{n-q}(X) & \to & H_{n-q}(X, X - A) & \to \end{array}$$

is sign-commutative (where D_U and D_X are the isomorphisms of the Poincaré Duality Theorem (26.6)). Apply the 5-lemma (14.7) (check that the 5-lemma still works when the diagrams are only sign-commutative). ■

(27.6) *Exercise.* Let (K, L) be a compact pair in X. Then there is a *relative Alexander duality*

$$\check{H}^q(K, L) \to H_{n-q}(X - L, X - K)$$

which in case $K = X$ yields an isomorphism

$$\check{H}^q(X, L) \to H_{n-q}(X - L)$$

Derive the relative Alexander duality by the following sequence of steps:

Step 1. Define

$$\check{H}^q(K, L) = \varinjlim_{(U, V)} H^q(U, V)$$

where (U, V) runs through the directed set of pairs of open sets containing (K, L). Derive the exact cohomology sequence

$$\to \check{H}^{q-1}(L) \to \check{H}^q(K, L) \to \check{H}^q(K) \to \check{H}^q(L) \to$$

by passing to the limit the exact cohomology sequences of the pairs (U, V).

Step 2. Define a cap product

$$H_n(X, X-K) \times H^q(U, V) \to H_{n-q}(X-L, X-K)$$

as follows: Let $z \in Z_n(X, X-K)$, $c \in Z^q(U, V)$. Let $\mathscr{V}$ be the open covering $\{V, X-K, U \cap (X-L)\}$ of X. By theorem (15.9), we may assume z is small of order $\mathscr{V}$. Since c annihilates $S_q(V)$, $z \cap c$ is a relative $(n-q)$-cycle on $(X-L, X-K)$.

This cap product is compatible with passage to smaller (U, V); hence, passage to the limit gives a cap product

$$H_n(X, X-K) \times \check{H}^q(K, L) \to H_{n-q}(X-L, X-K)$$

Step 3. The diagram

$$\begin{array}{ccccccccc}
\to & \check{H}^{q-1}(L) & \to & \check{H}^q(K, L) & \to & \check{H}^q(K) & \to & \check{H}^q(L) & \to \\
& \zeta_L \cap \downarrow & & \zeta_K \cap \downarrow & & \zeta_K \cap \downarrow & & \zeta_L \cap \downarrow & \\
\to & H_{n-q+1}(X, X-L) & \to & H_{n-q}(X-L, X-K) & \to & H_{n-q}(X, X-K) & \to & H_{n-q}(X, X-L) & \to
\end{array}$$

is sign-commutative. Apply the absolute Alexander duality and the 5-lemma. ∎

(27.7) *Remark.* We can similarly define a canonical homomorphism

$$\zeta_K \cap : H^q(X-L, X-K) \to \check{H}_{n-q}(K, L)$$

where the module on the right is the projective limit

$$\varprojlim_{(U, V)} H_{n-q}(U, V)$$

This is not in general an isomorphism: Let $X = \mathbf{P}^3$, $K = \mathbf{P}^2$, L empty, $R = \mathbf{Z}$. Then $H_1(K) = H_1(K) = \mathbf{Z}/2$. Now $X - K$ is an open 3-cell, so the exact cohomology sequence gives $H^2(X, X-K) = H^2(\mathbf{P}^3) = 0$. The trouble is caused by torsion in the homology groups. One can prove that this duality does hold when R is a field by an argument similar to the proof of (22.24), using the fact that projective limits commute with exact sequences of vector spaces.

(27.8) *Remark.* The Alexander duality can be generalized to the case of an arbitrary closed pair in a paracompact manifold, provided one uses Cech-Alexander cohomology with compact supports (see Spanier [52], Chap. 6, or Swan [56], Chap. XI).

(27.9) *Corollary. Let A be a compact submanifold of* $\mathbf{R}^n$. *Then*

$$H^q(A) \cong H^{*}_{n-q-1}(\mathbf{R}^n - A)$$

for all $q < n$ *and* $H^n(A) = 0$ *(hence* $\dim A < n$*).*

Proof: By (26.17), we may identify $\check{H}^q(A) = H^q(A)$. Regarding $\mathbf{R}^n$ as S^n—point, we have the isomorphisms

$$H^q(A) \underset{D_A}{\to} H_{n-q}(S^n, S^n - A) \leftarrow H_{n-q}(\mathbf{R}^n, \mathbf{R}^n - A) \underset{\partial}{\to} H^{*}_{n-q-1}(\mathbf{R}^n - A)$$ ■

(27.10) *General Separation Theorem. If A is a compact $(n-1)$-dimensional submanifold of* $\mathbf{R}^n$ *having k connected components, then the complement of A has $k+1$ connected components.*

Proof: Take coefficients in $\mathbf{Z}/2$. By (27.9) and Poincaré duality, $H_0(A) \cong H^{n-1}(A) \cong H^{*}_0(\mathbf{R}^n - A)$. ■

(27.11) *Theorem. A non-orientable compact n-dimensional manifold cannot be imbedded in* $\mathbf{R}^{n+1}$.

Proof: Otherwise, if k is the number of connected components of A, rank $H_n(A; \mathbf{Z}) = \text{rank } H^n(A; \mathbf{Z}) = \text{rank } H^{*}_0(\mathbf{R}^{n+1} - A; \mathbf{Z}) = k$ (since (27.9) holds for any coefficient ring!). But A non-orientable implies $H_n(A; \mathbf{Z})$ has rank less than k (22.28). ■

Thus the non-orientable surfaces U_h cannot be imbedded in 3-space.

(27.12) *Exercise.* If A is a compact n-dimensional manifold imbedded in compact connected manifold X of the same dimension, then $A = X$.

(27.13) *Exercise.* Apply Alexander Duality to get short proofs of Theorems 18.1 and 18.3.

(27.14) *Exercise.* Recall $X * Y$, the join of X, Y (17.18). Observe that S^n is homeomorphic to $S^p * S^q$ where $p + q = n - 1$. Embed S^p in S^n by $x \to (x, 0, y)$ where $y \in S^q$ is fixed. Use this embedding to interpret the duality isomorphism $H^p(S^p) \cong H_q(S^n - S^p)$ geometrically. The case $p = q = 1$ is pictured

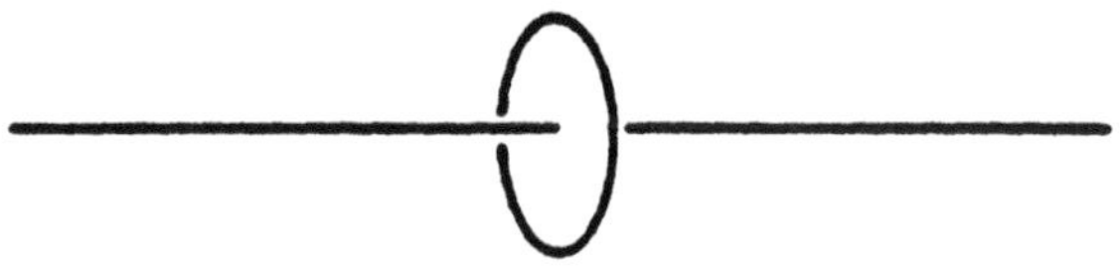

The following exercises are based on Steenrod and Epstein [87], p. 35.

(27.15) *Exercise.* Let M be a connected closed $(n - 1)$-manifold embedded in S^n. By (27.11), M is orientable and separates S^n into two open subsets with closures A and B such that $A \cup B = S^n$. Prove that no proper closed subset of M can separate S^n, hence $A \cap B = M$.

(27.16) *Exercise.* Use (27.15) to prove a theorem of Hopf. The inclusions $i:M \to A$ and $j:M \to B$ induce a representation of $H^q(M)$ as a direct sum

$$H^q(M) \cong i^*H^q(A) \oplus j^*H^q(B) \qquad 0 < q < n - 1$$

where i^*, j^* are monomorphisms ($i^* = H^q(i)$ an abbreviation). Show that $H^r(A) = 0 = H^r(B)$ for $r \geq n - 1$. Taking coefficients in a field F, prove

$$i^*H^q(A) \cong \mathrm{Hom}(j^*H^{n-q-1}(B), F) \qquad 0 < q < n - 1.$$

(27.17) *Exercise.* Use (27.16) to prove nonexistence of embeddings $\mathbf{P}^n \not\subset S^{n-1}$, $\mathbf{CP}^n \times S^m \not\subset S^{2n+m+1}$, $\mathbf{CP}^n \not\subset S^{2n-1}$, $\mathbf{HP}^n \not\subset S^{4n+1}$ provided $n \geq 2$.

(27.18) *Exercise.* Show that the topologist's sine curve (p. 56) without its lower arc is not tautly embedded in the plane.

28. Lefschetz Duality

An n-dimensional *manifold-with-boundary* is a space X locally homeomorphic to the Euclidean half-space

$$\mathbf{R}^n_+ = \{(x_1, \ldots, x_n) \in \mathbf{R}^n \mid x_n \geq 0\}$$

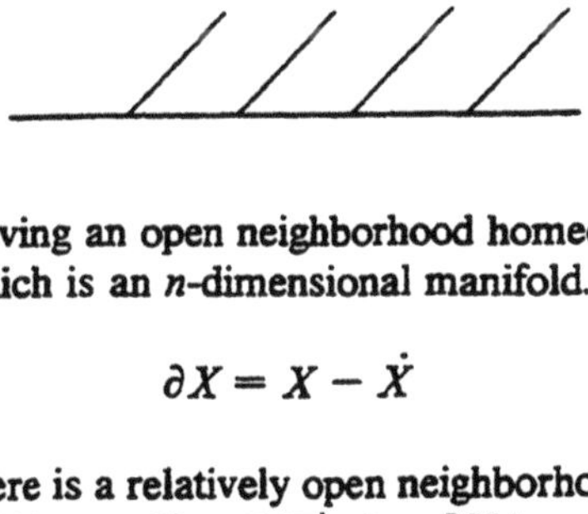

Those points in X having an open neighborhood homeomorphic to $\dot{E}^n$ form an open subset $\dot{X}$ which is an n-dimensional manifold. Let

$$\partial X = X - \dot{X}$$

For each $x \in \partial X$, there is a relatively open neighborhood homeomorphic to an open set in $\{x \in \mathbf{R}^n \mid x_n = 0\} \approx \mathbf{R}^{n-1}$; thus ∂X is an $(n-1)$-dimensional manifold.

(28.1) *Example* 1: The closed Möbius band. Its boundary is a circle.

(28.2) *Example* 2: The closed n-dimensional annulus

$$\{x \in \mathbf{R}^n; a \leq |x| \leq b\}$$

where $a > 0$. Its boundary is the disjoint union of two $(n-1)$-spheres.

(28.3) *Example* 3: E^n

(28.4) *Example* 4: $Y \times I$, where Y is any manifold without boundary.

(28.5) *Example* 5: Any open subspace of a manifold-with-boundary.

(28.6) *Note.* A complete classification of compact surfaces-with-boundary is given in Ahlfors and Sario [2], Chap. 1, or Massey [67].

Let V be open in X, so that $\dot{V} = V \cap \dot{X}$, $\partial V = V \cap \partial X$. Let $\Gamma(\dot{V})$ be the module of sections over $\dot{V}$ of the orientation sheaf of $\dot{X}$, $\Gamma(\partial V)$ the module of sections over ∂V of the orientation sheaf of ∂X (22.18 and 22.2).

(28.7) *Proposition. There are unique homomorphisms*

$$\partial_V : \Gamma(\dot{V}) \to \Gamma(\partial V)$$

which are compatible with restriction to smaller V and which take local orientations of $\dot{X}$ along $\dot{V}$ into local orientations of ∂X along ∂V.

(28.8) *Corollary. If $\dot{X}$ is R-orientable so is ∂X.*

In fact, (28.7) shows how an R-orientation of $\dot{X}$ induces one of ∂X.

Proof of the proposition: The question being local, it suffices to verify (28.7) for the opens in a basis. These are of two types:

Type 1. $V \subset \dot{X}$. Set $\partial_V = 0$.

Type 2. V homeomorphic to the half-ball $\dot{E}^n \cap \mathbf{R}^n_+$. In this case we have several natural isomorphisms.

(1) For any $x \in \dot{V}$, $X - \dot{V}$ is a strong deformation retract of $X - x$.

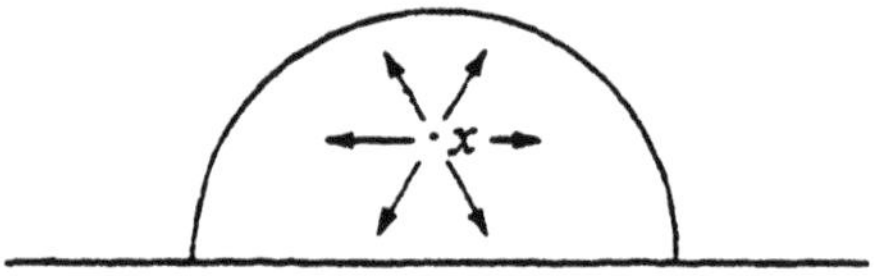

Hence $H_n(X, X - \dot{V}) \to H_n(X, X - x) \simeq H_n(\dot{X}, \dot{X} - x)$ is an isomorphism.

(2) Since $\dot{V}$ is an open n-cell, the evaluation homomorphism

$$\Gamma\dot{V} \to H_n(\dot{X}, \dot{X} - x)$$

is an isomorphism. Combining with (1) gives a commutative diagram of isomorphisms

$$\begin{array}{ccc} H_n(X, X - \dot{V}) & \simeq & \Gamma\dot{V} \\ \downarrow\wr & & \downarrow\wr \\ H_n(X, X - x) & \simeq & H_n(\dot{X}, \dot{X} - x) \end{array}$$

(3) Consider the exact homology sequence of the triple $(X, X - \dot V, X - V)$. We have $H_q(X, X - V) = H_q(X, X) = 0$ for all q, because $X - V$ is a strong deformation retract of X.

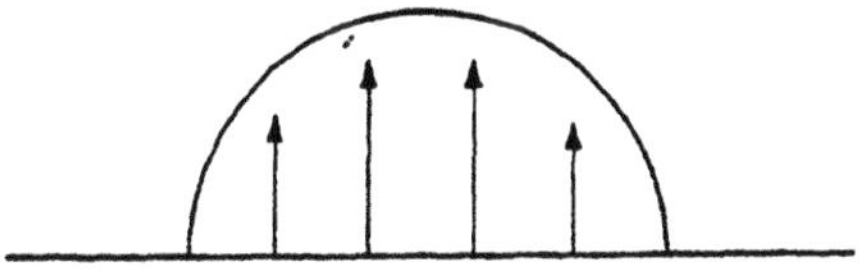

Hence $\partial : H_n(X, X - \dot V) \to H_{n-1}(X - \dot V, X - V)$ is an isomorphism.

(4) Suppose $x' \in \partial V$. Then $X - V$ is a strong deformation retract of $X - \dot V - x'$.

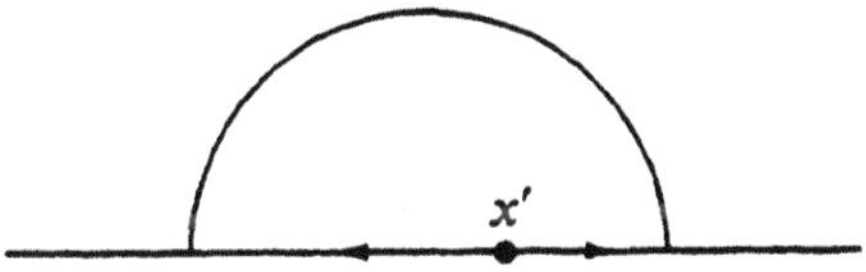

Similarly $\partial X - \partial V$ is a strong deformation retract of $\partial X - x'$.

(5) $X - \dot V - x'$ is relatively open in $X - \dot V$. The closure of $X - \dot V$ is $X - V \subset X - \dot V - x'$, hence $X - \dot V$ may be excised. Combining this isomorphism with those of (4) gives a commutative diagram of isomorphisms

$$\begin{array}{ccc} \Gamma \partial V & \xrightarrow{\ \sim\ } & H_{n-1}(X - \dot V, X - V) \\ \downarrow \wr & & \downarrow \wr \\ H_{n-1}(\partial X, \partial X - x') \xleftarrow{\ \simeq\ } H_{n-1}(\partial X, \partial X - \partial V) & \xrightarrow{\ \simeq\ } & H_{n-1}(X - \dot V, X - \dot V - x') \end{array}$$

(the left vertical isomorphism being the evaluation of sections at the point x' [∂V is an open $(n-1)$-cell in ∂x]).

(6) Combining (2), (3), and (5) gives an isomorphism

$$\partial_V : \Gamma \dot V \to \Gamma \partial V$$

which is clearly compatible with restriction to smaller V's of types 1 or 2. ■

(28.9) *Note.* The converse of (28.8) is false, as is shown by Example 1.

(28.10) We next imbed X as a closed subspace of an n-dimensional manifold without boundary in functorial way.

Let X_1, X_2 be two copies of X, and form the disjoint union

$$X_1 \amalg (\partial X \times I) \amalg X_2$$

Then identify any $x' \in \partial X_1$ with $(x', 0)$, and identify any $x' \in \partial X_2$ with $(x', 1)$. Let $2X$ be the quotient space. Then each X_i is mapped homeomorphically onto a closed subspace of $2X$. Clearly $2X$ is an n-dimensional manifold, compact if X is.

(28.11) *Examples*: 2(Möbius) = Klein bottle

2(Annulus of dim. 2) = Torus

$2E^n = S^n$

$2(S^1 \times I) =$ Torus

(28.12) *Exercise*. If $\dot{X}$ is R-oriented, then there is a unique R-orientation of $2X$ inducing the given R-orientation on $\dot{X}_1$ and $\dot{X}_2$.

Let $Y_1 = X_1 \cup (\partial X \times [0, 1))$, $Y_2 = (\partial X \times (0, 1]) \cup X_2$ (identifying with the images in $2X$). then X_i is a deformation retract of Y_i, $i = 1, 2$. And $Y_1 \cap Y_2 = \partial X \times (0, 1)$ is homotopically equivalent to ∂X. Since each Y_i is open in $2X$, $(2X, Y_1, Y_2)$ is an exact triad. Applying (20.8) to the Mayer-Vietoris sequence, we get

$$\chi(2X) = \chi(Y_1) + \chi(Y_2) - \chi(Y_1 \cap Y_2)$$

or

$$\chi(2X) = 2\chi(X) - \chi(\partial X)$$

where the Euler characteristic may be taken relative to any PID R.

(28.13) *Corollary. If X is compact and $\dot{X}$ is R-orientable* (*where R is a* PID), *then $\chi(\partial X; R)$ is even.*

Proof: If n is even, then ∂X is odd-dimensional R-orientable, so $\chi(\partial X; R) = 0$ (26.10). If n is odd, $\chi(2X; R) = 0$ for the same reason, whence $\chi(\partial X; R) = 2\chi(X; R)$. ■

(28.14) *Example*: $\mathbf{P}^{2n}$, $\mathbf{CP}^{2n}$, and $\mathbf{HP}^{2n}$ are not boundaries; ($\chi(\mathbf{P}^{2n}; \mathbf{Z}/2) = 1$, $\chi(\mathbf{CP}^{2n}) = \chi(\mathbf{HP}^{2n}) = 2n + 1$). The compact orientable surfaces T_g are all boundaries since they sit in 3-space. It's a deep theorem that all compact orientable 3-folds are boundaries (see Rohlin [47]).

(28.15) *Proposition. Let X be compact, $s \in \Gamma\dot{X}$ an R-orientation of $\dot{X}$.*

Then there is a unique homology class $\zeta \in H_n(X, \partial X)$ such that for any $x \in \dot{X}$,

$$s(x) = j_x^{\dot{X}}(\zeta)$$

Proof: There is a canonical homomorphism $j : H_n(X, \partial X) \to \Gamma\dot{X}$ given by

$$j(\alpha)(x) = j_x^{\dot{X}}(\alpha)$$

(verify $j(\alpha)$ is continuous). We claim j is an isomorphism: Identify $X = X_1$, so that X is a closed subspace of the open manifold $Y = Y_1$ and is a deformation retract of Y (28.10). Then we have the commutative diagram

$$\begin{array}{ccc}
H_n(X, X - \dot{X}) & \searrow^{j} & \\
\downarrow & & \Gamma\dot{X} \\
H_n(Y, Y - \dot{X}) & & \uparrow r \\
\downarrow & & \\
H_n(Y, Y - X) & \xrightarrow{j} & \Gamma X
\end{array}$$

where the left vertical arrows are isomorphisms ($Y - \dot{X}$ is homotopically equivalent to $Y - X$) and r is the restriction of sections over X of the orientation sheaf of Y to $\dot{X}$. Since X is compact, the bottom j is an isomorphism (22.24). Thus we are reduced to proving the following claim:

Sublemma. r is an isomorphism.

This follows from the fact that if $x' \in \partial X$, V a hemispherical neighborhood, W a coordinate neighborhood of x' in Y with $W \supset V$, then any section over V extends uniquely to W. ■

(28.16) *Corollary. (Same hypothesis.) Then $\partial\zeta \in H_{n-1}(\partial X)$ is the fundamental class for the induced orientation of ∂X.*

Proof: The proof of (28.7) shows that the diagram

$$\begin{array}{ccc}
H_n(X, \partial X) & \xrightarrow{j} & \Gamma(X) \\
\downarrow \partial & & \downarrow \partial_X \\
H_{n-1}(\partial X) & \xrightarrow{j} & \Gamma(\partial X)
\end{array}$$

is commutative. ■

(28.17) *Example*: Let X be the (closed) Möbius band, $\phi: I^2 \to X$ the quotient map, so that $\phi(0, t) = \phi(1, 1 - t)$ for all t. The loops

$$\sigma(s) = \phi(s, \tfrac{1}{2})$$

$$\tau(s) = \begin{cases} \phi(2s, 0) & s \leq \frac{1}{2} \\ \phi(2s - 1, 1) & s \geq \frac{1}{2} \end{cases}$$

represent the equator of X and the boundary of X respectively. Now the equator is a strong deformation retract of X, the deformation being given by

$$F_u(\phi(s, t)) = \phi(s, \tfrac{1}{2} + u(t - \tfrac{1}{2}))$$

Since $F_0\tau = \sigma^2$, we see that τ is homotopic to σ^2. It follows that $H_2(X) = 0$, while the inclusion $H_1(\partial X) \to H_1(X)$ becomes multiplication by 2 when we identify these modules with R. The exact homology sequence then shows that $H_2(X, \partial X) = 0$ when $R = \mathbf{Z}$, so that X is indeed non-orientable.

(28.18) *Lefschetz Duality Theorem. Let X be a compact manifold-with-boundary of dimension n, and let $\dot{X}$ be R-oriented. Let $\zeta \in H_n(X, \partial X)$ be the fundamental class. Then the diagram*

$$\begin{array}{ccccccccc}
\to & H^{q-1}(X) & \to & H^{q-1}(\partial X) & \xrightarrow{\delta} & H^q(X, \partial X) & \to & H^q(X) & \to \\
& \zeta\cap\downarrow & & (\partial\zeta)\cap\downarrow & & \zeta\cap\downarrow & & \downarrow & \\
\to & H_{n-q-1}(X, \partial X) & \xrightarrow[\partial]{} & H_{n-q}(\partial X) & \to & H_{n-q}(X) & \to & H_{n-q}(X, \partial X) & \to
\end{array}$$

is sign-commutative and the vertical arrows are isomorphisms.

Proof: Specifically, the left square is commutative only up to a factor of $(-1)^{q-1}$ (24.21), while the middle and right squares are actually commutative.

Note that the homomorphism $H^{q-1}(\partial X) \to H_{n-q}(\partial X)$ is the Poincaré Duality isomorphism (26.6). If we check the isomorphism $H^q(X) \to H_{n-q}(X, \partial X)$, the other isomorphism will follow from the 5-lemma (14.7).

Identify X with $X_1 \subset 2X$ (29.3). Then excision and deformation retraction give isomorphisms

$$H_{n-q}(X, \partial X) \cong H_{n-q}(\bar{Y}_1, \overline{Y_1 \cap Y_2}) \cong H_{n-q}(2X, 2X - X)$$

and under this isomorphism the fundamental class in $H_n(X, \partial X)$ (28.15) corresponds to the fundamental class in $H_n(2X, 2X - X)$ (28.12, 22.21, 22.24). But by Alexander Duality (27.5), cap product with the fundamental class gives an isomorphism

$$\check{H}^q(X) \xrightarrow{\approx} H_{n-q}(2X, 2X - X)$$

In this case, $\check{H}^q(X) = H^q(X)$ because X is a deformation retract of its neighborhoods $X \cup (\partial X \times [0, \varepsilon))$ in $2X$. ■

(28.19) *Exercise.* Let X be 3-dimensional with $H_1(X) = 0$. Then ∂X is a disjoint union of 2-spheres (show $H_1(\partial X) = 0$).

(28.20) *Exercise.* Suppose Y is a compact oriented manifold of dimension $4k$. Then there is a symmetric bilinear form β on $H^{2k}(Y)$ given by

$$\beta(\xi, \eta) = [\zeta, \xi \cup \eta]$$

where ζ is the fundamental homology class of dimension $4k$. Suppose Y is a boundary, $Y = \partial X$, where X is compact. Let A^{2k} be the image of $H^{2k}(X)$ under the inclusion $H^{2k}(X) \to H^{2k}(Y)$. Then A^{2k} is its own orthogonal complement under β.

In case R is the field of real numbers, this result can be expressed by saying that the *index* of the quadratic form induced by β is zero; this gives a necessary condition for a $4k$-dimensional manifold to bound. (See Hirzebruch [31], Section 8, for a formula for the index.)

(28.21) *Note.* The theory of bounding manifolds is called *cobordism*. Two compact n-dimensional manifolds are in the same cobordism class if their disjoint union is a boundary. Using the technique of (19.39) this can be shown to be an equivalence relation. The operation "disjoint union" defines an addition on the set $\mathcal{N}_n$ of n-dimensional cobordism classes making it into a vector space over $\mathbf{Z}/2$ (the zero class consists of the bounding manifolds). The topological product of manifolds induces a bilinear map

$$\mathcal{N}_p \times \mathcal{N}_q \to \mathcal{N}_{p+q}$$

by means of which the direct sum $\mathcal{N}$ of all the $\mathcal{N}_q$ becomes a graded commutative algebra over $\mathbf{Z}/2$. This algebra has been determined for the differentiable category, i.e., compact *differentiable* manifolds. In that case Thom showed that $\mathcal{N}$ is a free-commutative algebra generated by indeterminates x_q of degree q for all q not of the form $2^k - 1$; for q even, he showed that x_q is the class of projective space $\mathbf{P}^q$. The method of proof is to show that $\mathcal{N}_q$ is isomorphic to a certain homotopy group which one can compute (Thom [57]; also Milnor [41]). For q odd $\neq 2^k - 1$, the class x_q was computed by Dold [19] as follows: Write $q + 1 = 2^r(2s + 1)$. Then x_q is the class of $\mathbf{P}(2^r - 1, 2^r s)$, where in general $\mathbf{P}(m, n)$ is the manifold obtained from $S^m \times \mathbf{CP}^n$ by identifying (x, z) with $(-x, \bar{z})$. (See also Stong [72].)

There is also a cobordism theory for compact oriented differentiable manifolds. Define $-X$ to be X with the opposite orientation. Then X and Y are said to be in the same oriented cobordism class if the disjoint union of X and $-Y$ is an oriented boundary [comment after (28.8)]. Giving the topological product the product orientation (see Section 29), we obtain a graded skew-commutative ring

$$\Omega = \oplus_{q \geq 0} \Omega_q$$

The algebra $\Omega \oplus \mathbf{Q}$ over the rational numbers was determined by Thom; it is a free-skew-commutative algebra generated by the classes of even-dimensional complex projective spaces $\mathbf{CP}^{2k}$. Thus if q is not divisible by 4, Ω_q is a torsion group. The torsion part of Ω has been determined by Wall [58] and Milnor [40]. For small values of q we have

$$\Omega_q = \begin{cases} 0 & q = 2, 3 \\ \mathbf{Z} & q = 4 \\ \mathbf{Z}/2 & q = 5 \end{cases}$$

For any pair (Y, B), by considering all maps $(X, \partial X) \to (Y, B)$ for all X, a new homology theory for (Y, B) is constructed, called *bordism*. This theory satisfies all the Eilenberg-Steenrod axioms except that the bordism groups of a point are non-trivial (see Conner and Floyd [15]). A survey of recent developments is given by R. J. Milgram in Proc. Symp. Pure Math. Vol. 32 (1978) 79–89.

(28.22) *Note.* For a deeper study of the material in sections 26–30, see Dold [64]. For the generalization of the duality theorems to extraordinary homology theories, see Adams [74] or Gray [76].

(28.23) *Exercise.* If M is compact, ∂M is not a retract of M.

(28.24) *Exercise.* If M is the boundary of a contractible manifold W then M and $2W$ have the homology of spheres.

(28.25) *Exercise.* Suppose M is orientable and $\dim M = 2n + 1$. Let $i{:}\partial M \to M$ be inclusion. Let $K = \ker H_n(i){:}H_n(\partial M) \to H_n(M)$ where coefficients are in a field F. Prove $2 \dim K = \dim H_n(\partial M)$. If M is nonorientable, the result is true for $F = \mathbf{Z}/2\mathbf{Z}$.

(28.26) *Exercise.* ([79], p. 63). Let M be a compact 3-manifold such that ∂M contains no 2-spheres. Prove: if either (a) M is orientable and $\partial M \neq \emptyset$ or (b) M is nonorientable and ∂M (possibly empty) contains no projective planes, then $H_1(M; \mathbf{Z})$ is infinite.

Part IV
PRODUCTS AND LEFSCHETZ FIXED POINT THEOREM

Introduction to Part IV

We have a number of means at our disposal for systematically calculating the values of homology and cohomology *modules*. No such general devices are available for cup products. In practice two broad lines of attack are perceptible. On the one hand, special structure on the space in question may place restrictions on the possible cohomology *algebras*. For manifolds, the duality theorems impose strong restrictions. On the other hand, by working in an abstract setting, one can construct auxiliary products and determine many relations among them. These in turn can often be used to determine the cohomology algebra. This part develops some basic aspects of the formal procedure.

In chapter 29, cross products and slant products are introduced. Formal relations among these and the cup and cap products of chapter 24 are developed. Some relations, which hold on the level of cohomology, hold only up to chain homotopy on the underlying chain complex. It would be a considerable nuisance to have to make explicit such chain homotopies or even to give an argument like that for 24.8. Instead we use the acyclic model theorem. This is the most abstract piece of reasoning in the text, however *all* the vital details were already present in the proof of homotopy invariance (11.4).

Among the applications of the formal machinery is the description of the cohomology ring of a product of spaces in terms of the tensor product of the cohomology rings of the factors.

We have not tried to be complete in our elaboration of all the formal relations holding among the various products. The texts by Dold [64] and Spanier [52] provide much more information. The account in J. F. Adams [62] gives a good overview.

The *Thom complex*, introduced in chapter 30, has proved important in the study of manifolds and also in homotopy theory. The utility of the products introduced in chapter 29 reflects itself in the theorems of this section.

In chapter 31, a brief introduction to intersection theory is given. The Poincaré duality isomorphism receives a more geometric interpretation in the case where homology classes are represented by embedded submanifolds. Relaxing the manner in which homology is represented would lead to the geometric view of Poincaré duality afforded by the combinatorial approach [86]. We do not carry this out in general, but content ourselves with a discussion of the lens spaces.

29. Products

We have seen (7.11) that the nth homotopy group of a product of pointed spaces is canonically isomorphic to the product of the nth homotopy groups of the factors. However, the behavior of the homology modules with respect to products is more complicated. For example, $H_2(S^1 \times S^1) \cong R$, while $H_2(S^1) \times H_2(S^1) = 0$.

The analysis breaks into two parts. First is the geometric step relating $S(X \times Y)$ with $S(X) \otimes S(Y)$. Second is the algebraic step relating $H.(C \otimes C')$ with $H.(C) \otimes H.(C')$. In the case *when R is a field,* the *graded* homology module $H.(X \times Y)$ is isomorphic to the *tensor* product $H.(X) \otimes H.(Y)$.

Every singular n-simplex ω in $X \times Y$ is uniquely expressible in the form (σ, τ), where $\sigma = p_X\omega$, $\tau = p_Y\omega$ (p_X, p_Y the projections of $X \times Y$ on X, Y). For each integer p with $0 \leq p \leq n$, we can associate to (σ, τ) the element $\sigma\lambda_p \otimes \tau\rho_{n-p}$ of the tensor product $S_p(X) \otimes S_{n-p}(Y)$ (front p-face of σ tensored with the back $(n-p)$-face of τ; see Section 24). Extending by linearity gives a homomorphism $S_n(X \times Y) \to S_p(X) \otimes S_{n-p}(Y)$. We can then add all of these homomorphisms: Let

$$[S(X) \otimes S(Y)]_n = \bigoplus_{p=0}^{n} S_p(X) \otimes S_{n-p}(Y)$$

and define *the Alexander-Whitney homomorphism*

$$A: S_n(X \times Y) \to [S(X) \otimes S(Y)]_n$$

by

$$A(\sigma, \tau) = \sum_{p=0}^{n} \sigma\lambda_p \otimes \tau\rho_{n-p}.$$

(29.1) *Lemma. The Alexander-Whitney homomorphism is functorial in* (X, Y).

Proof. This means: Given maps $f:X \to X'$, $g:Y \to Y'$, they induce the map $f \times g: X \times Y \to X' \times Y'$; then the diagram

$$\begin{array}{ccc} S_n(X \times Y) & \xrightarrow{A} & [S(X) \otimes S(Y)]_n \\ {\scriptstyle S_n(f \times g)}\big\downarrow & & \big\downarrow{\scriptstyle \bigoplus_p S_p(f) \otimes S_{n-p}(g)} \\ S_n(X' \times Y') & \xrightarrow{A} & [S(X') \otimes S(Y')]_n \end{array}$$

is commutative. The proof is immediate from the definitions. ■

(29.2) We now make the sequence of modules $[S(X) \otimes S(Y)]_n$ into an algebraic chain complex by defining a boundary operator. In general, let C, C' be algebraic chain complexes, and set

$$[C \otimes C']_n = \bigoplus_{p=0}^{n} C_p \otimes C'_{n-p}$$

as before. Then $\partial:[C \otimes C']_n \to [C \otimes C']_{n-1}$ is defined by

$$\partial(z \otimes z') = \partial z \otimes z' + (-1)^p z \otimes \partial z'$$

for $z \in C_p$, $z' \in C'_{n-p}$, and extended by linearity. Clearly $\partial\partial = 0$.

(29.3) *Lemma. A is a chain homomorphism, i.e.,*

$$A\partial = \partial A$$

Proof: The proof is by a computation essentially identical to that in the proof of (24.2). ■

We can form the homology modules $H_n(C \otimes C')$ as usual by taking the kernel of ∂ modulo the image of the preceding ∂. (29.3) implies that A induces a functorial homomorphism

$$\bar{A} : H_n(X \times Y) \to H_n(S(X) \otimes S(Y))$$

(29.4) *Eilenberg-Zilber Theorem. $\bar{A}$ is an isomorphism. In fact, A is a* *chain equivalence of the chain complex $S(X \times Y)$ with the chain complex $S(X) \otimes S(Y)$.*

To say that A is a *chain equivalence* means that there is a chain homomorphism $B:S(X) \otimes S(Y) \to S(X \times Y)$ such that AB and BA are both chain homotopic to the identity. The construction of B and the chain homotopies is by the method of acyclic models—see Appendix (29.23A).

We next relate the homology of the algebraic complex $S(X) \otimes S(Y)$ to the tensor product of the homology of X with the homology of Y. If z is a p-cycle on X, w an $(n-p)$-cycle on Y, then $z \otimes w$ is a cycle in $[S(X) \otimes S(Y)]_n$, and we get a well-defined bilinear pairing

$$H_p(X) \times H_{n-p}(Y) \to H_n(S(X) \otimes S(Y))$$

by $(\bar{z}, \bar{w}) \to \overline{z \otimes w}$. Hence there is a unique homomorphism

$$H_p(X) \otimes H_{n-p}(Y) \to H_n(S(X) \otimes S(Y))$$

sending $\bar{z} \otimes \bar{w}$ into $\overline{z \otimes w}$. If, as before, we set

$$[H(X) \otimes H(Y)]_n = \bigoplus_{p=0}^{n} H_p(X) \otimes H_{n-p}(Y)$$

we obtain a homomorphism

$$i:[H(X) \otimes H(Y)]_n \to H_n(S(X) \otimes S(Y))$$

given by

$$i\left(\sum_{p=0}^{n} \bar{z}_p \otimes \bar{w}_{n-p}\right) = \sum_{p=0}^{n} \overline{z_p \otimes w_{n-p}}$$

Clearly i is functorial in the pair (X, Y). The image of an element $\zeta \otimes \omega$ under the composite homomorphism

$$\bar{B}i:H_p(X) \otimes H_q(Y) \to H_{p+q}(X \times Y)$$

will be denoted $\zeta \times \omega$.

When R is a PID, we will show that i is a monomorphism and we will determine its cokernel. For later applications, we will treat a more general situation: Replace $S(X)$ and $S(Y)$ by arbitrary algebraic chain complexes C, C'. Moreover, we take C and C' to be indexed by all the integers, not just the non-negative ones; this allows us to simultaneously treat algebraic cochain complexes, setting $C_{-n} = C^n$ to convert a cochain complex into a chain complex. Let

$$[C \otimes C']_n = \bigoplus_{p+q=n} C_p \otimes C'_q$$

where now p and q may take on negative values, and define the boundary operator for the complex $C \otimes C'$ as in (29.2). We obtain the canonical chain homomorphism

$$i{:}H(C) \otimes H(C') \to H(C \otimes C')$$

as before. We determine the kernel and cokernel of i under the assumptions that R is a PID and C is free [i.e., all C_n are free modules, satisfied by $C_n = S_n(X)$].

Let Z, B be the complexes of cycles and boundaries of C. Let B^- be the complex given by

$$B_q^- = B_{q-1}$$

Then we have an exact sequence of chain complexes

(29.5) $$0 \to Z \to C \xrightarrow{\partial} B^- \to 0$$

Since B^- is free by our assumptions, this sequence splits, hence tensoring with C' gives another exact sequence of chain complexes

$$0 \to Z \otimes C' \to C \otimes C' \to B^- \otimes C' \to 0$$

By the usual argument, this sequence induces an infinite exact homology sequence

(29.6) $$\to H_n(Z \otimes C') \to H_n(C \otimes C') \to H_n(B^- \otimes C')$$
$$\to H_{n-1}(Z \otimes C') \to$$

Now Z is a free complex whose boundary operator is identically zero. In this case $Z = H(Z)$, and we obtain the following result:

Sublemma. $i{:}Z \otimes H(C') \to H(Z \otimes C')$ is an isomorphism.

Proof: The boundary operator on $Z \otimes C'$ is given by

$$\partial(z \otimes w) = (-1)^p z \otimes \partial w$$

hence, the proof of the sublemma comes down to showing that the sequence

$$0 \to Z_p \otimes B'_q \to Z_p \otimes Z'_q \to Z_p \otimes H_q(C') \to 0$$

is exact, which follows from the fact that Z_p is free. ■

Similarly, $i{:}B^- \otimes H(C') \to H(B^- \otimes C')$ is an isomorphism. Substitute these isomorphic modules in the exact sequence (29.6), obtaining

$$\to [B \otimes H(C')]_n \to [Z \otimes H(C')]_n \to H_n(C \otimes C')$$
$$\to [B \otimes H(C')]_{n-1} \to$$

In other words, we have the exact sequence

(29.7) $$0 \to \text{Coker}\,\phi_n \to H_n(C \otimes C') \to \text{Ker}\,\phi_{n-1} \to 0$$

where $\phi_n{:}[B \otimes H(C')]_n \to [Z \otimes H(C')]_n$ is the connecting homomorphism.
To determine Coker ϕ_n and Ker ϕ_{n-1}, we use the exact sequence

(29.8) $$0 \to B_p \xrightarrow{j_p} Z_p \to H_p(C) \to 0$$

which need not split. Tensoring with $H_{n-p}(C')$ gives an exact sequence

(29.9) $$B_p \otimes H_{n-p}(C') \xrightarrow{j_p \otimes 1} Z_p \otimes H_{n-p}(C')$$
$$\to H_p(C) \otimes H_{n-p}(C') \to 0$$

where the left arrow needn't be a monomorphism; in fact, since (29.8) is a free resolution of $H_p(C)(Z_p$ is free), its kernel is by definition

$$\text{Tor}(H_p(C), H_{n-p}(C')).$$

Adding these sequences for all p gives the exact sequence (since) $\phi_n = \oplus_p j_p \otimes 1$).

$$0 \to \bigoplus_p \text{Tor}(H_p(C), H_{n-p}(C')) \to [B \otimes H(C')]_n \xrightarrow{\phi_n}$$
$$\xrightarrow{\phi_n} [Z \otimes H(C')]_n \to [H(C) \otimes H(C')]_n \to 0$$

which gives us Ker ϕ_n and Coker ϕ_n. We state the result.

(29.10) *The Künneth Exact Sequence. Assume R is a* PID *and C is free. For all n, we have the exact sequence*

$$0 \to \bigoplus_p H_p(C) \otimes H_{n-p}(C') \to H_n(C \otimes C')$$
$$\to \bigoplus_p \text{Tor}(H_p(C), H_{n-p-1}(C')) \to 0$$

This sequence actually splits (but non-canonically): For sequence (29.5) splits, so we have a projection $\pi: C \to Z$; similarly, we have $\pi': C' \to Z'$. Composing these with the quotient homomorphisms $\psi: C \to H(C)$, $\psi': C' \to H(C')$, hence a homomorphism

$$H(\psi \otimes \psi'): \mathrm{H}(C \otimes C') \to H(H(C) \otimes H(C')) = H(C) \otimes H(C')$$

which is a projection making $H(C) \otimes H(C')$ a direct summand.

In the geometric case $C = S(X)$, $C' = S(Y)$, we can use the Eilenberg-Zilber theorem to substitute for $H_n(S(X) \otimes S(Y))$ and obtain the homology of $X \times Y$.

(29.11) *The Künneth Formula. Assume R is a* PID. *Then*

$$H_n(X \times Y) \cong \bigoplus_{p=0}^{n} H_p(X) \otimes H_{n-p}(Y) \oplus \bigoplus_{p=0}^{n} \mathrm{Tor}(H_p(X), H_{n-p-1}(Y))$$

(29.11.1) *Corollary. If all the homology modules of* Y *(or of* X*) in dimensions* $< n$ *are free (e.g., if* R *is a field), then*

$$H_n(X \times Y) \cong \bigoplus_{p=0}^{n} H_p(X) \otimes H_{n-p}(Y)$$

Proof: In that case (29.9) is exact with a zero on the left, hence $\mathrm{Tor}(H_p(X), H_{n-p-1}(Y)) = 0$. ■

(29.11.2) *Corollary.* (*R a* PID.) *If the Euler characteristics* $\chi(X; R)$, $\chi(Y; R)$ *are defined, then* $\chi(X \times Y; R)$ *is defined and*

$$\chi(X \times Y; R) = \chi(X; R)\chi(Y; R)$$

As special case of (29.10), let $C = S(X, \mathbf{Z})$, and let C' be the chain complex given by $C'_0 = R$, $C'_n = 0$ for $n \neq 0$, where R is any commutative ring. In this case

$$[C \otimes C']_n = S_n(X; \mathbf{Z}) \otimes R = S_n(X; R)$$

so that we obtain

(29.12) *Universal Coefficient Theorem. The sequence*

$$0 \to H_n(X; \mathbf{Z}) \otimes R \to H_n(X; R) \to \mathrm{Tor}(H_{n-1}(X; \mathbf{Z}), R) \to 0$$

is split exact.

For example, if $R = \mathbf{Z}/2$, $\mathrm{Tor}(H_{n-1}(X; \mathbf{Z}), \mathbf{Z}/2)$ is the subgroup of $H_{n-1}(X; \mathbf{Z})$ of elements of order 2, while $H_n(X; \mathbf{Z}) \otimes \mathbf{Z}/2$ is the cokernel of multiplication by 2 on $H_n(X; \mathbf{Z})$.

(29.13) If the homology modules are finitely generated, we can determine the Tor's more explicitly in terms of the cyclic components. By an argument completely analogous to that in (23.23), one can show that $\mathrm{Tor}(M, N)$ is bi-additive in (M, N). Since the free summands don't contribute anything (29.12), we are reduced to computing $\mathrm{Tor}(R/a, R/b)$, where a and b are non-zero. Just as in proof of (23.21), we can choose any free resolution to compute Tor, so we use

$$0 \to R \xrightarrow{a^*} R \to R/a \to 0$$

where a^* is multiplication by a. Tensoring with R/b gives

$$0 \to \mathrm{Tor}(R/a, R/b) \to R/b \xrightarrow{a^*} R/b \to R/d \to 0$$

where d is the greatest common divisor of a and b. If m is the least common multiple of a and b, the kernel of multiplication by a on R/b is $(m/a)R/bR \cong R/d$. Thus if the torsion submodules of M, N have the respective decompositions

$$\bigoplus_i R/a_i, \quad \bigoplus_j R/b_j$$

and $d_{ij} = gcd(a_i, b_j)$, then

$$\mathrm{Tor}(M, N) \cong \bigoplus_{i,j} R/d_{ij}$$

(29.14) *Example*: Consider the manifolds $\mathbf{P}^3 \times S^2$, $\mathbf{P}^2 \times S^3$. Since the homology of a sphere is free, we obtain (29.11.1)

$$H_n(\mathbf{P}^2 \times S^3; \mathbf{Z}) \cong \begin{cases} \mathbf{Z} & n = 0 \\ \mathbf{Z}/2 & n = 1 \\ 0 & n = 2 \\ \mathbf{Z} & n = 3 \\ \mathbf{Z}/2 & n = 4 \\ 0 & n \geq 5 \end{cases}$$

$$H_n(\mathbf{P}^3 \times S^2; \mathbf{Z}) \cong \begin{cases} \mathbf{Z} & n = 0 \\ \mathbf{Z}/2 & n = 1 \\ \mathbf{Z} & n = 2 \\ \mathbf{Z} \oplus \mathbf{Z}/2 & n = 3 \\ 0 & n = 4 \\ \mathbf{Z} & n = 5 \\ 0 & n = 6 \end{cases}$$

Thus these manifolds are not homotopically equivalent. However,

$$\pi_1(\mathbf{P}^2 \times S^3) \cong \mathbf{Z}/2 \cong \pi_1(\mathbf{P}^3 \times S^2)$$

and for $n > 1$ we have (7.12)

$$\pi_n(\mathbf{P}^2 \times S^3) \cong \pi_n(S^2) \times \pi_n(S^3) \cong \pi_n(S^2 \times \mathbf{P}^3)$$

so that all the homotopy groups are the same. (A theorem of Whitehead gives a sufficient condition for two spaces with the same homotopy groups to have the same homology groups. See Spanier [52], Chap. 7, Section 5.)

(29.15) *Example*: Consider the 6-dimensional manifolds $S^2 \times S^4$ and $\mathbf{CP}^3$. By (29.11.1) and (19.21), they both have the same homology. However, their homotopy groups are different, e.g.,

$$\pi_4(S^2 \times S^4) \cong \pi_4(S^2) \times \pi_4(S^4)$$

contains the subgroup $1 \times \pi_4(S^4) \cong \mathbf{Z}$, whereas the exact sequence of the fibre space $S^7 \to \mathbf{CP}^3$ shows

$$\pi_4(\mathbf{CP}^3) \cong \pi_4(S^7) = 0$$

(See Spanier [52], Chap. 7, Section 2.)

(29.16) *Remark*. In the relative case, (29.10) yields the exact sequence

$$0 \to \bigoplus_{p=0}^{n} H_p(X, A) \otimes H_{n-p}(Y, B) \to H_n(S(X, A) \otimes S(Y, B)) \to$$

$$\to \bigoplus_{p=0}^{n-1} \operatorname{Tor}(H_p(X, A), H_{n-p-1}(Y, B)) \to 0$$

In case one of A, B is empty, say B, then an easy argument shows that the Alexander-Whitney homomorphism induces an equivalence of chain complex $S(X \times Y)/S(A \times Y)$ with chain complex $S(X, A) \otimes S(Y)$, hence an isomorphism

$$H_n(S(X, A) \otimes S(Y)) \cong H_n(X \times Y, A \times Y)$$

for all n. On the other hand, in the case where $(X \times Y, A \times Y, X \times B)$ is an exact triad (e.g., if A and B are open), then joining the previous case to the relative Mayer-Vietoris sequence via the five lemma yields the isomorphism

$$H_n(S(X, A) \otimes S(Y, B)) \cong H_n(X \times Y, A \times Y \cup X \times B)$$

for all n.

(29.17) *Cup and cross products*. The basis for using the previous material to study cup products is their description (due to Lefschetz) as compositions. The cup product $c \cup d$ of cochains $c \in S^p(X)$, $d \in S^q(X)$ is the composite

$$S(X) \xrightarrow{S(\Delta)} S(X \times X) \xrightarrow{A} S(X) \otimes S(X) \xrightarrow{c \otimes d} R \otimes R \xrightarrow{m} R$$

where $\Delta: X \to X \times X$ is the diagonal and m is multiplication in R. This is readily checked. On singular n-simplices σ $(n = p + q)$, $S(\Delta)(\sigma) = (\sigma, \sigma)$ and

$$m(c \otimes d)A(\sigma, \sigma) = m(c \otimes d)(\sum_{i+j=n} \sigma\lambda_i \otimes \sigma\rho_j)$$

$$= [\sigma\lambda_p, c][\sigma\rho_q, d].$$

By isolating the steps, formal relations are found which are useful in calculations.

Consider spaces X, Y and cochains $c \in S^p(X)$, $d \in S^q(Y)$.

(29.17.1) *Definition*. The *cross* (or *exterior*) *product* $c \times d \in S^{p+q}(X \times Y)$ is the cochain given by the composite

$$S(X \times Y) \xrightarrow{A} S(X) \otimes S(Y) \xrightarrow{c \otimes d} R \otimes R \xrightarrow{m} R.$$

More precisely, consider the cochain complex $[S(X) \otimes S(Y)]^*$ whose component of degree n is the dual of $[S(X) \otimes S(Y)]_n$. There is a canonical homomorphism of cochain complexes

$$r: S^{\cdot}(X) \otimes S^{\cdot}(Y) \to [S(X) \otimes S(Y)]^*$$

defined as follows: Given $c \in S^p(X)$, $d \in S^q(Y)$, $r(c \otimes d)$ is the linear form on $S(X) \otimes S(Y)$ whose value on $z \otimes w \in S_{p'}(X) \otimes S_{q'}(Y)$ is given by

$$[z \otimes w, r(c \otimes d)] = \begin{cases} [z, c][w, d] & \text{if } p' = p \text{ and } q' = q, \\ 0 & \text{otherwise.} \end{cases}$$

Then $c \times d = {}^t\!Ar(c \otimes d)$.

(29.17.2) *Lemma.* $\delta(c \times d) = \delta c \times d + (-1)^p c \times \delta d$.

Proof. Routine using (29.2) and (29.3) ∎

As usual, the cross product of cocycles is a cocycle and the addition of a coboundary varies the cross product of cocycles by a coboundary. Hence we receive a pairing

$$H^q(X) \otimes H^q(Y) \xrightarrow{\times} H^{p+q}(X \times Y).$$

called the *cohomology cross product.* The basic relation between cup products and cross products is

$$c \cup d = H^{\cdot}(\Delta)(c \times d)$$

for $c, d \in H^{\cdot}(X)$. This is often used as a definition for the cup product.
A useful relation is given by

(29.17.3) *Proposition.* $c \times d = S^{\cdot}(p_X)(c) \cup S^{\cdot}(p_Y)(d)$ *where* p_X, p_Y *are the projections of* $X \times Y$ *on* X, Y.

Proof. For any singular $(p + q)$-simplex (σ, τ) in $X \times Y$,

$$\begin{aligned} [(\sigma, \tau), {}^t\!Ar(c \otimes d)] &= [A(\sigma, \tau), r(c \otimes d)] \\ &= [\sigma\lambda_p \otimes \tau\rho_q, r(c \otimes d)] \\ &= [\sigma\lambda_p, c][\tau\rho_q, d], \\ [(\sigma, \tau), S^{\cdot}(p_X)c \cup S^{\cdot}(p_Y)d] &= [p_X(\sigma, \tau)\lambda_p, c][p_Y(\sigma, \tau)\rho_q, d] \\ &= [\sigma\lambda_p, c][\tau\rho_q, d]. \end{aligned}$$ ∎

Corollary. The cross products for homology and cohomology are related by the formula

$$(29.17.4) \qquad [\zeta \times \omega, \xi \times \eta] = [\zeta, \xi][\omega, \eta].$$

For some applications it is useful to generalize (29.17.3) to

(29.17.5) *Proposition. Let* $a, b \in H^{\cdot}(X)$, $c, d \in H^{\cdot}(Y)$. *Then* $(a \cup b) \times (c \cup d) = (-1)^{|b||c|}(a \times c) \cup (b \times d)$ *where* $|x|$ *denotes dimension.*

Proof. Unlike (29.17.3) this relation does not hold for cochains. We begin the proof here. It will be completed by an appeal to the acyclic model theorem (29.25). The following diagram contains routes to define the various products.

$$\begin{array}{ccccc}
 & & S(X \times Y) & \xrightarrow{A} & S(X) \otimes S(Y) \\
 & \swarrow^{S(\Delta_{X \times Y})} \quad ① & \downarrow S(\Delta_X \times \Delta_Y) & ② & \downarrow S(\Delta_X) \otimes S(\Delta_Y) \\
S(X \times Y \times X \times Y) & \xrightarrow{S(1 \times i \times 1)} & S(X \times X \times Y \times Y) & \xrightarrow{A} & S(X \times X) \otimes S(Y \times Y) \\
\downarrow A & & ③ & & \downarrow A \otimes A \\
S(X \times Y) \otimes S(X \times Y) & \xrightarrow{A \otimes A} & S(X) \otimes S(Y) \otimes S(X) \otimes S(Y) & \xrightarrow{1 \otimes T \otimes 1} & S(X) \otimes S(X) \otimes S(Y) \otimes S(Y)
\end{array}$$

where i and T are defined in (29.27). Triangle ① is induced from commuting maps while ② is the functorial property of A (29.1). Diagram ③ commutes up to chain homotopy, a fact proved in (29.28). From this the result follows. ■

Next we place (29.17.5) in an algebraic context. Form the graded module $H^{\cdot}(X) \otimes H^{\cdot}(Y)$, whose component of degree n is

$$[H^{\cdot}(X) \otimes H^{\cdot}(Y)]^n = \bigoplus_{p+q=n} H^p(X) \otimes H^q(Y).$$

We make this into a graded R-algebra with multiplication

$$(a_p \otimes b_q)(c_r \otimes d_s) = (-1)^{qr}(a_p \cup c_r) \otimes (b_q \cup d_s).$$

(With the same definition of multiplication in $S^{\cdot}(X) \otimes S^{\cdot}(Y)$, the coboundary operator becomes a graded-algebra derivation; this accounts for the choice of sign.) Note that this multiplication is in general noncommutative, even though each of the factors may be commutative.

The graded algebra homomorphism

$$H^{\cdot}(X) \otimes H^{\cdot}(Y) \xrightarrow{\times} H^{\cdot}(X \times Y)$$

is functorial in (X, Y) and by (29.17.3) is given by

$$a \otimes b \to a \times b = H^{\cdot}(p_X)(a) \cup H^{\cdot}(p_Y)(b)$$

for $a \in H^{\cdot}(X)$, $b \in H^{\cdot}(Y)$ (p_X, p_Y being the projections of $X \times Y$ on X, Y).

(29.17.6) *Exercise* (R a PID). Assume all the homology modules of X are finitely generated. Then there is an exact sequence

$$0 \to [H^{\cdot}(X) \otimes H^{\cdot}(Y)]^n \xrightarrow{\times} H^n(X \times Y) \to \bigoplus_{p-q=n+1} \mathrm{Tor}(H^p(X), H^q(Y)) \to 0$$

so that if, in addition, the homology modules of X or Y are all torsion-free, the cross product is an isomorphism. (Show $S(X)$ is chain homotopically equivalent to a subcomplex C such that each C_p is free finitely generated. Then $C^p = C_p^*$ is free for all p and $C^{\cdot}$ is cochain homotopically equivalent to $S^{\cdot}(X)$. Convert the cochain complexes $C^{\cdot}$, $S^{\cdot}(Y)$ into chain complexes by negating the indices, and apply the Künneth exact sequence (29.10). Finally, use the fact that C_p is free finitely generated to note that $r: C^p \otimes S^q(X) \to [C_p \otimes S_q(Y)]^*$ is an isomorphism.)

In particular, if one of $H_{\cdot}(X)$, $H_{\cdot}(Y)$ is torsion-free (or coefficients are in a field), we have a complete description of the cohomology ring $H^{\cdot}(X \times Y)$ as the tensor product of the rings $H^{\cdot}(X)$, $H^{\cdot}(Y)$.

Exercise. Given another space W and maps $f: W \to X$, $g: W \to Y$, they induce a map $(f, g): W \to X \times Y$. For $\xi \in H^{\cdot}(X)$, $\eta \in H^{\cdot}(Y)$, we have

(29.17.7) $\quad H^{\cdot}((f, g))(\xi \times \eta) = H^{\cdot}(f)(\xi) \cup H^{\cdot}(g)(\eta).$

In particular, if $\Delta: X \to X \times X$ is the diagonal map $\Delta(x) = (x, x)$,

$$H^{\cdot}(\Delta)(\xi \times \eta) = \xi \cup \eta.$$

There is also a formula relating cap and cross products, namely

(29.17.8) $\quad (\zeta \times \omega) \cap (\xi \times \eta) = (-1)^{s(p-r)}(\zeta \cap \xi) \times (\omega \cap \eta)$

for $\zeta \in H_p(X)$, $\omega \in H_q(Y)$, $\xi \in H^r(X)$, $\eta \in H^s(Y)$. This formula is not obvious. The analogous formula on the chain-cochain level fails to hold, because AB is only chain homotopic to the identity, not equal to the identity, and $'Ar$ is an algebra homomorphism only on the cohomology, not on the cochain level. The proof is by the same techniques as in the Eilenberg-Zilber theorem (see Appendix (29.23A).

(29.18) *Example.* The cohomology ring is a finer invariant than the homology (or cohomology) groups. Consider the space $S^1 \times S^2$ and the space $S^1 \vee S^2 \vee S^3$ obtained by joining S^1, S^2, and S^3 at a common point P. By the Künneth formula and formulas (19.16–19.18)), these spaces have the same homology groups (**Z** in dimensions 0 to 3, 0 in higher dimensions), hence the same cohomology groups (23.28). By (27.17.6), if a is a generator of $H^1(S^1 \times S^2)$, b a generator of $H^2(S^1 \times S^2)$, then $a \cup b$ generates $H^3(S^1 \times S^2)$.

Define the map f: $S^1 \vee S^2 \vee S^3 \to S^1 \vee S^2$ to be the identity on $S^1 \vee S^2$ and the constant map onto P on S^3. Then $H^q(f)$ is a group isomorphism for $q = 0, 1, 2$. (Since $H_q(f)$ is, exercise.) Let a' generate $H^1(S^1 \vee S^2 \vee S^3)$, b' generate $H^2(S^1 \vee S^2 \vee S^3)$, and put

$$a' = H^1(f)(a''), \qquad b' = H^2(f)(b'').$$

Then $a'' \cup b'' \in H^3(S^1 \vee S^2) = 0$, so that $a' \cup b' = H^3(f)(a'' \cup b'') = 0$. Thus $H^{\cdot}(S^1 \times S^2)$ and $H^{\cdot}(S^1 \vee S^2 \vee S^3)$ are not isomorphic as rings.

(29.19) Another useful operation on $X \times Y$ is *the slant product* (division by chains), a bilinear pairing

$$H^{p+q}(X \times Y) \times H_p(X) \to H^q(Y)$$

$$(\gamma, \alpha) \to \gamma/\alpha$$

which satisfies the formula

$$[\beta, \gamma/\alpha] = [\alpha \times \beta, \gamma]$$

for $\beta \in H_q(Y)$.

First suppose $c \in [S(X) \otimes S(Y)]^*_{p-q}$, $w \in S_p(X)$. Define c/w to be the q-cochain on Y given by

$$[z, c/w] = [w \otimes z, c]$$

for all $z \in S_q(Y)$. Then

$$\begin{aligned} [z, \delta(c/w)] &= [\partial z, c/w] \\ &= [w \otimes \partial z, c] \\ &= (-1)^p[\partial(w \otimes z) - \partial w \otimes z, c] \\ &= (-1)^p[z, \delta c/w - c/\partial w] \end{aligned}$$

so that $\delta(c/w) = (-1)^p(\delta c/w - c/\partial w)$. Hence passage to the quotient gives a slant product

$$H^{p+q}([S(X) \otimes S(Y)]^*) \times H_p(X) \to H^q(Y)$$

For $\gamma \in H^{p+q}(X \times Y)$, $\alpha \in H_p(X)$ we then define

$$\gamma/\alpha = \overline{{}'A^{-1}}\gamma/\alpha$$

Note that in the special case $p = q = 0$, $\gamma = 1$, we get

$$1/\alpha = [\alpha, 1]1 \in H^0(Y) \tag{29.20}$$

for all $\alpha \in H_0(X)$.

The basic formula relating all the different products is

$$\{(\xi \times \eta) \cup \gamma\}/\alpha = (-1)^{r(s-q)}\eta \cup \{\gamma/\alpha \cap \xi\} \tag{29.21}$$

for $\gamma \in H^p(X \times Y)$, $\xi \in H^q(X)$, $\eta \in H^r(Y)$, $\alpha \in H_s(X)$. To make this plausible, suppose $\beta \in H_{p-q-r-s}(Y)$. Then

$$\begin{aligned}
[\beta, \eta \cup \{\gamma/\alpha \cap \xi\}] &= [\beta \cap \eta, \gamma/\alpha \cap \xi] \\
&= [(\alpha \cap \xi) \times (\beta \cap \eta), \gamma] \\
&= (-1)^{r(s-q)}[(\alpha \times \beta) \cap (\xi \times \eta), \gamma] (29.17.8) \\
&= (-1)^{r(s-q)}[\alpha \times \beta, (\xi \times \eta) \cup \gamma] \\
&= (-1)^{r(s-q)}[\beta, \{(\xi \times \eta) \cup \gamma\}/\alpha].
\end{aligned}$$

(For the proof see (29.30).)

In the special case $q = s$, $\gamma = 1$, we get

$$(\xi \times \eta)/\alpha = [\alpha, \xi]\eta \tag{29.22}$$

using (29.20).

Given maps $f: X \to X'$, $g: Y \to Y'$, the slant product satisfies the formula

$$H^{p+q}(f \times g)(\gamma')/\alpha = H^q(f)(\gamma'/H_p(g)(\alpha)). \tag{29.23}$$

(Exercise.)

Appendix (29.23A) Acyclic Models

In this appendix we treat the Eilenberg-Mac Lane method of acyclic models (see [22], or Mac Lane [38], Chapter VIII) and apply it to the theory of products.

We have seen that to an ordered pair (X, Y) of topological spaces we can associate

$$F(X, Y) = S(X \times Y),$$

$$F'(X, Y) = S(X) \otimes S(Y)$$

which are *augmented chain complexes*: An *augmentation* of a chain complex C of R is an epimorphism $\varepsilon: C_0 \to R$ such that the composite

$$C_1 \xrightarrow{\partial} C_0 \xrightarrow{\varepsilon} R$$

is zero; the augmentation of $S(X \times Y)$ is ∂^* (9.7), while that of $S(X) \otimes S(Y)$ is

$$\partial^* \otimes \partial^*: S_0(X) \otimes S_0(Y) \to R \otimes R \cong R$$

We can make a category $\mathscr{P}$ out of the ordered pairs (X, Y) by defining a morphism $(X, Y) \to (X', Y')$ to be a map of the form $f \times g$, where $f: X \to X'$, $g: Y \to Y'$. We can make a category $\mathscr{C}$ out of the augmented chain complexes by defining a morphism $h: C \to C'$ to be a chain homomorphism which preserves the augmentations, i.e., $\varepsilon' h = \varepsilon$. Then F is clearly a functor from $\mathscr{P}$ to $\mathscr{C}$, and F' is also ($F'(f \times g) = S(f) \otimes S(g): S(X) \otimes S(Y) \to S(X') \otimes S(Y')$).

Moreover, Alexander-Whitney gives us a morphism (natural transformation) of functors

$$A: F \to F'$$

as follows from (29.1), (29.3), and the fact that

$$A: S_0(X \times Y) \to S_0(X) \otimes S_0(Y)$$

preserves the augmentations (exercise). The Eilenberg-Zilber theorem can be rephrased to state that A is a *chain equivalence* of functors: This means that there exists a morphism of functors

$$B: F' \to F$$

such that AB and BA are chain homotopic to the identity morphisms. In general, if $\mathscr{A}$ is any category, $F, F': \mathscr{A} \to \mathscr{C}$ functors ($\mathscr{C}$ as above), $\Phi, \Psi: F \to F'$ morphisms of functors, we say Φ and Ψ are *chain homotopic* if for every object X in $\mathscr{A}$, the chain homomorphisms $\Phi(X), \Psi(X): F(X) \to F'(X)$ are chain homotopic by means of a chain homotopy which is functorial in X.

The technique of acyclic models gives us a sufficient condition for the two functors F, F' with values in $\mathscr{C}$ to be chain equivalent. In fact, the condition is so strong it implies that any morphism of functors $F \to F'$ is a chain equivalence.

Given a category $\mathscr{A}$ with a distinguished set of objects $\mathscr{M}$. We say that the pair $(\mathscr{A}, \mathscr{M})$ is a *category with models*, the *models* being the objects in $\mathscr{M}$. Let $F: \mathscr{A} \to \mathscr{C}$ be any functor. For each $q \geq 0$ we have a functor F_q from $\mathscr{A}$ to the category of R-modules which assigns to any X in $\mathscr{A}$ the qth component of the chain complex $F(X)$. We say F_q has a *basis* (relative to $\mathscr{M}$) if there is an indexed family $\{d_j \in F_q(M_j)\}_{j \in J}$, where the M_j are some of the models, such that for every X in $\mathscr{A}$, $F_q(X)$ is the free R-module spanned by all the elements

$$\{F_q(u)(d_j)\}_{j \in J}, \qquad \text{all } u: M_j \to X$$

We say F is *free* if all the F_q's have bases.

(29.24) *Example*: Let $F, F': \mathscr{S} \to \mathscr{C}$ be the two functors of the Eilenberg-Zilber theorem. As models for $\mathscr{S}$ we take all the ordered pairs (Δ^p, Δ^q) (where Δ^p is the standard geometric p-simplex). Then F and F' are both free: *Let* $d_q: \Delta^q \to \Delta^q \times \Delta^q$ be the diagonal map, so that $d_q \in S_q(\Delta^q \times \Delta^q)$. For any ordered pair (X, Y), we know that the singular q-simplexes σ form a basis for the free R-modules $S_q(X \times Y)$. But the formula

$$\sigma = [(p_X\sigma) \times (p_Y\sigma)] \circ d_q$$

(where $p_X: X \times Y \to X$, $p_Y: X \times Y \to Y$ are the projections) shows that the singleton $\{d_q\}$ is a basis for F_q. Thus F is free relative to the chosen models. On the other hand, a basis for $F'_n(X, Y)$ consists of all $\sigma \otimes \tau$, σ a singular p-simplex in X, τ a singular q-simplex in Y, $p + q = n$. But the formula (see 9.10)

$$\sigma \otimes \tau = F'_n(\sigma \times \tau)(\delta_p \otimes \delta_a)$$

(where $\delta_p \in S_p(\Delta^p)$, $\delta_q \in S_q(\Delta^q)$ are the identity maps) shows that the family $\{\delta_p \otimes \delta_q\}_{p-q=n}$ is a basis for F'_n. Thus F' is free.

If C is an augmented chain complex with augmentation $\varepsilon: C_0 \to R$, we can form the reduced chain complex C given by

$$\tilde{C}_q = \begin{cases} C_q & q > 0 \\ \text{Kernel } \varepsilon & q = 0 \end{cases}$$

The homology modules of $\tilde{C}$ are called the *reduced* (*augmented*) *homology modules* of C. A functor $F:\mathscr{A} \to \mathscr{C}$ will be called *acyclic* (relative to $\mathscr{M}$) if for every model $M \in \mathscr{M}$, all the augmented homology modules of the chain complex $F(M)$ are zero.

(29.24) *Example* (continued): In the Eilenberg-Zilber situation, $H_n^*(\Delta^p \times \Delta^q) = 0$ for all n, p, and q, since $\Delta^p \times \Delta^q$ is contractible. Thus F is acyclic. We also have $H_n^*[S(\Delta^p) \otimes S(\Delta^q)] = 0$ for all n, p, and q: Let R be the chain complex which has R in dimension 0 and 0 in all other dimensions, with zero boundary operator. Then the chain homomorphism $\partial^*:S(\Delta^p) \to R$ is a chain equivalence since Δ^p is contractible. Hence

$$\partial^* \otimes \partial^*:S(\Delta^p) \otimes S(\Delta^q) \to R \otimes R \simeq R$$

is a chain equivalence, from which the assertion follows. Thus F' is also acyclic.

We can now state our main result.

(29.25) *Theorem on Acyclic Models. Let $(\mathscr{A},\mathscr{M})$ be a category with models, $\mathscr{C}$ the category of augmented chain complexes over R. Let F, $F':\mathscr{A} \to \mathscr{C}$ be functors such that F is free and F' is acyclic. Then there exists a morphism of functors $\Phi:F \to F'$ unique up to natural chain homotopy.*

(29.26) *Corollary. If both F and F' are free and acyclic, then F and F' are chain equivalent, and in fact any morphism of functors $F \to F'$ is a chain equivalence.*

Proof of Corollary: By the theorem there are morphisms $\Phi:F \to F'$ and $\Psi:F' \to F$. Since by the theorem there are unique morphisms $F \to F$ and $F' \to F'$ up to chain homotopy, $\Phi\Psi$ and $\Psi\Phi$ must be chain homotopic to the respective identity morphisms. ∎

Using (29.24) we see that the Eilenberg-Zilber theorem follows at once.

(29.27) *Exercise.* An explicit chain homomorphism

$$B:S(X) \otimes S(Y) \to S(X \times Y)$$

which is functorial in (X, Y) is *the shuffle homomorphism given by*

$$B\,((\sigma \otimes \tau) = \sum \pm (D_{j_q} \ldots D_{j_1}\sigma, D_{i_p} \ldots D_{i_1}\tau)$$

where σ is a singular p-simplex, τ a singular q-simplex, D_k is the kth degeneracy operator (proof of (24.8), part 2), the sum is over all permutations $(i_1 \ldots i_p j_1 \ldots j_q)$ of $(0 \ldots p + q - 1)$ such that $i_1 < i_2 < \cdots < i_p$, $j_1 < j_2 < \cdots < j_q$, and the sign is the signature of the permutation.

Before proving (29.25) we give some other applications.

For any ordered pair (X, Y), define a homomorphism

$$T: S(X) \otimes S(Y) \to S(Y) \otimes S(X)$$

by

$$T(z \otimes w) = (-1)^{pq} w \otimes z$$

for $z \in S_p(X)$, $w \in S_q(Y)$. Then T is a chain homomorphism:

$$\partial T(z \otimes w) = (-1)^{pq}[\partial w \otimes z + (-1)^q w \otimes \partial z]$$

$$T\partial(z \otimes w) = T[\partial z \otimes w + (-1)^p z \otimes \partial w]$$

$$= (-1)^{pq-q} w \otimes \partial z + (-1)^{pq} \partial w \otimes z$$

Clearly T preserves augmentation and is functorial in (X, Y).

Consider the diagram

$$\begin{array}{ccc} S(X \times Y) & \xrightarrow{S(i)} & S(Y \times X) \\ {\scriptstyle A}\downarrow & & \downarrow{\scriptstyle A} \\ S(X) \otimes S(Y) & \xrightarrow[T]{} & S(Y) \otimes S(X) \end{array}$$

where i is the interchange homeomorphism $X \times Y \to Y \times X$.

(29.28) *Proposition. This diagram is chain homotopy commutative, i.e., $AS(i)$ is chain homotopic to TA.*

The proof follows immediately from the uniqueness part of the acyclic model theorem. ■

(29.29) *Corollary. If $\zeta \in H_p(X)$, $\omega \in H_q(Y)$ then*

$$H_{p+q}(i)(\zeta \times \omega) = (-1)^{pq} \omega \times \zeta$$

(29.30) *Exercise.* Prove formulas (29.17.8) and (29.21) by the same method, i.e., deduce them from the chain homotopy commutativity of suitable diagrams, the latter being proved by referring to the theorem on acyclic models.

Proof of (29.25): Let $\{d_j \in F_0(M_j)\}_{j \in J_0}$ be a basis for F_0. Since F' is acyclic, the augmentation induces an isomorphism

$$\varepsilon': H_0(F'(M_j)) \simeq R$$

for all $j \in J_0$. Hence there is a unique $z_j \in H_0(F'(M_j))$ such that

$$\varepsilon'(z_j) = \varepsilon(d_j)$$

for all $j \in J_0$.

There is a unique morphism of functors $\phi: H_0(F) \to H_0(F')$ preserving augmentation, namely the one sending the class of a linear combination

$$\sum \nu_{uj} F_0(u)(d_j) \in F_0(X)$$

($\nu_{uj} \in R$, $u: M_j \to X$) to the element

$$\sum \nu_{uj} H_0(F')(u)(z_j) \in H_0(F'(X))$$

We will construct a morphism $\Phi: F \to F'$ so as to give $\phi = H_0(\Phi)$. If Ψ is any morphism $F \to F'$, then by uniqueness $\phi = H_0(\Psi)$, and we will simultaneously construct the chain homotopy $D: \Phi \simeq \Psi$.

Thus for every object $X \in \mathscr{Q}$ we must define a chain homomorphism $\Phi(X): F(X) \to F'(X)$ (resp. a chain homotopy $D(X): F(X) \to F'(X)$) such that for any $h: X \to Y$ we have

$$\Phi(Y)F(h) = F'(h)\Phi(X)$$

$$(\text{resp. } D(Y)F(h) = F'(h)D(X))$$

For $q \geq 0$, let $\{d_j \in F_q(M_j)\}_{j \in J_q}$ be a basis for F_q. Once we know the values $\Phi_q(M_j)(d_j)$ (resp. $D_q(M_j)(d_j)$) then for any X we are forced to define

$$\Phi_q(X)F_q(u)(d_j) = F'_q(u)\Phi_q(M_j)(d_j)$$

$$(\text{resp. } D_q(X)F_q(u)(d_j) = F'_q(u)D_q(M_j)(d_j))$$

which determines $\Phi_q(X)$ (resp. $D_q(X)$). These values are not arbitrary since we want

$$\partial\Phi_q(X) = \Phi_{q-1}(X)\partial$$

$$(\text{resp. } \partial D_q(X) = \Phi_q(X) - \Psi_q(X) - D_{q-1}(X)\partial)$$

We use induction on q. For $q = 0$, define $\Phi_0(M_j)(d_j)$ to be any representative in $F'(M_j)$ of $z_j \in H_0(F'(M_j))$ (resp. define $D_0(M_j)(d_j)$ to be any element of $F'_1(M_j)$ whose boundary is the element $\Phi_0(M_j)$ $(d_j) - \Psi_0(M_j)(d_j)$ of kernel of the augmentation). Then for $w \in F_0(X)$ the homology class of $\Phi_0(X)(w)$ is obtained by applying $\phi(X)$ to the class of w. In particular, for $j \in J_1$, $\Phi_0(M_j)(\partial d_j)$ is a boundary in $F'_0(M_j)$, so define $\Phi_1(M_j)(d_j)$ to be any element such that

$$\partial\Phi_1(M_j)(d_j) = \Phi_0(M_j)(\partial d_j)$$

Assume $q > 1$ and Φ_p defined for $p < q$ so as to commute with ∂. Note that $\Phi_{q-1}(M_j)(d_j)$ is a cycle; since $H_{q-1}(F'(M_j)) = 0$, we can define $\Phi_q(M_j)(d_j)$ to be any element such that

$$\partial\Phi_q(M_j)(d_j) = \Phi_{q-1}(M_j)(d_j)$$

[Assume $q > 0$ and D_p defined for $p < q$ so as to be a chain homotopy. Note that for every $j \in J_q$, the element

$$\Phi_q(M_j)(d_j) - \Psi_q(M_j)(d_j) - D_{q-1}(M_j)(\partial d_j)$$

is a cycle. Since $H_q(F'(M_j)) = 0$, we can define $D_q(M_j)(d_j)$ to be any element whose boundary is this cycle.]. ∎

(29.31) *Remark*. We can view our results somewhat more categorically if we notice that $S(X)$ carries all the structure for the homology and cohomology of the space X. Thus far we have regarded $S(X)$ only as an augmented chain complex. To obtain the multiplicative properties, consider the augmentation preserving chain homomorphism

$$S(X) \xrightarrow{S(d)} S(X \times X) \xrightarrow{A} S(X) \otimes S(X)$$

where d is the diagonal map and A is Alexander-Whitney; let $m = AS(d)$. Then m is the *comultiplication* for a structure of *coalgebra* on $C = S(X)$, i.e., the diagram

$$
\begin{array}{ccc}
C \otimes C \otimes C & \xleftarrow{m \otimes id} & C \otimes C \\
{\scriptstyle id \otimes m}\uparrow & & \uparrow{\scriptstyle m} \\
C \otimes C & \xleftarrow{m} & C
\end{array}
$$

is commutative (this diagram represents the dual of the associative law; to prove it is commutative, use the explicit formula

$$m(\sigma) = \sum_{p+q=n} \sigma\lambda_p \otimes \sigma\rho_q$$

for any singular n-simplex σ). Moreover the augmentation $\partial^*:S(X) \to R$ is a *co-unit*, i.e., the composites

$$C \xrightarrow{m} C \otimes C \xrightarrow{id \otimes \partial^*} C \otimes R \simeq C$$

$$C \xrightarrow{m} C \otimes C \xrightarrow{\partial^* \otimes id} R \otimes C \simeq C$$

are both equal to the identity automorphism of C.

Let $\mathscr{G}$ be the category whose objects are those augmented chain complexes having a comultiplication making them into a coalgebra for which the augmentation is a co-unit. A morphism in the category $\mathscr{G}$ will be an augmentation preserving chain homomorphism $h:C \to C'$ such that the diagram

$$
\begin{array}{ccc}
C \otimes C & \xrightarrow{h \otimes h} & C' \otimes C' \\
m\uparrow & & \uparrow m' \\
C & \xrightarrow{h} & C'
\end{array}
$$

is commutative. Then the functor $X \to S(X)$ actually takes its values in the category $\mathscr{G}$ (if $f:X \to Y$ then (29.1) implies that $S(f)$ is a morphism in $\mathscr{G}$).

However, the category $\mathscr{G}$ is not appropriate for products, since it does not have them! Indeed, given $C, C' \in \mathscr{G}$, the tensor product $C \otimes C'$ does have the natural comultiplication

$$C \otimes C' \xrightarrow{m \otimes m'} C \otimes C \otimes C' \otimes C' \xrightarrow{id \otimes T \otimes id} C \otimes C' \otimes C \otimes C'$$

where T is the interchange operator defined in (29.28). This makes $C \otimes C'$

into a coalgebra (exercise). There are projections $p: C \otimes C' \to C, p': C \otimes C' \to C'$ given by

$$\begin{array}{ccc} & C \otimes C' & \\ {}^{id \otimes \varepsilon'}\swarrow & & \searrow^{\varepsilon \otimes id} \\ C \otimes R & & R \otimes C' \\ \downarrow\wr & & \downarrow\wr \\ C & & C' \end{array}$$

which are morphisms in $\mathscr{D}$ (exercise). Suppose $h: C'' \to C, h': C'' \to C'$ are morphisms in $\mathscr{D}$. If there exists a morphism $(h, h'): C'' \to C \otimes C'$ such that $p(h, h') = h, p'(h, h') = h'$, then it is unique:

$$\begin{aligned} (h, h') &= (p \otimes p')(id \otimes T \times id)(m \otimes m')(h, h') \\ &= (p \otimes p')[(h, h') \otimes (h, h')]m'' \\ &= (h \otimes h')m'' \end{aligned}$$

Now $h \otimes h'$ is a morphism in $\mathscr{D}$, but m'' is not, unless the diagram

$$\begin{array}{ccc} C'' \otimes C'' & \xrightarrow{T} & C'' \otimes C'' \\ & {}^{m''}\searrow \quad \swarrow^{m''} & \\ & C'' & \end{array}$$

is commutative, i.e.. C'' is skew-commutative (exercise). Thus $C \otimes C'$ has the universal property of a product only relative to the subcategory of skew-commutative objects. We have seen (24.7) that the functor $X \to S(X)$ does not have values in this subcategory.

Hence we consider the category $\bar{\mathscr{D}}$ whose objects are the same as those in $\mathscr{D}$ but whose morphisms are the chain homotopy classes of morphisms in $\mathscr{D}$, and consider the full subcategory $\bar{\mathscr{D}}_c$ of skew-commutative objects ($C \in \bar{\mathscr{D}}_c$ if and only if Tm is chain homotopic to m). Then the category $\bar{\mathscr{D}}_c$ does have products, and $X \to S(X)$ can be regarded as a functor with values in $\bar{\mathscr{D}}_c$ (since the functor $X \to S(X) \otimes S(X)$ is acyclic on the models Δ^q, and the functor $X \to S(X)$ is free on those models, the acyclic model theorem implies Tm chain homotopic to m for $S(X)$—gives another proof of (24.8)). The Eilenberg-Zilber theorem then states that this functor with values in $\bar{\mathscr{D}}_c$ preserves products (up to isomorphism).

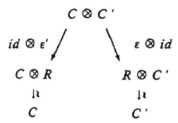

(29.32) *Exercise.* The Alexander-Whitney map $S(X \times Y) \to S(X) \otimes S(Y)$ is a morphism in the category $\mathscr{D}_c$.

(29.33) *Relative cup products.* Let $(X \times Y, A \times Y, X \times B)$ be an exact triad. We derive the relative Eilenberg-Zilber theorem and use it to construct relative cup products. From exactness of the triad, we obtain a chain equivalence

$$S(X \times Y)/S(A \times Y) + S(X \times B) \to S(X \times Y)/S(A \times Y \cup X \times B).$$

Using the Eilenberg-Zilber theorem (29.4) we have chain equivalences

$$S(X) \otimes S(Y) \to S(X \times Y),$$

$$S(A) \otimes S(Y) \to S(A \times Y),$$

$$S(X) \otimes S(B) \to S(X \times B)$$

which combine, using functoriality, to produce a chain equivalence

$$S(X) \otimes S(Y)/S(A) \otimes S(Y) + S(X) \otimes S(B) \to$$

$$S(X \times Y)/S(A \times Y) + S(X \times B)$$

Combining these chain equivalences with the chain isomorphism

$$S(X) \otimes S(Y)/S(A) \otimes S(Y) + S(X) \otimes S(B) \cong S(X)/S(A) \otimes S(Y)/S(B)$$

we obtain

(29.34) *Relative Eilenberg-Zilber Theorem. If $(X \times Y, A \times Y, X \times B)$ is an exact triad, there is a chain equivalence of $S(X, A) \otimes S(Y, B)$ with $S(X \times Y, A \times Y \cup X \times B)$.*

(29.35) *Corollary. There is a pairing, the relative cross product,*

$$H^p(X, A) \otimes H^q(Y, B) \xrightarrow{\times} H^{p+q}(X \times Y, A \times Y \cup X \times B)$$

compatible with the cross product in the absolute case and satisfying the analogues of functoriality, associativity, and behavior under the interchange map (29.29) *which hold for the cross product.*

The proof mimics the acyclic model method used to prove these properties for the cross product.

In case $X = Y$, the diagonal induces

$$(X, A \cup B) \to (X \times X, A \times X \cup X \times B)$$

and hence a pairing

$$H^p(X, A) \otimes H^q(X, B) \to H^{p+q}(X, A \cup B).$$

(29.36) *Proposition. If $X = \cup_{i=1}^n U_i$ where U_i are open and contractible in X, then all n-fold cup products of elements of positive dimension are* 0 *in* $H^\cdot(X)$.

Proof. The inclusion $U_i \to X$ induces the zero map $H^\cdot(X) \to H^\cdot(U_i)$ in positive dimensions, since U_i is contractible in X. By the long exact sequence, there are epimorphisms $H^q(X, U_i) \to H^q(X)$, $q > 0$. Given $x_1, \ldots, x_n$, let $y_i \in H^\cdot(X, U_i)$ be the preimage of x_i. Then the relative cup product $y_1 \cup \ldots \cup y_n \in H^\cdot(X, \cup_{i=1}^n U_i) = 0$ in positive dimensions. By naturality, $x_1 \cup \ldots \cup x_n = 0$. ∎

In particular, cup products in a suspension are 0.

(29.37) *Remark*. An elegant application of relative cup products to Hopf invariants is given in Steenrod and Epstein [87], p. 12–15. The reader is encouraged to look at it.

(29.38) *Exercise.* Show the cohomology rings of $\mathbf{CP}^{n(n+1)/2}$ and $S^2 \times S^4 \times \ldots \times S^{2n}$ are not isomorphic.

(29.39) *Excercise*. Prove $S^n \vee S^m$ is not a retract of $S^n \times S^m$.

(29.40) *Exercise*. Prove that if the diagonal $\Delta: X \to X \times X$ factors up to homotopy through $X \vee X$, then cup products of elements of positive degree are 0 in $H^\cdot(X)$.

$$\begin{array}{ccc} X & \xrightarrow{\Delta} & X \times X \\ & \searrow & \simeq \; \cup \\ & & X \vee X. \end{array}$$

Show such a factorization exists if X is a suspension.

(29.41) *Exercise.* Prove if X is a closed manifold with diagonal factoring as in (29.40), then X has the homology of a sphere.

(29.42) *Exercise.* Let $f: X \to Y$. Adapt the proof of (29.36) to the situation $Cf = CX \cup_f Y$ to show if n-fold cup products are 0 in $H^{\cdot}(Y)$. then $(n + 1)$-fold cup products are 0 in $H^{\cdot}(Cf)$ where Cf is the mapping cone (21.15).

(29.43) *Exercise.* Use the method of derived functors applied to the functor "tensoring with R" to derive (29.12).

(29.44) *Exercise.* For $R = \mathbf{Z}/n\mathbf{Z}$, derive (29.12) directly from the long exact sequence in homology induced by the coefficient sequence $0 \to \mathbf{Z} \xrightarrow{n} \mathbf{Z} \to \mathbf{Z}/n\mathbf{Z} \to 0$. Suggestion: Identify $H_q(X; \mathbf{Z}/n\mathbf{Z})$ with

$$\text{Coker}\,\{H_q(X; \mathbf{Z}) \xrightarrow{n} H_q(X; \mathbf{Z})\}$$

and $\text{Tor}(H_{q-1}(X; \mathbf{Z}), \mathbf{Z}/n\mathbf{Z})$ with

$$\text{Ker}\,\{H_{q-1}(X; \mathbf{Z}) \xrightarrow{n} H_{q-1}(X; \mathbf{Z})\}.$$

(29.45) *Exercise.* Apply the Universal Coefficient Theorem to prove that if X is an n-dimensional manifold and $A \subset X$ a closed connected subspace, then the torsion subgroup of $H_{n-1}(X, X - A; \mathbf{Z})$ is of order two if A is compact and X non-orientable along A, and is zero otherwise. (Hint: (22.24) applies.) In particular, taking $X = A$ when X is compact connected, we see that $H_{n-1}(X; \mathbf{Z})$ has torsion of order two if X is non-orientable and is torsion-free if X is orientable.

30. Thom Class and Lefschetz Fixed Point Theorem

Let X be an orientable n-manifold without boundary. We construct a cohomology class $\mu \in H^n(X \times X, X \times X - \Delta)$ called the Thom class. This class has intimate connections with the fundamental class ζ and is used to construct an inverse to the Poincaré duality isomorphism.

Let X be an n-dimensional manifold (without boundary). In Section 22 we have defined the R-orientation sheaf $X^0 \to X$ of X whose fibre over a point x is the local homology module $H_n(X, X - x)$. Given an R-orientation of X, i.e., given a section $s{:}X \to X^0$ such that for each x, $s(x)$ generates $H_n(X, X - x)$.

Consider now the dual sheaf $X^{0*} \to X$, whose fibre over x is the local cohomology module $H^n(X, X - x)$. The R-orientation s determines a global section $s^*{:}X \to X^{0*}$ which is characterized by

$$[s(x), s^*(x)] = 1$$

for all $x \in X$.

Suppose U is open in X. Denote by $\Gamma^* U$ the module of sections over U of the dual sheaf. If Δ is the diagonal of $X \times X$,

$$U_u i_x{:}(X, X - x) \to (X \times U, X \times U - \Delta)$$

will be the map given by

$$U_u i_x(x') = (x', x), \qquad x' \in X$$

where $x \in U$. Recall that

$$j_x^U:(X, X - U) \to (X, X - x)$$

is the inclusion map.

(30.1) *Theorem. Let X be an R-oriented n-dimensional manifold, U an open subspace. Then*

$$H^q(X \times U, X \times U - \Delta) = 0 \quad \text{all } q < n$$

and there is a unique isomorphism $\phi: H^n(X \times U, X \times U - \Delta) \to \Gamma^* U$ such that

$$\phi(\beta)(x) = H^n({}_U i_x)(\beta)$$

for all $\beta \in H^n(X \times U, X \times U - \Delta)$, $x \in U$.

(30.2) *Corollary. There is a unique cohomology class $\mu = \mu_X$ in $H^n(X \times X, X \times X - \Delta)$ such that for all $x \in X$,*

$$s^*(x) = H^n({}_X i_x)(\mu)$$

This class μ will be called *the Thom class* of the given R-orientation of X.

(30.3) *Corollary. Suppose X is compact. Let $\zeta \in H_n(X)$ be the fundamental class of the R-orientation. Let*

$$H^n(j): H^n(X \times X, X \times X - \Delta) \to H^n(X \times X)$$

be the homomorphism induced by inclusion, and let $\mu' = H^n(j)(\mu)$. Then

$$\mu'/\zeta = 1$$

Proof: For any $x \in X$, consider the commutative diagram

$$\begin{array}{ccc} (X, X - x) & \xrightarrow{i_x} & (X \times X, X \times X - \Delta) \\ {\scriptstyle j_x}\uparrow & & \uparrow{\scriptstyle j} \\ X & \xrightarrow{i_x} & X \times X \end{array}$$

where $i_x = {}_X i_x$ and the vertical arrows are inclusions. Then

$$\begin{aligned} 1 &= [s(x), H^n(i_x)\mu] = [H_n(j_x)\zeta, H^n(i_x)\mu] = [\zeta, H^n(i_x j_x)\mu] \\ &= [\zeta, H^n(j i_x)\mu] = [H_n(i_x)\zeta, \mu']. \end{aligned}$$

If $\bar{x}$ is the homology class of x(x is a 0-cycle), the $H_n(i_x)\zeta = \zeta \times \bar{x}$ (since $X \approx X \times x$), so by (29.19)

$$1 = [\bar{x}, \mu'/\zeta]$$

for all x, which proves the corollary.

■

Proof of Theorem (*30.1*): For any $\beta \in H^n(X \times U, X \times U - \Delta)$, define a set-theoretic section $\phi(\beta)$: $U \to X^{0*}$ by

$$\phi(\beta)(x) = H^n({}_U i_x)(\beta) \qquad x \in U$$

If $\Gamma' U$ denotes the module of set-theoretic sections over U, then for $V \subset U$ we have the commutative diagram

$$\begin{array}{ccc} H^n(X \times U, X \times U - \Delta) & \to & H^n(X \times V, X \times V - \Delta) \\ \phi \downarrow & & \downarrow \phi \\ \Gamma' U & \longrightarrow & \Gamma' V \end{array}$$

Thus to verify that the homomorphism ϕ actually takes its values in the module $\Gamma^* U$ of (continuous) sections, it suffices to consider the following special case:

Case 1. U is contained in a coordinate neighborhood U' and U is contractible (an open n-cell, say).

For each $x \in U$ we have the commutative diagram

$$\begin{array}{ccc} H^q(X \times U, X \times U - \Delta) & \cong & H^q(U' \times U, U' \times U - \Delta) \\ {}_U i_x \downarrow & & \downarrow {}_U i_x \\ H^q(X, X - x) & \xrightarrow{\sim} & H^q(U', U' - x) \end{array}$$

where the horizontal isomorphisms are excisions. Thus we may assume $X = U' \approx \mathbf{R}^n$. In that case we have a homeomorphism

$$f:(\mathbf{R}^n \times U, (\mathbf{R}^n - 0) \times U) \cong (\mathbf{R}^n \times U, \mathbf{R}^n \times U - \Delta)$$

given by $f(y, x) = (y + x, x)$, and for each $x \in U$ a commutative diagram

$$\begin{array}{ccc}
(\mathbf{R}^n \times U, (\mathbf{R}^n - 0) \times U) & \xrightarrow{\mathcal{L}} & (\mathbf{R}^n \times U, \mathbf{R}^n \times U - \Delta) \\
\uparrow \scriptstyle{id \times j_x} & & \uparrow \scriptstyle{{}_U i_x} \\
(\mathbf{R}^n \times 0, (\mathbf{R}^n - 0) \times 0) & & \\
\uparrow & & \\
(\mathbf{R}^n, \mathbf{R}^n - 0) & \underset{f_x}{\xrightarrow{\approx}} & (\mathbf{R}^n, \mathbf{R}^n - x)
\end{array}$$

where $f_x(y) = x + y$, and j_x is the map of the point 0 onto x. We may assume $0 \in U$.

Sublemma 1. The map $s' \to s'(0)$ *is an isomorphism of* Γ^*U *onto* $H^n(\mathbf{R}^n, \mathbf{R}^n - 0)$.

Proof: This follows from the fact that $H^n(j_0^U)$ is an isomorphism (22.4). ■

Sublemma 2. If $s' \in \Gamma^*U, x \in U$, *then*

$$s'(x) = H^n(f_{-x})(s'(0))$$

Proof: Let $\alpha \in H^n(\mathbf{R}^n, \mathbf{R}^n - U)$ be the unique class such that $\alpha = H^n(j_x^U)(s'(x))$ for all $x \in U$. Now j_x^U and $f_x j_0^U$ are homotopic maps $(\mathbf{R}^n, \mathbf{R}^n - U) \to (\mathbf{R}^n, \mathbf{R}^n - x)$ (at time t the map is $f_{tx} j_0^U$), whence

$$\begin{aligned}
s'(x) &= H^n(j_x^U)^{-1}(\alpha) \\
&= H^n(f_x)^{-1} H^n(j_0^U)^{-1} H^n(j_0^U)(s'(0)) \\
&= H^n(f_{-x})(s'(0))
\end{aligned}$$ ■

Using these sublemmas, the continuity of $\phi(\beta)$ is equivalent to the formula

$$H^n({}_U i_x)(\beta) = H^n(f_{-x}) H^n({}_U i_0)(\beta)$$

which is due to the fact that ${}_U i_x$ is homotopic to ${}_U i_0 f_{-x}$.

Sublemma 3. $H^q({}_U i_0)$ *is an isomorphism for all* q.
The theorem in case 1 follows from sublemmas 1 and 3.

Using the comutative diagram above, we must show that

$$H^q(id \times j_0): H^q(\mathbf{R}^n \times U, (\mathbf{R}^n - 0) \times U) \to H^q(\mathbf{R}^n \times 0, (\mathbf{R}^n - 0) \times 0)$$

is an isomorphism, which follows from the contractibility of U. ■

Case 2. If the theorem holds for opens U, V and $B = U \cap V$, then it holds for the open $Y = U \cup V$.

For brevity, let $U' = X \times U - \Delta$, $V' = X \times V - \Delta$, $B' = X \times B - \Delta$, $Y' = X \times Y - \Delta$.

Sublemma 4. There exists an exact sequence

$$\to H^q(X \times Y, Y') \xrightarrow{i} H^q(X \times U, U') \oplus H^q(X \times V, V') \xrightarrow{j} H^q(X \times B, B') \xrightarrow{k} H^{q+1}(X \times Y, Y') \to$$

where i is induced by the chain homomorphism $z \to (z, z)$, j by the chain homomorphism $(z, w) \to z - w$, and k is a connecting homomorphism.

Granting this sublemma, we see by the case assumption that $H^q(X \times Y, Y') = 0$ for $q < n$. Moreover, the cummutative diagram

$$\begin{array}{ccccccc} 0 \to & H^n(X \times Y, Y') & \xrightarrow{i} & H^n(X \times U, U') \oplus H^n(X \times V, V') & \xrightarrow{j} & H^n(X \times B, B') \\ & \downarrow \phi & & \downarrow \phi \oplus \phi & & \downarrow \phi \\ 0 \to & \Gamma^* Y & \xrightarrow{i} & \Gamma^* U \oplus \Gamma^* V & \xrightarrow{j} & \Gamma^* B \end{array}$$

and the 5-lemma imply that ϕ is an isomorphism for Y.

Proof of sublemma: To derive the exact sequence of sublemma 4, consider the monomorphism of chain complexes

$$i : \frac{S(X \times B)}{S(B')} \to \frac{S(X \times U)}{S(U')} \oplus \frac{S(X \times V)}{S(V')}$$

given by $i(\bar{z}) = (\bar{z}, \bar{z})$, and the chain epimorphism

$$j : \frac{S(X \times U)}{S(U')} \oplus \frac{S(X \times V)}{S(V')} \to \frac{S(X \times U) + S(X \times V)}{S(U') + S(V')}$$

given by $j(\bar{z}, \bar{w}) = \overline{z - w}$. Clearly $ji = 0$. Suppose $j(\bar{z}, \bar{w}) = 0$. If $z = \Sigma \nu_h \sigma_h$, let v be the sum of those $\nu_h \sigma_h$ such that $|\sigma_h|$ meets Δ; then v is equal to the chain defined in the same way using w instead of z, since $z - w \in S(U') + S(V')$, so that $i(\bar{v}) = (\bar{z}, \bar{w})$. Thus we have an exact sequence of chain complexes. Since all these complexes are free, dualizing gives an exact sequence of cochain complexes, hence an infinite exact cohomology sequence

$$\to H^q(C/C') \xrightarrow{i} H^q(X \times U, U') \oplus H^q(X \times V, V')$$
$$\xrightarrow{j} H^q(X \times B, B') \xrightarrow{k} H^{q-1}(C/C') \to$$

where $C = S(X \times U) + S(X \times V)$, $C' = S(U') + S(V')$. Since $X \times Y = X \times U \cup X \times V$, the inclusion of C into $S(X \times Y)$ is a chain homotopy equivalence. This holds similarly for the inclusion $C' \to S(Y')$, and by passage to the quotient, for the inclusion $C/C' \to S(X \times Y)/S(Y')$. Thus we can replace $H^q(C/C')$ by $H^q(X \times Y, Y')$ in the above exact sequence. ∎

Case 3. If $U = \cup_i U_i$, where (U_i) is a totally ordered family of open subspaces for each of which the theorem holds, then the theorem holds for U.

The restriction homomorphism $\Gamma^* U \to \Gamma^* U_i$ give a representation of $\Gamma^* U$ as the projective limit (25.9)

$$\Gamma^* U = \varprojlim_i \Gamma^* U_i$$

Thus we need only verify that the canonical homomorphism

$$\theta: H^q(X \times U, U') \to \varprojlim_i H^q(X \times U_i, U_i')$$

is an isomorphism for all $q \leq n$. If C stands for either S, Z, or B, we have

$$C_q(X \times U, U') = \bigcup_i C_q(X \times U_i, U_i')$$

since each chain has compact support and the U_i are totally ordered. Taking annihilators, we see that

$$C^q(X \times U, U') = \varprojlim_i C^q(X \times U_i, U_i')$$

for $C = S$ or Z. We verify it for $C = B$: If $c = \delta d$, with

$$d \in S^{q-1}(X \times U, U'),$$

then $c \mid X \times U_i = \delta(d \mid X \times U_i)$ for all i. Conversely, suppose for each i there is $d_i \in S^{q-1}(X \times U_i, U_i')$ such that $c \mid X \times U_i = \delta d_i$. We claim that after modifying each d_i by a coboundary, we can get the various d_i to path together to give $d \in S^{q-1}(X \times U, U')$ such that $c = \delta d$. If $U_i \subset U_j$, then $d_i - (d_j | X \times U_i)$ is a cocycle, hence by the theorem for U_i, a coboundary. If $q \leq 1$, this means that $d_i = d_j | X \times U_i)$, so no modification is needed. Assume now $q > 1$ and the claim is true for $q - 1$. Clearly we can assume that the index set is an ordinal number, so that we can perform our modifications by transfinite induction. Suppose j is an index such that the d_i patch together for $i < j$. If j has a predecessor i, and

$$d_i - (d_j | X \times U_i) = e_{ij}$$

we can extend e_{ij} to an element of $S^{q-2}(X \times U_j, U'_j)$ (because

$$S_{q-2}(X \times U_i, U'_i)$$

is a direct summand in $S_{q-2}(X \times U_j, U'_j)$), and replace d_j *by* $d_j + e_{ij}$. If j is a limit ordinal, then we can assume the e_{ij} patch together for all $i < j$ (by inductive assumption on q), giving an $e_j \in S^{q-2}(X \times U_j, U'_j)$, and we replace d_j by $d_j + e_j$. Thus

$$B^q(X \times U, U') = \varprojlim_i B^q(X \times U_i, U'_i)$$

for all $q \leq n$. It follows that θ is an isomorphism for $q \leq n$.

Case 4. U is any open set contained in a coordinate neighborhood.

We enumerate the dense set of points in U having rational coordinates and let V_j be the largest open convex set contained in U about the jth point. Let

$$U_1 = V_1,$$

$$U_i = U_{i-1} \cup V_i, \qquad i > 1.$$

By Cases 1 and 2 and induction on i, the theorem is true for all the U_i. By Case 3, the theorem is true for their union U.

General Case. Apply Zorn's lemma, using the previous cases. ■

(30.4) *Note.* Let $p: X \times X \to X$ be projection on the second factor. If $U \subset X$ is a coordinate neighborhood of x, then, by Case 1, $p^{-1}(U) \cap (X \times X - \Delta)$ is homeomorphic to $(X - \{x\}) \times U$. Hence the pair $(p^{-1}(U), p^{-1}(U) \cap (X \times X - \Delta))$ is homeomorphic to $(X, X - \{x\}) \times U$, a local product structure for p analogous to covering projections. Furthermore $H^n({}_Xi_x)(\mu)$ generates $H^n(X, X - \{x\})$. These are the ingredients used to prove the *Thom Isomorphism Theorem*: *the map*

$$\Phi: H^q(X) \to H^{q+n}(X \times X, X \times X - \Delta)$$

given by $\Phi(c) = p^*(c) \cup \mu$ *is an isomorphism.* The relative cup product is used. We leave the proof as a project for the interested reader. The theorem is true for coordinate neighborhoods by the Künneth Theorem. The general case is obtained by Mayer-Vietoris methods embodied in sublemma 4. The theorem has important extensions and applications to fibre bundles and characteristic classes. Further developments are in [64], [52], [65], [84].

One reason for working with the relative cohomology module $H(X \times X, X \times X - \Delta)$ is the following commutativity lemma.

(30.5) *Lemma. If* $\gamma \in H^p(X \times X, X \times X - \Delta)$, $\eta \in H^q(X)$, *then*

$$H^p(j)(\gamma) \cup (\eta \times 1) = H^p(j)(\gamma) \cup (1 \times \eta)$$

where $j: X \times X \to (X \times X, X \times X - \Delta)$ *is the inclusion.*

Proof. By (26.17.7), there is an open neighborhood V of Δ in $X \times X$ and retraction of $r: V \to \Delta$ such that $ir \simeq k$, where $i: \Delta \to X \times X$, $k: V \to X \times X$ are the inclusion maps. Denote the inclusion

$$(V, V - \Delta) \to (X \times X, X \times X - \Delta)$$

by k', and note that k' is an excision. The following diagram is commutative:

$$\begin{array}{ccccc}
 & & \beta & \to & \gamma \cup \beta \\
 & H^q(i) & H^q(X \times X) & \to & H^{p+q}(X \times X, X \times X - \Delta) \\
H^q(\Delta) & \swarrow & \downarrow H^q(k) & & \downarrow \wr \; H^{p+q}(k') \\
 & \searrow H^q(r) & H^q(V) & \to & H^{p+q}(V, V - \Delta) \\
 & & \xi & \to & H^p(k')(\gamma) \cup \xi
\end{array}$$

where we are using the mixed cup products of absolute and relative cohomology.

Let $p_i: X \times X \to X$, $i = 1, 2$, be the projections. Then $1 \times \eta = H^0(p_1)(1) \cup H^q(p_2)(\eta) = H^q(p_2)(\eta)$, $\eta \times 1 = H^q(p_1)(\eta)$. Let $p: \Delta \to X$ be the common restriction of p_1 and p_2 to the diagonal. From the diagram we get

$$\begin{aligned}
\gamma \cup H^q(p_i)(\eta) &= H^{p+q}(k')^{-1}(H^p(k')(\gamma) \cup H^q(p_i ir)(\eta)) \\
&= H^{p+q}(k')^{-1}(H^p(k')(\gamma) \cup H^q(pr)(\eta))
\end{aligned}$$

for both $i = 1$ and $i = 2$. By definition of the mixed cup product,

$$H^p(j)(\gamma) \cup H^q(p_i)(\eta) = H^{p+q}(\gamma \cup H^q(p_i)(\eta))$$

which proves the lemma. ■

In case X is compact, we can use the Thom class μ (or rather its image μ' in $H^n(X \times X)$) to make explicit the inverse to the Poincaré duality isomorphism.

(30.6) *Theorem. Let X be a compact R-oriented n-dimensional manifold with fundamental class $\zeta \in H_n(X)$. Then for any $p \leq n$, the inverse to the Poincaré duality isomorphism $H^p(X) \cong H_{n-p}(X)$ is given by*

$$\alpha \to (-1)^{pn}\mu'/\alpha, \qquad \alpha \in H_{n-p}(X).$$

Proof. If $\eta \in H^p(X)$, then $\zeta \cap \eta$ is its image in $H_{n-p}(X)$, and

$$\begin{aligned}
\mu'/\zeta \cap \eta &= 1 \cup \{\mu'/\zeta \cap \eta\} \\
&= \{(\eta \times 1) \cup \mu'\}/\zeta && (29.21) \\
&= \{(1 \times \eta) \cup \mu'\}/\zeta && (30.5) \\
&= (-1)^{pn}(-1)^0\eta \cup \{\mu'/\zeta \cap 1\} && (29.21) \\
&= (-1)^{pn}\eta \cup (\mu'/\zeta) \\
&= (-1)^{pn}\eta \cup 1 && (30.2) \\
&= (-1)^{pn}\eta.
\end{aligned}$$ ■

Suppose we have a map $f: X \to Y$, where Y is another compact R-oriented manifold, of dimension m. We define the cohomology class μ_f of the graph of f by

$$\mu_f = H^m(f \times id)(\mu'_Y) \in H^m(X \times Y)$$

where $\mu'_Y \in H^m(Y \times Y)$ is the image of the Thom class of Y. The class μ_f completely determines the homomorphism induced by f on the cohomology.

(30.7) *Proposition. For any $\eta \in H^p(Y)$,*

$$H^p(f)(\eta) = (-1)^{pm}\mu_f/\zeta_Y \cap \eta$$

where $\zeta_Y \in H_m(Y)$ is the fundamental class of Y.

Proof. $\mu_f/\zeta_Y \cap \eta = H^p(f)(\mu'_Y/\zeta_Y \cap \eta)$ (29.23)

$= (-1)^{pm} H^p(f)(\eta).$ (30.6)

■

Assume now $Y = X$. We seek sufficient conditions for f to have a fixed point.

(30.8) *Proposition. If $\mu_f \neq 0$ then f has a fixed point.*

Proof. If f has no fixed point, then we have a factorization

$$\begin{array}{ccc} & & X \times X - \Delta \\ & \swarrow & \downarrow i \\ X \times X & \xrightarrow[f \times id]{} & X \times X \end{array}$$

where i is the inclusion. Since $H^n(i)H^n(j) = 0$ and $\mu'_X = H^n(j)(\mu_X)$, $\mu_f = H^n(f \times id)(\mu'_X) = 0$. ■

To obtain a numerical criterion, consider *the Lefschetz class* $L_f = H^n(f, id)(\mu') = H^n(\Delta)(\mu_f) \in H^n(X)$, and define *the Lefschetz number* by

$$\Lambda_f = [\zeta, L_f] = [\zeta, H^n(f, id)\mu'].$$

By (30.8), $\Lambda_f \neq 0$ implies f has a fixed point.

We compute Λ_f when R is a field: Let $\{\sigma_i\}$ be a basis for $H^{\cdot}(X)$, where i runs through a finite set; let q_i be the integer such that $\alpha_i \in H^{q_i}(X)$. By the Künneth formula (29.17.6), the $\alpha_i \times \alpha_j$ form a basis for $H^{\cdot}(X \times X)$, so that

$$\mu' = \sum_{i,j} c_{ij}\, \alpha_i \times \alpha_j$$

(where $c_{ij} = 0$ when $q_i + q_j \neq n$). Let

$$H^{\cdot}(f)(\alpha_i) = \sum_k a_{ki}\alpha_k$$

(where $a_{ki} = 0$ when $q_k \neq q_i$), and let

$$y_{kj} = [\zeta \cap \alpha_k, \alpha_j] = [\zeta, \alpha_k \cup \alpha_j]$$

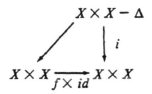

(so that $y_{kj} = 0$ when $q_k + q_j \neq n$, and $y_{jk} = (-1)^{q_k(n-q_k)} y_{kj}$ when $q_k + q_j = n$). Then

$$\Lambda_f = \sum_{i,j} c_{ij}[\zeta, H^{\cdot}(f, id)(\alpha_i \times \alpha_j)]$$

$$= \sum_{i,j} c_{ij}[\zeta, H^{\cdot}(f)(\alpha_i) \cup \alpha_j] \qquad (29.17.7)$$

$$= \sum_{i,j,k} c_{ij} a_{ki}[\zeta, \alpha_k \cup \alpha_j]$$

$$= \sum_{i,j,k} c_{ij} a_{ki} y_{kj} = \sum_{i,k} a_{ki} \left(\sum_j c_{ij} y_{kj} \right).$$

On the other hand,

$$(-1)^{nq_k} \alpha_k = \mu'/\zeta \cap \alpha_k \qquad (30.6)$$

$$= \left(\sum_{j,i} c_{ji} \alpha_j \times \alpha_i \right) / \zeta \cap \alpha_k$$

$$= \sum_{j,i} c_{ji}[\zeta \cap \alpha_k, \alpha_j] \alpha_i \qquad (29.22)$$

$$= (-1)^{q_k(n-q_k)} \sum_i \left(\sum_j c_{ji} y_{jk} \right) \alpha_i$$

so that

$$\sum_j c_{ji} y_{jk} = (-1)^{q_k} \delta_{ik}.$$

Since the right inverse of a matrix is also its left inverse,

$$\sum_j c_{ij} y_{kj} = (-1)^{q_k} \delta_{ik}$$

Thus

$$\Lambda_f = \sum_k (-1)^{q_k} a_{kk}$$

We summarize these results:

(30.9) *Lefschetz Fixed Point Theorem. Let X be a compact R-oriented manifold, where R is a field. If $f: X \to X$ is any map, then the Lefschetz number of f is given by*

$$\Lambda_f = \sum_q (-1)^q \text{ Trace } H^q(f)$$

If $\Lambda_f \neq 0$, then f has a fixed point.

(30.10) *Corollary. The Lefschetz number of the identity map is equal to the Euler characteristic $\chi(X; R)$ of X, i.e.,*

$$\chi(X; R) = [\zeta, H^n(d)\mu']$$

where $d: X \to X \times X$ is the diagonal map.

(30.11) *Exercise.* If $R = \mathbf{Q}$, then Λ_f is actually an integer.

(30.12) *Corollary. If $f: S^n \to S^n$ has degree $d \neq (-1)^{n+1}$, then f has a fixed point.* (The antipodal map (16.4) shows the hypothesis is necessary.)
Proof: Trace $H^n(f) = d$. Trace $H^0(f) = 1$, and $\Lambda_f = 1 + (-1)^n d$. ■

(30.13) *Exercise.* Using (26.12) or (26.14), prove that if n is even every map of $\mathbf{CP}^n$ or $\mathbf{HP}^n$ into itself has a fixed point (see [63]).

(30.14) *Note.* In case f has only finitely many fixed points $x_1, \ldots x_r$, there is a purely local way of assigning an index $I(x_j) \in \mathbf{Z}$ to each point so that the Lefschetz Fixed Point Formula holds:

$$\Lambda_f = \sum_j I(x_j)$$

(see Brown [11, 63]). The Lefschetz formula has very interesting analogues in differential geometry (Bott and Atiyah [8]) and in algebraic geometry (Weil [60]).

(30.15) *Exercise.* Let X, Y be R-oriented manifolds of dimensions n, m respectively, not necessarily compact. Then for each point $(x, y) \in X \times Y$, we have a canonical isomorphism (29.16)

$$\times: H_n(X, X - x) \otimes H_m(Y, Y - y)$$
$$\to H_{m+n}(X \times Y, X \times Y - (x, y))$$

(since all other terms in the Künneth formula are zero). Then there is a unique R-orientation of $X \times Y$ such that the local R-orientation at (x, y) is $\zeta_x \times \zeta_y$, where ζ_x (resp. ζ_y) is the local R-orientation of X at x (resp. of Y at y); this is called *the product R-orientation* of $X \times Y$. In case X and Y are compact, then

$$\zeta_{X \times Y} = \zeta_X \times \zeta_Y$$

where ζ denotes the fundamental class for the R-orientation.

Suppose X is compact R-oriented and $\mu_{X \times X}$ is the Thom class for the manifold $X \times X$ in the product R-orientation, $\mu'_{X \times X}$ its image in $H^{2n}(X \times X \times X \times X)$. What is the relation between $\mu'_{X \times X}$ and $\mu'_X \times \mu'_X$? Let μ''_X be the Poincaré dual of $H_n(\Delta)(\zeta_X)$, i.e.,

$$(\zeta_X \times \zeta_X) \cap \mu''_X = H_n(\Delta)(\zeta_X).$$

Prove $\mu'_X = \mu''_X$. (see 30.17)

(30.16) *Project.* Generalize all the results of Section 30 to compact manifolds with boundary; in particular, determine the inverse isomorphism to the Lefschetz Duality Theorem and prove a Lefschetz fixed point theorem.

(30.17) *Remark.* Let X be compact of dimension n. Take coefficients in a field. We construct an explicit representation of μ', the restriction of the Thom class to $H^n(X \times X)$. Given a basis $\{b_i\}$ for $H^{\cdot}(X)$, construct a dual basis $\{b_i^*\}$ for $H^{\cdot}(X)$ by requiring

$$[\zeta, b_j^* \cup b_i] = \begin{cases} 1 & i = j, \\ 0 & i \neq j. \end{cases}$$

This is possible by Poincaré Duality. We use the slant product and (29.21) to show

(30.18) *Proposition.* $\mu' = \Sigma_i (-1)^{|b_i|} b_i^* \times b_i$ *where* $|b_i| = \dim b_i$.

Proof. Using (29.17.6), write $\mu' = \Sigma \alpha_{rs} b_r^* \times b_s$. Then for $a \in H^k(X)$, by (30.5)

$$(a \times 1) \cup \mu' = (1 \times a) \cup \mu'.$$

Using (29.21) and (30.3) we obtain

$$\{(1 \times a) \cup \mu'\}/\zeta = (-1)^{nk} a \cup \{\mu'/\zeta\} = (-1)^{nk} a.$$

Substituting for μ' and using (29.17.5), we have

$$(-1)^{nk} a = \{(a \times 1) \cup \mu'\}/\zeta = \Sigma \alpha_{rs} \{(a \cup b_r^*) \times b_s\}/\zeta.$$

Substituting $a = b_r$ and using (29.22) yields

$$(-1)^{n|b_r|}b_r = \Sigma\alpha_{rs}(-1)^{|b_r^*||b_r|}b_s.$$

Since $\{b_s\}$ are independent, we obtain

$$\alpha_{rs} = \begin{cases} (-1)^{|b_r|} & r = s, \\ 0 & r \neq s. \end{cases}$$ ■

The appearence of (30.18) (signs, whether b_i^* is on the right or left) is sensitive to the conventions used when defining cap products.

Prove $\mu_x' = \mu_x''$. Suggestion: Write $\mu_x'' = \Sigma\alpha_{rs}b_r^* \times b_s$ and evaluate $\zeta \dot{\times} \zeta \cap \mu_x''$ on $b_i \times b_j^*$.

(30.19) *Exercise.* Show that $H^n(i)\mu' = (-1)^n\mu'$ where i:$X \times X \to X \times X$ is the interchange map and $n =$ dim X.

(30.20) *Exercise.* Show the existence of a class μ satisfying (30.3) implies X is orientable.

(30.21) *Exercise.* Derive (30.10) from (30.18). This is easy but worth knowing.

(30.22) *Exercise.* Let $f: X \to Y$ as in (30.9) and define g:$X \to X \times Y$ by $g(x) = (x, f(x))$. Let $\Gamma = H(g)(\zeta_X)$. Prove $\zeta_X \times \zeta_Y \cap \mu_f = \Gamma$. Suggestion: Take coefficients in a field and let w satisfy $\zeta_X \times \zeta_Y \cap w = \Gamma$. Write $w = \Sigma\lambda_{r,s}b_r^* \times c_s$ and $\Gamma = H(1 \times f)(\zeta_X \times \zeta_Y \cap \mu_X')$ by (30.15). Then evaluate on $b_r \times c_s^*$ and use (30.18) to compute μ_f. *Remark.* The identification of μ' and μ_f as Poincaré duals of the diagonal $H(\Delta)(\zeta_X)$ and the graph Γ respectively provides geometric insight. For example the formula $\Lambda_f = [\zeta \times \zeta, \mu' \cup \mu_f]$ follows from (30.15). In the next chapter, we can interpret this formula as the "intersection number" of the diagonal and the graph.

31. *Intersection numbers and cup products.*

Consider, once again, the torus. A basis for H_1 is represented by a pair of embedded circles intersecting in a point. The cup product of the dual basis elements is nonzero. Another example is the orientable surface of genus 2 with a basis for H_1 represented in the picture.

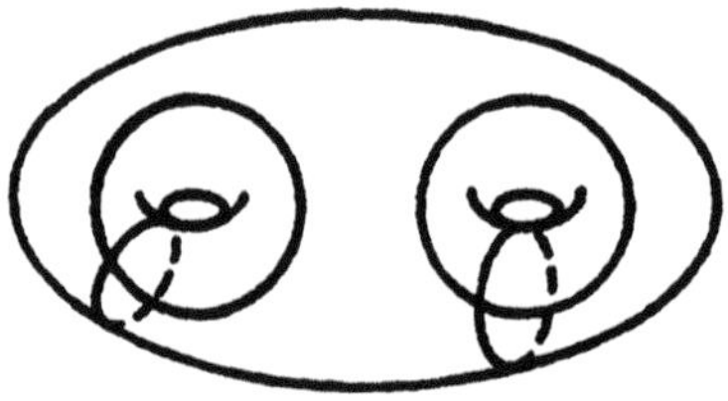

Using (24.30) to calculate, the nonzero and zero values among cup products of different elements in the dual basis for H^1 correspond to nonempty or empty intersections among the representative circles. We use the Thom class to develop the relation between intersection numbers and cup products. Our account is introductory. A fuller account is given in Dold [64] and a geometric treatment in Seifert and Threlfall [86]. The basic ideas for this treatment come from Dold's book and Milnor [82]. The basic paper is Thom [57]. Throughout coefficients are either a field or the integers.

Let V be a closed n-manifold with submanifolds M, N of dimensions r, s respectively.

Basic assumption, $r + s = n$ and $M \cap N$ (if nonempty) is a finite set of points.

(31.1) *Definition.* M intersects N *transversally* at a point x if there is a coordinate neighborhood U of x and a homeomorphism $(U, U \cap M, U \cap N) \cong (\mathbf{R}^{r+s}, \mathbf{R}^r \times \{0\}, \{0\} \times \mathbf{R}^s)$. Necessarily x corresponds to the origin. We say $M \cap N$ is *transverse*, if it is transverse at each point.

Examples. The intersection of either the graph of $y = x$ or $y = x^3$ and the x-axis is transverse at the origin. A nontransverse intersection is exhibited by the graph of $y = x^2$ and the x-axis.

If each of the manifolds M, N, V is oriented (or coefficients are $\mathbf{Z}/2\mathbf{Z}$) we can assign an *intersection number* ± 1 to each transverse intersection as follows. Let ζ_M^x, ζ_N^x and ζ_V^x be the local orientations at $x \in M \cap N \subset V$ induced by the given global orientations. Let U be a coordinate neighborhood of x as in (22.1). The inclusion $(U, U - x) \subset (V, V - x)$ is an excision. By transversality we have a homeomorphism

$$(U, U - x) \cong (\mathbf{R}^{r+s}, \mathbf{R}^{r+s} - \{0\}) = (\mathbf{R}^r, \mathbf{R}^r - \{0\}) \times (\mathbf{R}^s, \mathbf{R}^s - \{0\}).$$

Now form the commutative diagram

$$\begin{array}{ccc} H_r(\mathbf{R}^r, \mathbf{R}^r - \{0\}) \otimes H_s(\mathbf{R}^s, \mathbf{R}^s - \{0\}) & \xrightarrow{\times} & H_{r+s}(\mathbf{R}^{r+s}, \mathbf{R}^{r+s} - \{0\}) \\ \downarrow & & \downarrow \\ H_r(M \cap U, M \cap U - x) \otimes H_s(N \cap U, N \cap U - x) & \xrightarrow{\times} & H_{r+s}(U, U - x) \\ & & \downarrow \\ & & H_{r+s}(V, V - x) \end{array}$$

where the horizontal maps are cross products and the vertical maps are excisions. By (30.15)

$$\zeta_M^x \times \zeta_N^x = \varepsilon \zeta_V^x \quad \text{where } \varepsilon = \pm 1.$$

(31.2) *Definition. The intersection number of M and N at x is ε.* We write $M \cdot N = \Sigma_i \varepsilon_i$ where ε_i is the intersection number of M and N at each transverse intersection x_i. $M \cdot N$ is not defined here if $M \cap N$ fails to be transverse.

(31.3) *Lemmas. The intersection number is independent of coordinate neighborhood U.*

We leave the proof as an exercise. ■

(31.4) *Lemma.* $M \cdot N = (-1)^{rs} N \cdot M$.

Proof. This follows from the corresponding property for cross products (29.29). ■

Remark. A useful geometric way to think of the local intersection number ε is in terms of equivalence classes of frames. If we juxtapose a local orientation frame for M at x with one for N at x, we obtain a frame for V at x. The value of ε is ± 1 according as this frame is (is not) equivalent to the orientation of V.

(31.5) *Lemma. Let* $k:(V, V-x) \to (V\times V, V\times V-\Delta)$ *be the inclusion. Then* $M \cdot N = \Sigma_i[\zeta_M^{x_i} \times \zeta_N^{x_i}, H^n(k)\mu]$, where μ is the Thom class and $\zeta_M^{x_i} \times \zeta_N^{x_i}$ is regarded as an element of $H_n(V, V-x)$.

Proof. By (30.2), $[\zeta_V^{x_i}, H^n(k)\mu] = 1$. ■

As things stand $M \cdot N$ is not well related to $H.(V)$. For example, if $V = S^2$, M and N a pair of intersecting arcs, we can have $M \cdot N \neq 0$.

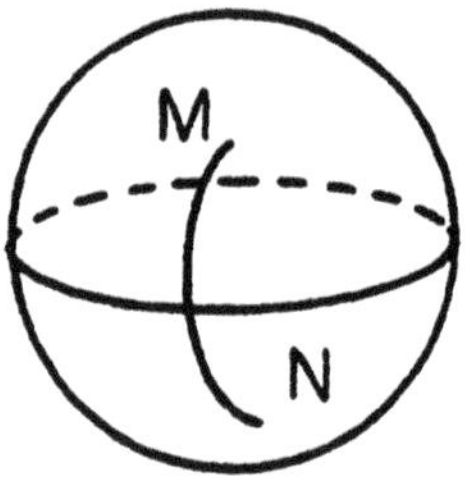

(31.6) *Definition.* A class $\alpha \in H_r(V)$ is *represented* by a submanifold M if $\alpha = H_r(i)(\zeta_M)$ where $i:M \to V$ is inclusion.

The connection between algebra and geometry is supplied by

(31.7) *Proposition. Let* $\alpha \in H_r(V)$, $\beta \in H_s(V)$ *be represented by submanifolds* M, N *such that* $M \cap N$ *is transverse. Then* $[\alpha \times \beta, \mu'] = M \cdot N$ *where* μ' *is the restriction of the Thom class.*

Proof. We first set up notation. Let $\{U_i\}$ be coordinate neighborhoods of the transverse intersections at $\{x_i\} = M \cap N$. Write $\Delta | M \cap N \subset V\times V$ for the set $\{(x, x) | x \in M \cap N\}$. We have a commutative diagram

$$\begin{array}{ccc} M\times N & \to & (M\times N,\ M\times N-\Delta \mid M\cap N) \\ \downarrow & & \downarrow l \\ V\times V & \underset{j}{\to} & (V\times V,\ V\times V-\Delta). \end{array}$$

Write the pair of open sets $(M\cap U_i, N\cap U_i)=(A_i, B_i)$. By naturality of cross products, we have the commutative diagram

$$\begin{array}{ccccc} H_r(M)\otimes H_s(N) & \overset{\times}{\to} & H_{r+s}(M\times N) \to & & H_{r+s}(M\times N,\ M\times N-\Delta \mid M\cap N) \\ \downarrow & & \downarrow & & \downarrow H(l) \\ H_r(V)\otimes H_s(V) & \overset{\times}{\to} & H_{r+s}(V\times V) & \underset{H(j)}{\to} & H_{r+s}(V\times V,\ V\times V-\Delta). \end{array}$$

Using excision, we factor $H(l)$ into a composite

$$H_n(M\times N, M\times N-\Delta \mid M\cap N)\simeq \sum_i H_n(A_i\times B_i, A_i\times B_i-(x_i, x_i))$$

$$\simeq \sum_i H_n(U_i, U_i-x_i)\simeq \sum_i H_n(V, V-x_i)\overset{H(k)}{\to} H_n(V\times V, V\times V-\Delta).$$

Here the second isomorphism is from transversality, and the others are excisions. Now we calculate,

$$[\alpha\times\beta, \mu']=[\alpha\times\beta, H^{\cdot}(j)\mu]=[H_{\cdot}(j)(\alpha\times\beta), \mu]$$

$$=[H_{\cdot}(l)(\zeta_M\times\zeta_N), \mu]=\sum_i[\zeta_M^{x_i}\times\zeta_N^{x_i}, H^{\cdot}(k)\mu]$$

$$=M\cdot N.$$ ■

We apply (31.7) to calculate cup products in $H^{\cdot}(V)$.

(31.8) *Corollary. If* $a=\mu'/\alpha$, $b=\mu'/\beta$, *then* $[\zeta_V, a\cup b]=(-1)^s M\cdot N$. *Recall* $\beta\in H_s(V)$.

Proof. By (30.6), $\zeta\cap a=(-1)^{ns}\alpha$. Hence

$$[\zeta, a\cup b]=[\zeta\cap a, b]=(-1)^{ns}[\alpha, \mu'/\beta]=(-1)^{ns}[\beta\times\alpha, \mu']$$

$$=(-1)^s[\alpha\times\beta, \mu']=(-1)^s M\cdot N.$$ ■

Remark. The annoying sign is a consequence of our conventions for cap products.

Remark. The possibility that $\alpha = \beta$ is allowed. But if $\alpha = \beta$ we still require different submanifolds representing the classes, and they must intersect transversally or not at all.

Remark. The reader should examine the opening examples in light of (31.8). In particular note the different usages of the word "dual" to describe bases in H_1, H^1 related by algebraic duality and Poincaré duality.

Example. The cup product structure for the mod 2 cohomology of $\mathbf{P}^3$ is read off a suitable picture. Think of $\mathbf{P}^3$ as a solid 3-ball with antipodal points on the boundary identified. There are embeddings $\mathbf{P}^2 \subset \mathbf{P}^3$ from an equatorial 2-disc and $\mathbf{P}^1 \subset \mathbf{P}^3$ from a diameter. Their intersection detects the nonzero cup product $H^1 \otimes H^2 \to H^3$.

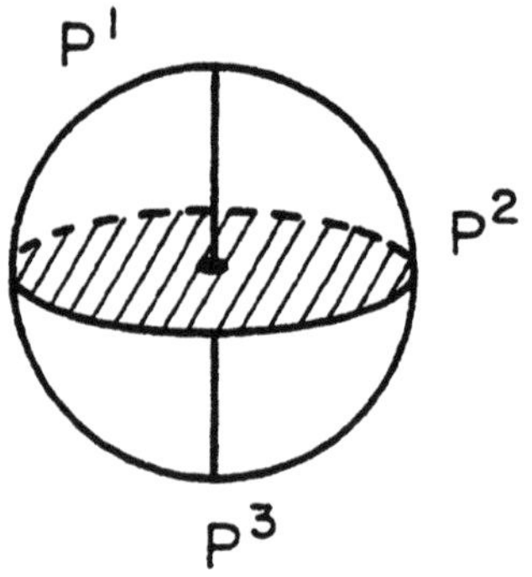

Exercise. Use (31.8) to calculate the cohomology rings of compact surfaces, and $\mathbf{P}^n$, $\mathbf{CP}^n$, $\mathbf{HP}^n$.

Exercise. Let $f: X^n \to Y^n$ be a map of closed manifolds with deg $f = k$. Show $k\chi(Y) = [\Gamma, \mu_f]$ where Γ is defined in (30.22). Use (31.6) and (30.22) to obtain the equation $k\chi(Y) = \Gamma \cdot \Gamma$.

Note. The reader wishing further development of intersection theory is referred to Dold [64], VIII, §11, p. 314–325. Here we give a brief outline of the connection between the material here and Dold's formulation.

Let $e: M' \to V^n$ be an embedding with $e(M)$ sufficiently well behaved that $\check{H}^q(M) \cong H^q(M)$ (27.1). Then there is an Alexander duality isomorphism $H^{n-q}(M) \to H_q(V, V - M)$ (27.6). Combined with Poincaré duality, we obtain the *transfer* isomorphism

$$H_q(V,\ V-M) \to H^{n-q}(M) \xrightarrow{\zeta_M} H_{q-s}(M)$$

where $n = r + s$. Under this isomorphism, $1 \in H_0(M)$ corresponds to a class $\nu_M \in H_s(V,\ V-M)$ called the *transverse class of M in V*. Using universal coefficients, we obtain the *Thom class of M in V*, $\tau_M \in H^s(V,\ V-M)$ such that $[\nu_M, \tau_M] = 1$. In the case of the diagonal $\Delta \subset V \times V$, we have $\tau_\Delta \equiv \mu \in H^n(V \times V,\ V \times V - \Delta)$ as defined in chapter 30.

Now suppose M^r, $N^s \subset V$ intersect transversally in a single point. It is proved that under the cup product pairing

$$H^s(V,\ V-M) \otimes H^r(V,\ V-N) \to H^n(V,\ V-M \cap N)$$

we have $\tau_M \cup \tau_N = \pm\ \tau_{M\cap N}$. The sign is the intersection number. The requirement $r + s = n$ is unnecessary and the nonconnected situation develops in the expected way. Furthermore, if $\mu_M \in H^s(V)$ is the pull-back of τ_M, one can prove $\zeta_V \cap \mu_M = e_*(\zeta_M)$. This is a generalization of (30.15).

Our (31.7) results in treating the case $M \times N$, $\Delta \subset V \times V$. The formalism is:

$$\zeta_{V\times V} \cap \mu_{M\times N} = e_*(\zeta_M \times \zeta_N) = \alpha \times \beta.$$

$$\zeta_{V\times V} \cap \mu_\Delta = e_*(\zeta_\Delta) \qquad (30.15).$$

Then writing (31.7) in this notation, we have

$$\begin{aligned}[\alpha \times \beta, \mu'] &= [\zeta_{V\times V} \cap \mu_{M\times N}, \mu_\Delta] \\ &= [\zeta_{V\times V}, \mu_{M\times N} \cup \mu_\Delta] \\ &= [\zeta_{V\times V}, \tau_{M\times N} \cup \tau_\Delta] \quad \text{by naturality of cup product,} \\ &= [\zeta_{V\times V}, \pm\ \tau_{(M\times N)\cap\Delta}].\end{aligned}$$

An extension of intersection theory (lengthy example). The local character of intersection theoretic calculation permits substantial generalization. We give a simple extension in the case of lens spaces $L(p, q)$ and refine earlier calculations. Our goal is to prove

(31.9) *Proposition.* Let p be prime. *There is a generator* $x \in H^1(L(p, q); \mathbf{Z}/p\mathbf{Z})$ *such that* $x \cup \beta x = qz$ *where* $z \in H^3(L(p, q); \mathbf{Z}/p\mathbf{Z})$ *satisfies* $[\zeta, z] = 1$, ζ *is the mod p reduction of the integral fundamental class and* β *is the Bockstein homomorphism associated with the short exact sequence* $0 \to \mathbf{Z}/p\mathbf{Z} \to \mathbf{Z}/p^2\mathbf{Z} \to \mathbf{Z}/p\mathbf{Z} \to 0$.

Before proving this, we first recall the main features of the spaces $L(p, q)$ and introduce the models to be used. Next we discuss the extension of (31.7) to these models. After this, the proof will be calculation.

Recall $L(p, q) = (E_1^2 \times S^1) \cup_h (E_2^2 \times S^1)$ where h is a homeomorphism of $\partial(E_i^2 \times S^1) = S^1 \times S^1$ such that $h| \partial E_1^2 \times \{pt\}$ is a (p, q)-torus knot on $\partial(E_2^2 \times S^1)$. The integers (p, q) are coprime. For models we form $E^2 \times S^1$ by identifying $(x, 0)$ with $(x, 1)$ in $E^2 \times I$. We write $M = \partial E^2 \times \{\frac{1}{2}\}$ for the circle whose image under h is the (p, q) torus knot. For suitable h, $h(M)$ is represented by p helicial arcs on $\partial E_2^2 \times I$ with the top of the arc displaced from the bottom by rotation through $2\pi q/p$ radians.

Here is a picture for $L(5, 2)$.

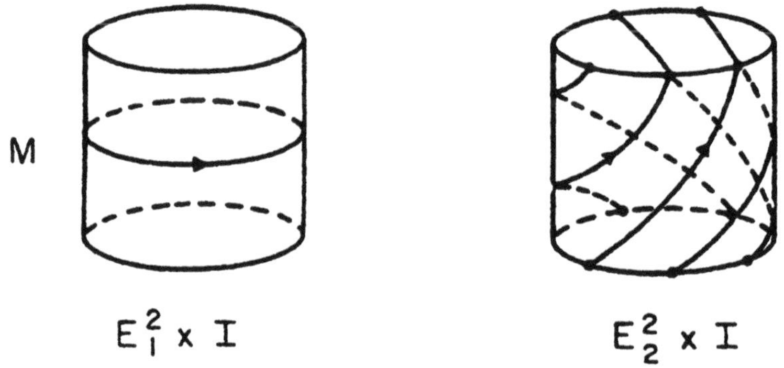

We orient $L(p, q)$ with the right hand orientation on each block $E_i^2 \times I$ and take h to be orientation reversing, and choose h to map M in the indicated direction.

We have calculated the integral and mod p cohomology groups by Mayer-Vietoris sequences (combine (21.28) and (23.28)). Abbreviating $L(p, q)$ to L, the results are $H^q(L; \mathbf{Z}/p\mathbf{Z}) \cong \mathbf{Z}/p\mathbf{Z}$, $0 \leq q \leq 3$ and with integer coefficients $H^3 \cong \mathbf{Z}$, $H^2 \cong \mathbf{Z}/p\mathbf{Z}$, $H^1 = 0$. For the mod p cohomology ring, recall $\beta: H^1 \to H^2$ is an isomorphism (23.37). By Poincaré duality, the cup product pairing $H^1 \otimes H^2 \to H^3$ is nondegenerate. Hence if $x \in H^1(L; \mathbf{Z}/p\mathbf{Z})$ is a generator, then βx and $x \cup \beta x$ generate H^2 and H^3. Of course the integral cohomology ring is trivial. Consequently, these cohomology rings are independent of q.

The effect of the $2\pi q/p$ twist shows up when we use intersection theory. We turn to the appropriate modifications of (31.7).

Clearly, nonzero elements of $H_2(L; \mathbf{Z}/p\mathbf{Z})$ cannot be represented by submanifolds (why? note $H_2(L; \mathbf{Z}) = 0$). Instead we shall represent H_2 by an embedded "pseudo-projective plane" P. P is obtained by attaching a 2-cell along its boundary to a circle by a map of degree p

$$P = S^1 \cup_p e^2.$$

The abstract pseudo-projective plane P may be visualized as a regular p-sided polygon with identifications on the boundary. We embed P in L by describing the pieces in the blocks $E_i^2 \times I$. One piece is the disc $E_1^2 \times \{\frac{1}{2}\}$ bounding M. The other piece consists of flanges from the p helicial arcs to the core axis of $E_2^2 \times I$. The pieces are pictured in figure 10. A Mayer-Vietoris calculation shows the inclusion $P \to L$ induces isomorphisms of H_1, H_2 with $\mathbf{Z}/p\mathbf{Z}$ coefficients. We use P to represent H_2 in the sense of (31.6).

Next we consider orientation. While P is not a manifold, it has local euclidean structure at points x in the interior of the polygon described above. For such points and coefficients $\mathbf{Z}/p\mathbf{Z}$ there is an excision $(\mathbf{R}^2, \mathbf{R}^2 - \{0\}) \to (P, P - x)$ and an isomorphism

$$H_2(P; \mathbf{Z}/p\mathbf{Z}) \to H_2(P, P - x; \mathbf{Z}/p\mathbf{Z}).$$

(Prove this and also see what happens when integer coefficients are used.)

Thus we can identify a generator of $H_2(P, P - x)$ with an orientation and, with mod p coefficients, pull this class back to $H_2(P; \mathbf{Z}/p\mathbf{Z})$. In figure 10 an orientation of the preimage of P is indicated along with the corresponding orientation of P in L. The orientation of P depends on the choice of h.

The key step in the proof of (31.7) is the factorization of the map $H(l)$ (in the notation of (31.7)). For this only local transversality is required. We shall use the extension of (31.7) to the current situation without formal statement.

Next we look for a generator of H_1 having transverse intersection with P. The Mayer-Vietoris calculations shows the core axis of $E_2^2 \times I$ represents a generator of H_1 after identification of its end points. The core axis is oriented

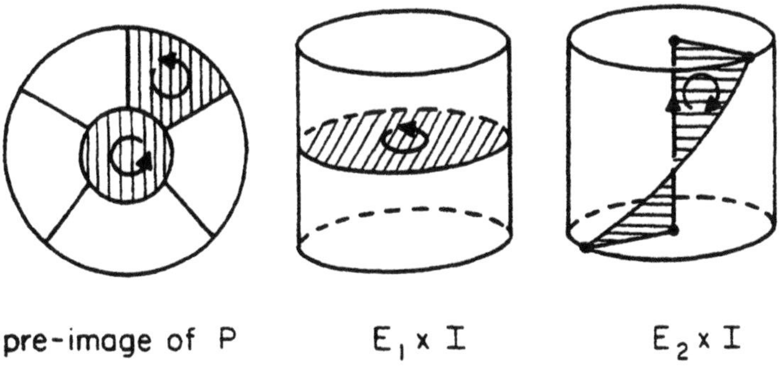

Figure 10.

upward as indicated in figure 10. A small parallel displacement of the core axis gives a line C intersecting P transversally in q-points. The intersection number at each of these points is -1, since the orientation of C juxtaposed with the orientation of P is left-handed.

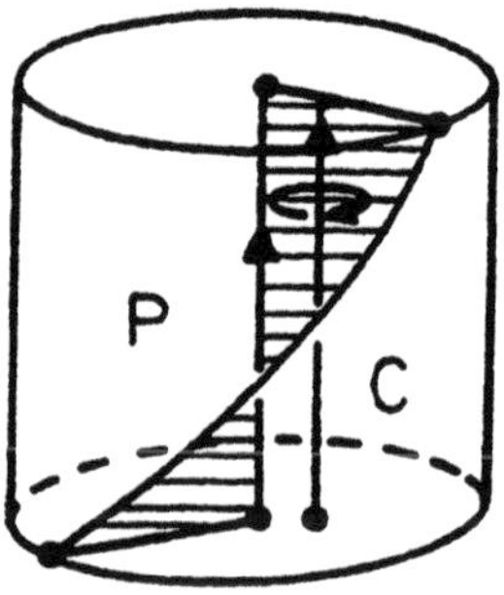

We summarize this discussion:

1. Let $\gamma \in H_1(L; \mathbf{Z}/p\mathbf{Z})$ be the class represented by C, a small parallel displacement of the core axis of $E_2 \times I$ with the upward orientation.

2. Let $\pi \in H_2(L; \mathbf{Z}/p\mathbf{Z})$ be the image of the generator $g \in H_2(P; \mathbf{Z}/p\mathbf{Z})$ pulling back an orientation of $(P, P - x)$ where $x \in C \cap P$. Then

3. $[\gamma \times \pi, \mu'] = -q$ where $\mu' \in H^3(L \times L; \mathbf{Z}/p\mathbf{Z})$ is the restriction of the Thom class and

4. $\beta\pi = \gamma$ where $\beta{:}H_2 \to H_1$ is the Bockstein.

Proof. Item 3 is proved in the preceding discussion. For item 4, consider the short exact sequence of chain complexes

$$0 \to S(L) \otimes \mathbf{Z}/p\mathbf{Z} \xrightarrow{p} S(L) \otimes \mathbf{Z}/p^2\mathbf{Z} \to S(L) \otimes \mathbf{Z}/p\mathbf{Z} \to 0.$$

Then a chain representing π pulls back to a chain z whose boundary is divisible by p. Using the model P, one sees $p^{-1}\partial(z)$ is a chain representing γ. ■

We can now complete the proof of (31.9). Let $y \in H^1(L; \mathbf{Z}/p\mathbf{Z})$ satisfy $[\gamma, y] = 1$. Then $[\pi, \beta y] = [\beta\pi, y] = [\gamma, y] = 1$. Since $-q = [\gamma \times \pi, \mu'] = [\pi, \mu'/\gamma]$, we have $\mu'/\gamma = -q\beta y$. Similarly $\mu'/\pi = -qy$. Hence by (31.8)

$$q^2[\zeta, y \cup \beta y] = (-1)(-q) = q$$

so $y \cup \beta y = q'z$ where $qq' \equiv 1 \mod p$ and z satisfies $[\zeta, z] = 1$. Set $x = qy$ to obtain the desired formula $x \cup \beta x = qz$. ■

(31.10) *Corollary. If $L(p, q)$ is homotopy equivalent to $L(p, q')$, then $\pm q' \equiv qm^2 \mod p$.*

Proof. Let $f: L(p, q) \to L(p, q')$ be a homotopy equivalence. Then $H(f)\zeta = \pm\zeta'$ since ζ is the reduction of an integral class. Let x, x' be the classes in H^1 of $L(p, q), L(p, q')$ given by (31.8). Then $H^{\cdot}(f)x' = mx$ for some mod p integer m. Calculating

$$\begin{aligned} \pm q' &= [\pm\zeta', x' \cup \beta x'] = [\zeta, H^{\cdot}(f)(x' \cup \beta x')] \\ &= m^2[\zeta, x \cup \beta x] = m^2 q \end{aligned}$$

where the equations hold in $\mathbf{Z}/p\mathbf{Z}$. ■

Remark. The homeomorphisms among lens spaces (21.27) allow the rephrasing of the necessary condition for homotopy equivalence; $\pm qq'$ is a quadratic residue mod p.

Remark. Examples can be found among the lens spaces which are not homotopy equivalent, but have isomorphic cohomology rings and isomorphic homotopy groups. For example $L(5, 1)$ and $L(5, 2)$ are not homotopy equivalent by (31.10). We have seen their cohomology rings and fundamental group are isomorphic. Since the universal covering space of each is S^3, their higher homotopy groups are isomorphic as well.

Note. Perhaps the most important examples are pairs of lens spaces which are homotopy equivalent but not homeomorphic. For arbitrary spaces, such examples are common place, but for closed manifolds the issue is much more subtle. Recall that there are no such examples for $n \leq 2$, and the solution of the generalized Poincaré conjecture asserts that a manifold of the homotopy type of S^n is homeomorphic to S^n for $n \geq 5$ (see note (19.12)). The lens space examples arise because the sufficient condition for homeomorphism, $L(p, q) \cong L(p, q')$ if $\pm q' \equiv \pm q^{-1} \mod p$, is also necessary (due to Moise and Reidemeister using work of Franz) while the necessary condition for homotopy equivalence (31.10) is also sufficient (due to J. H. C. Whitehead). Thus e.g., $L(7, 1)$ and $L(7, 2)$ are homotopy equivalent but not homeomorphic. These matters and more are discussed in detail in M. Cohen [78].

Exercise. The core axis of $E_1 \times I$ *oriented downward* gives another generator $\alpha \in H_1(L(p, q); \mathbf{Z}/p\mathbf{Z})$, after identification of endpoints. Show, both by intersection theory and Mayer-Vietoris sequences, $\gamma = q\alpha$ and $[\alpha, x] = 1$ where x is from (31.9).

TABLE OF SYMBOLS

TABLE OF SYMBOLS

Standard Symbols

S^n	n-sphere
$\mathbf{R}$	field of real numbers
$\mathbf{Z}$	the integers
$\mathbf{Q}$, $\mathbf{C}$, $\mathbf{H}$	rational numbers, complex numbers, quaternions
$\mathbf{I}$	closed unit interval
$\cong$	isomorphic
$\xrightarrow{\approx}$	isomorphism
$\approx$	homeomorphic
$\overline{U}$	closure of U
$\dot{A}$	interior of A
ϕ	empty set
M^*	the dual $\mathrm{Hom}_R(M, R)$ of the moduel M
t_f	transpose of the homomorphism f

BIBLIOGRAPHY

BIBLIOGRAPHY

1. J. F. Adams, "Vector fields on spheres," *Ann. Math.*, **75**, 603–632 (1962).
2. L. Ahlfors and L. Sario, *Riemann Surfaces*, 2nd printing (corrected), Princeton Univ. Press, Princeton, New Jersey (1965).
3. E. Artin, *Vorlesungen über Algebraische Topologie*, Hamburg (1964).
4. E. Artin and J. Tate, *Class Field Theory*, Harvard notes.
5. M. Artin, *Grothendieck Topologies*, Harvard notes (1962).
6. M. Atiyah, *K-Theory*, Harvard notes (1964).
7. R. L. Bishop and R. J. Crittenden, *Geometry of Manifolds*, Academic Press, New York (1964).
8. R. Bott and M. Atiyah, "A Lefschetz fixed point formula for elliptic differential operators," *Bull. Amer. Math. Soc.*, **72**, 245–250 (1966).
9. M. Brown, "A proof of the generalized Schoenflies theorem," *Bull. Amer. Math. Soc.*, **66**, 74–76 (1960).
10. M. Brown, "Locally flat imbeddings of topological manifolds," *Ann. Math.*, **75**, 331–341 (1962).
11. R. F. Brown "On the Lefschetz fixed point theorem," *Amer. J. Math.*, **87**, 1–10 (1965).
12. H. Cartan, *Séminaire*, 1948-present, Institut Henri Poincaré, Paris.
13. C. Chevalley, *Algebraic Functions of One Variable*, Amer. Math. Soc., N. Y. (1951).
14. C. Chevalley, *Lie Groups*, 2nd printing, Princeton Univ. Press, Princeton, New Jersey (1960).
15. P. E. Conner, Jr., and E. E. Floyd, *Differentiable Periodic Maps*, Academic Press, New York (1964).
16. R. Crowell and R. Fox, *Knot Theory*, Ginn, Boston, Massachusetts (1963).
17. G. de Rham, *Variétés Différentiables*, Hermann, Paris (1960).
18. P. Dolbeault, "Formes différentielles et cohomologie sur une variété analytique complexe, I and II," *Ann. Math.*, **64**, 83–130 (1956); *Ann. Math.*, **65**, 282–330 (1957).
19. A. Dold, "Erzeugende der Thomschen Algebra," *Math. Zeit.*, **65**, 25–35 (1956).

20. J. Dugundji, *Topology*, Allyn and Bacon, Boston, Massachusetts (1960).
21. S. Eilenberg, "Singular homology theory," *Ann. Math.*, **45**, 407–447 (1944).
22. S. Eilenberg and S. MacLane, "Acyclic Models," *Amer. J. Math* **75**, 189–199 (1953).
23. S. Eilenberg and N. Steenrod, *Foundations of Algebraic Topology*, Princeton Univ. Press, Princeton, New Jersey (1952).
24. E. Fadell, "Generalized normal bundles for locally flat imbeddings," *Trans. Amer. Math. Soc.*, **114**, 488–513.
25. I. M. Gelfand, R. A. Minlos, and Z. Ya. Shapiro, *Representations of Rotation and Lorentz Groups*, Macmillan, N. Y. (1963).
26. S. Goldberg, *Curvature and Homology*, Academic Press, N.Y. (1962).
27. A. Grothendieck, "Cohomology theory of abstract algebraic varieties," *Proc. Intern. Congr. Math. Edinburgh*, 103–118 (1958).
28. P. Halmos, *Finite Dimensional Vector Spaces*, Van Nostrand, Princeton, New Jersey (1958).
29. O. Hanner, "Some theorems on absolute neighborhood retracts," *Arkiv für Mat.*, **2**, 315–360 (1952).
30. P. Hilton and S. Wylie, *Homology Theory*, Cambridge University Press, London and New York (1960).
31. F. Hirzebruch, *New Topological Methods in Algebraic Geometry*, Springer, Berlin (1965).
32. J. Hocking and G. Young, *Topology*, Addison-Wesley, Reading, Massachusetts (1960).
33. S. T. Hu, *Homotopy Theory*, Academic Press, New York (1959).
34. J. Kelley, *General Topology*, Van Nostrand, Princeton, New Jersey (1955).
35. S. Lang, *Algebra*, Addison-Wesley, Reading, Massachusetts (1964).
36. R. Lashof, "Problems in differential and algebraic topology," *Ann. Math.*, 81, 565–591 (1965).
37. S. Lefschetz, *Algebraic Topology*, Amer. Math. Soc., N.Y. (1942).
38. S. MacLane, *Homology*, Academic Press, New York (1963).
39. B. Mazur, "On embeddings of spheres," *Bull. Amer. Math. Soc.*, **65**, 59–65 (1959).
40. J. Milnor, "On the cobordism ring and a complex analogue, I" *Amer. J. Math.*, **82**, 505–521 (1960).
41. J. Milnor, *Lectures on Characteristic Classes*, Princeton notes (1957).
42. J. Milnor, *Lectures on Differential Topology*, Princeton notes (1958).
43. J. Milnor, *Morse Theory*, Princeton Univ. Press, Princeton, New Jersey, 1963.
43a. J. Milnor, *Topology from the differentiable viewpoint*, University of Virginia Press, Charlottesville (1965).

44. J. Munkres, *Elementary Differential Topology*, Princeton Univ. Press, Princeton, N.J. (1963).
44a. S. P. Novikov, "New ideas in algebraic topology," *Russian Math. Surveys*, **20**, 37–62 (1965).
45. J. H. Poincaré, *Analysis situs*, J. de l'École Polytechnique, Paris (1895).
46. D. Puppe, *Topologie II*, Bonn, 1960.
47. V. A. Rohlin, "New results in the theory of 4-dimensional manifolds," (Russian) *Doklady* **84**, 221–224 (1952).
48. J. Schwartz, "De Rham's theorem for arbitrary spaces," *Amer. J. Math.*, **77**, 29–44 (1955).
49. J.-P. Serre, *Corps Locaux*, Hermann, Paris (1962).
50. J.-P. Serre, *Cohomologie Galoisienne*, Springer, Berlin (1964).
51. S. Smale, "A survey of recent results in differential topology," *Bull. Amer. Math. Soc.*, **69**, 131–145 (1963).
51a. S. Smale, "On the structure of 5-manifolds," *Ann. Math.*, **75**, 38–46, (1962).
52. E. Spanier, *Algebraic Topology*, McGraw-Hill (1966).
53. M. Spivak, *Calculus on Manifolds*, Benjamin, New York (1965).
54. G. Springer, *Introduction to Riemann Surfaces*, Addison-Wesley, Reading, Massachusetts (1957).
54a. J. Stallings, "Polyhedral homotopy-spheres," *Bull. Amer. Math. Soc.* **66**, 485–488 (1960).
55. N. Steenrod, *Topology of Fibre Bundles*, 5th printing, Princeton Univ. Press., Princeton, New Jersey (1965).
56. R. Swan, *Theory of Sheaves*, Univ. of Chicago Press, Chicago, Illinois (1964).
57. R. Thom, "Quelques propriétés globales des variétés différentiables," *Comment. Math. Helv.*, **28**, 17–86 (1954).
58. C. T. C. Wall, "Determination of the cobordism ring," *Ann. Math.*, **72**, 292–311 (1960).
58a. C. T. C. Wall, "Classification of (n − 1)-connected 2n-manifolds," *Ann. Math.* **75**, 163–189 (1962).
59. A. Wallace, *Algebraic Topology*, Macmillan (Pergamon), New York (1963).
60. A. Weil, "Abstract versus classical algebraic geometry," *Proc. Int. Congr. Math.*, Amsterdam, III, (1954).
61. J. H. C. Whitehead, "Combinatorial Homotopy, I," *Bull. Amer. Math. Soc.*, **55**, 213–245 (1949).
62. J. F. Adams, *Algebraic Topology—A Student's Guide*, Cambridge University Press, N.Y. and London, 1972.
63. R. F. Brown, *The Lefschetz Fixed Point Theorem*, Scott, Foresman and Co., Glenview, Illinois, 1971.
64. A. Dold, *Lectures on Algebraic Topology*, Springer-Verlag, N.Y.,

Heidelberg, Berlin, 1972.

65. D. Husemoller, *Fibre Bundles*, Springer-Verlag, N.Y., 1975.
66. A. T. Lundell and S. Weingram, *The topology of CW-Complexes*, Van Nostrand, N.Y., 1969.
67. W. S. Massey, *Algebraic Topology: An Introduction*, Harcourt-Brace, N.Y., 1967.
68. F. W. Warner, *Foundations of Differentiable Manifolds and Lie Groups*, Scott-Foresman, Glenview, Illinois, 1971.
69. R. E. Mosher and M. C. Tangora, *Cohomology Operations and Applications in Homotopy Theory*, Harper, N.Y., 1968.
70. J. F. P. Hudson, *Piecewise Linear Topology*, Benjamin. Reading, Mass., 1969.
71. V. Guillemin and A. Pollack, *Differential Topology*, Prentice-Hall, Englewood Cliffs, N.J., 1974.
72. R. E. Strong, *Notes on Cobordism Theory*, Princeton University Press, 1968.
73. C. T. C. Wall, *Surgery on Compact Manifolds*, Academic, N.Y., 1970.
74. J. F. Adams, *Stable Homotopy and Generalized Homology*, U. of Chicago, 1974.
75. J. W. Vick, *Homology Theory*, Academic, N.Y., 1973.
76. B. Gray, *Homotopy Theory—An Introduction to Algebraic Topology*, Academic, N.Y., 1975.
77. H. Cartan and S. Eilenberg, *Homological Algebra*, Princeton Univ. Press, 1956.
78. M. Cohen, *A Course in Simple-Homotopy Theory*, Graduate Texts in Mathematics, **10**, Springer-Verlag, 1973.
79. J. Hempel, *3-Manifolds*, Annals of Mathematics Studies, **86**, Princeton Univ. Press, 1978.
80. F. Klein, *On Riemann's Theory of Algebraic Functions and Their Integrals*, Dover Publications, 1963.
81. F. Klein, *Lectures on Icosehedron and the Solution of Equations of the Fifth Degree*, Dover Publications.
82. J. Milnor, *Lectures on the h-Cobordism Theorem*, Princeton Mathematical Notes, Princeton Univ. Press, 1965.
83. J. Milnor, "On Manifolds Homeomorphic to the 7-Sphere," *Annals Math.* **64** (1956), 399–405.
84. J. Milnor and J. Stasheff, *Characteristic Classes*, Annals of Math. Studies, **76**, Princeton Univ. Press, 1975.
85. D. Rolfsen, *Knots and Links*, Mathematics Lecture Series, **7**, Publish or Perish Inc., 1976.
86. H. Seifert and W. Threlfall, *Lehrbuch der Topologie*, Leipzig, Teubner, 1934, Chelsea Publishing Company, 1947.

87. N. Steenrod and D. B. A. Epstein, *Cohomology Operations*, Annals of Math. Studies, **50**, Princeton University Press, 1962.
88. G. Whitehead, *Elements of Homotopy Theory*, Graduate Texts in Mathematics, **61**, Springer-Verlag, 1978.
89. M. Newman, *Integral Matrices*, Academic Press, New York and London, 1972.
90. H. Bass (ed.), *Algebraic K-Theory I, II, III*. Proceedings of the Conference held at Seattle Reasearch Center of the Battelle Institute. Springer Verlag Lecture Notes in Mathematics, **341–343**, 1973.
91. R. S. Palais, Seminar on the Atiyah-Singer Index Theorem. Annals of Mathematics Studies, **57**, Princeton Univ. Press, 1965.
92. P. Shanahan, The Atiyah-Singer Index Theory: An Introduction. Lecture Notes in Mathematics, Springer Verlag, 1978.
93. R. Bott, On topological obstructions to integrability, *International Congress of Mathematicians*, Nice 1970, **1**, Gauthier-Villars, 1971.

INDEX

INDEX

Milton Keynes UK
Ingram Content Group UK Ltd.
UKHW050256161024
449569UK00042B/1733